Bio-Inspired Data-driven Distributed Energy in Robotics and Enabling Technologies

This book begins by introducing bio-inspired data-driven computation techniques, discussing bio-inspired swarm models, and highlighting the development of interactive bio-inspired energy harvesting systems to drive transportation infrastructure. It further covers important topics such as efficient control systems for distributed and hybrid renewable energy sources, and smart energy management systems for developing intelligent systems.

This book:

- Presents data-driven intelligent heuristics for improving and advancing environmental sustainability in both eco-cities and smart cities.
- Discusses various efficient control systems for distributed and hybrid renewable energy sources and enhances the scope of smart energy management systems for developing even intelligent systems.
- Showcases how distributed energy systems improve the data-driven robots in the Internet of Medical Things.
- Highlights practical approaches to optimize power generation, reduce costs through efficient energy, and reduce greenhouse gas emissions to the possible minimum.
- Covers bio-inspired swarm models, smart data-driven sensing to combat environmental issues, and futuristic data-driven enabled schemes in blockchain-fog-cloud-assisted medical energy ecosystem.

This book is primarily written for graduate students and academic researchers in diverse fields, including electrical engineering, electronics and communications engineering, computer science and engineering, and environmental engineering.

Intelligent Data-Driven Systems and Artificial Intelligence
Series Editor: Harish Garg

Cognitive Machine Intelligence: Applications, Challenges, and Related Technologies
Inam Ullah Khan, Salma El Hajjami, Mariya Ouaissa, Salwa Belqziz and Tarandeep Kaur Bhatia

Artificial Intelligence and Internet of Things based Augmented Trends for Data Driven Systems
Anshu Singla, Sarvesh Tanwar, Pao-Ann Hsiung

Modelling of Virtual Worlds Using the Internet of Things
Edited by Simar Preet Singh and Arun Solanki

Data-Driven Technologies and Artificial Intelligence in Supply Chain
Tools and Techniques
Mahesh Chand, Vineet Jain and Puneeta Ajmera

Bio-Inspired Data-driven Distributed Energy in Robotics and Enabling Technologies
Edited by Abhishek Kumar, Hemant Kumar Saini, Ashutosh Kumar Dubey and Vicente García-Díaz

For more information about this series, please visit: www.routledge.com/Intelligent-Data-Driven-Systems-and-Artificial-Intelligence/book-series/CRCIDDSAAI

Bio-Inspired Data-driven Distributed Energy in Robotics and Enabling Technologies

Edited by
Abhishek Kumar, Hemant Kumar Saini,
Ashutosh Kumar Dubey, and
Vicente García-Díaz

CRC Press
Taylor & Francis Group
Boca Raton London New York

CRC Press is an imprint of the
Taylor & Francis Group, an **informa** business

Designed cover image: fixed image template

MATLAB® and Simulink® are trademarks of The MathWorks, Inc. and are used with permission. The MathWorks does not warrant the accuracy of the text or exercises in this book. This book's use or discussion of MATLAB® or Simulink® software or related products does not constitute endorsement or sponsorship by The MathWorks of a particular pedagogical approach or particular use of the MATLAB® and Simulink® software.

First edition published 2025
by CRC Press
2385 NW Executive Center Drive, Suite 320, Boca Raton FL 33431

and by CRC Press
4 Park Square, Milton Park, Abingdon, Oxon, OX14 4RN

CRC Press is an imprint of Taylor & Francis Group, LLC

© 2025 selection and editorial matter, Abhishek Kumar, Hemant Kumar Saini, Ashutosh Kumar Dubey, and Vicente García-Díaz; individual chapters, the contributors

ISBN: 9781032640631 (hbk)
ISBN: 9781032869414 (pbk)
ISBN: 9781003530077 (ebk)

DOI: 10.1201/9781003530077

Typeset in Sabon
by Newgen Publishing UK

Contents

About the editors

Abhishek Kumar is working as an Associate Professor at Chandigarh University, India and is currently a Post-Doctoral Fellow in the Ingenium Research Group Lab at Universidad De Castilla-La Mancha, Ciudad Real, Spain. He has more than 170 publications in reputed, peer-reviewed national and international journals and conferences.

Hemant Kumar Saini is currently working as an Assistant Professor in the Computer Science & Engineering Department at Chandigarh University, Punjab, India. He had been honored by the Governor of Rajasthan for his research works in 2015. He is a TEQIP-III coordinator. Saini has delivered many expert talks in ATAL-sponsored Faculty Development Programs and keynote talks in many national and international conferences.

Ashutosh Kumar Dubey is an Associate Professor in the Department of Computer Science at Chitkara University School of Engineering and Technology, Himachal Pradesh, India. He is a postdoctoral fellow at the Ingenium Research Group Lab, Universidad de Castilla-La Mancha, Ciudad Real, Spain.

Vicente García-Díaz is an Associate Professor in the Department of Computer Science at the University of Oviedo, Spain. He is a software engineer and has a PhD in Computer Science. He has a Master's degree in Occupational Risk Prevention and the qualification of University Expert in Blockchain Application Development. He is part of the editorial and advisory board of several indexed journals and conferences and has been editor of several special issues in books and indexed journals. He has supervised 100+ academic projects and published 100+ research papers in journals, conferences, and books. His teaching interests are primarily in the design and analysis of algorithms and the design of domain-specific languages. His current research interests include decision support systems, health informatics, and eLearning.

Contributors

Taslima Ahmed
IIMT College of Engineering, Greater Noida, Uttar Pradesh, India

Firos A
Rajiv Gandhi University, Arunachal Pradesh, India

Monica Bhutani
Bharati Vidyapeeth's College of Engineering, New Delhi, India

Veer P. Gangwar
Lovely Professional University, Phagwara, India

Ganesh Gupta
Sharda University, Noida, Uttar Pradesh, India

Monica Gupta
Bharati Vidyapeeth's College of Engineering, New Delhi, India

Shika Gupta
Chandigarh University, Punjab

Inzimam Ul Hassan
Chandigarh University, Punjab, India

Agha Imran Husain
Manav Rachna University, Faridabad, Haryana, India

Zaiba Ishrat
Meerut Institute of Engineering & Technology, Meerut, Uttar Pradesh, India

Tarun Jain
Bharati Vidyapeeth's College of Engineering, New Delhi, India

Christian Kaunert
Dublin City University (Ireland) & University of South Wales (UK)

Aakansha Khanna
Chandigarh University, Punjab, India

Bhavani Krishna
Dayananda Sagar College of Engineering, Bengaluru, Karnataka, India

Abhishek Kumar
Ingenium Research Group Lab, Universidad De Castilla-La Mancha, Spain

Naveen Kumar
Mahatma Jyotiba Phule Rohilkhand University, Bareilly, Uttar Pradesh And Department of Mathematics, National Institute of Technology Kurukshetra, Haryana, India

Babar Ali Kunwar
Meerut Institute of Engineering & Technology, Meerut, Uttar Pradesh, India

Aavaig Malhotra
Bharati Vidyapeeth's College of Engineering, New Delhi, India

Shyam Shankar Menon
Hewlett Packard India, Bengaluru Karnataka, India

Sriniwas Mishra
IIMT College of Engineering, Greater Noida, Uttar Pradesh, India

Dipra Mitra
Amity University, Jharkhand, India

Gajendra Nagaraj
Dayananda Sagar University, Bengaluru, Karnataka, India

Padma Nandanan
St Joseph's University, Bengaluru, Karnataka, India

Srinivas Kumar Palvadi
SRK Institute of Technology, Vijayawada, Andhra Pradesh, India

Gurpreet Singh Panesar
Chandigarh University, Punjab, India

Hariharan R
St Joseph's University, Bengaluru, Karnataka, India

Manju Rani
Gurugram University, Gurugram, Haryana, India

Ruchika
K.M. Govt. (P.G) College, Narwana, Haryana, India

Channi Sachdeva
Lovely Professional University, Phagwara, Punjab, India

Hemant Kumar Saini
Parul University, Vadodra, Gujarat and
  Chandigarh University, Gharuan, Punjab, India

Anu Priya Sharma
Manav Rachna University, Faridabad, Haryana, India

Tej Prakash Sharma
Bharati Vidyapeeth's College of Engineering, New Delhi, India

Bhupinder Singh
Sharda University, Greater Noida, Uttar Pradesh, India

Gurwinder Singh
Chandigarh University, Mohali, Punjab, India

Monika Singh
Chandigarh University, Mohali, Punjab, India

Akansha Solanki
Bharati Vidyapeeth's College of Engineering, New Delhi, India

Komal Vig
Sharda University, Greater Noida, Uttar Pradesh, India

Data-driven aircrafts transforming the manufacturing world with new idea of bio-inspired interactions

Hemant Kumar Saini and Abhishek Kumar

1.1 INTRODUCTION

Recently, with the advancements in visual recognition within neuroscience, researchers are increasingly inclined toward bioinspired approaches. However, the challenge lies in the adaptability of these approaches due to preprocessing burdens. To address this issue, a novel preprocessing technique using data-driven mechanisms has emerged in the realm of bioinspired methods, providing more effective and accurate vision recognition.

Previous studies on data-driven manufacturing have highlighted its potential to revolutionize traditional engineering design processes. However, ongoing advancements in these fields continuously push the limits of what can be achieved, leaving vast areas relatively unexplored. This research seeks to investigate the potential of using a combination of spatial data sets as a gateway for innovative, data-driven designs (Arad et al. 2013). The process commences with diverse data acquisition tools and processing techniques, progresses through computational analysis and optimization designs, and culminates in digital manufacturing using contemporary artifacts (Shukla et al. 2017). These data sets facilitate the development of Industry 5.0 models, enabling the initiation of a reverse engineering process for various applications within the manufacturing world (Xu et al.2021).

Architects consistently encounter design challenges that demand efficient solutions. Drawing inspiration from bioinspired principles, coupled with data sets collected from users, can anticipate trends and provide solutions tailored to specific conditions. This approach facilitates informed decision-making. Modern architects frequently adopt a strategy of mimicking nature, relying on biological structures that optimize efforts for desired outcomes, honed over the course of evolution.

The application of data-driven artifacts is prevalent in various fields, including pest control, vision monitoring, coal mining, chemical industry, gas plants, and X-ray laboratories. This involves extracting insights to identify essential aspects of manufacturing plants, both visually and functionally. This transformative approach to designing, building, and utilizing space

DOI: 10.1201/9781003530077-1

relies on readily available and accessible information, fostering diverse types of engagement. The integration of artifacts with bioinspired mechanisms enhances decision-making capabilities.

In parallel, emerging technologies like the Internet of Things (IoT) in healthcare services and intelligent towns necessitate efficient data processing. However, existing processing approaches face challenges, viz., waiting acknowledgments and delays due to double-data transfers—first from computing systems to the cloud and then from the cloud to IoT applications. The expanding volume and variety of data collected by IoT devices, particularly in smart cities, pose challenges in terms of effective decision-making and data analytics (Sfar et al. 2017).

To address these issues, a bioinspired, algorithm-based, data-driven analytics paradigm is introduced interfacing between sensor devices and the computing resources. The approach by Cao et al.(2021)efficiently processes user-driven data from diverse environments, providing solutions to the bottlenecks experienced by computational systems, which receive raw data from IoT artifacts. The upcoming sections elaborate on successfully implemented bioinspired paradigms.

1.2 BIOINSPIRED ALGORITHMS

1. Raff et al. (2020) explored the existing bioinspired algorithms applicable to various aspects of data-driven applications. These algorithms, with diverse targets, can be broadly classified into three groups: ecological, swarm-based, and evolutionary.
2. Genetic Algorithms (GA): These algorithms are used basically for the enhancement and best utilization of bandwidth where the data is segregated from the service resources, which need the separate computing part. Bonyadi and Michalewicz (2017) proposed one of the differential equations to improve the exploitation searching.
3. Simulated Annealing (SA): Colorni et al. (1992) introduced an optimization technique based on the SA algorithm. He proposed an impactful model, which is used to select only those attributes that impact the complete training, thus improving the accuracy.
4. Artificial Bee Colony (ABC) Algorithm: This algorithm identifies the optimal set and optimizes for varying dataset sizes, reducing implementation instance and civilizing precision. In a MapReduce-based Hadoop environment, Google Maps utilizes the ABC algorithm, yielding more effective outcomes in terms of execution time.
5. Cat Swarm Optimization (CSO) Algorithm: Lin et al. (2015) selected characteristics during classification and utilized term frequency-inverse occurrence to enhance feature selection accuracy. This algorithm finds applications, particularly in medical chemical therapy using chemical robots.

6. Swarm Intelligence (SI) Algorithm: This approach, designed by Cheng et al. (2016), tackles financial weight post evils. The SI algorithm enhances the accuracy of data processing by efficiently handling high-dimensional data.

7. Ant Colony Optimization (ACO) Algorithm: Banerjee and Badr (2018) tailored this approach for small aircraft mobile data, especially relevant during times like the coronavirus disease (COVID) pandemic, using rough set techniques. The ACO algorithm aids in selecting optimal features for informed decisions, contributing to the effective management of social networks, such as tweets and advertisements.

1.3 CASE STUDY OF PEST INFESTATION

Unmanned aerial vehicles (UAVs) have become increasingly popular in the agriculture industry in the past few decades. UAV-based actuation is utilized to release biological control agents and spray pesticides. Taking wind speed and UAV flight characteristics into consideration to maximize the precise distribution of pesticides, biological control agents is a major problem in such UAV-based actuation. The density distribution patterns of vermiculite dispensed from a hovering UAV are predicted using a data-driven framework as a function of the dispenser setting, wind speed, and UAV movement status. The vermiculite distribution pattern was properly predicted by the SI learning system, which was used to both train and test data. The architecture and algorithm are easily adaptable to additional precision pest management scenarios including various UAVs, dispensers, insecticides, and crop types. Furthermore, because of this model's straightforward analytical structure, it can be integrated into the controller's design to maximize the autonomous UAV's delivery of the appropriate number of predatory mites to several target sites.

1.4 OPEN CHALLENGES AND RESEARCH DIRECTIONS

The extensive activity documented in the literature serves as a clear testament to recent technological advancements achieved through the application of bioinspired computation to Big Data. Certainly, bioinspired computation has been applied across diverse domains for data-centric applications, including, but not limited to, energy (Andreou et al. 2016; Wang, Liu, et al. 2020), transport and mobility (Del Ser J 2016), health (Munir et al. 2019), industry (Yan et al. 2018; Diez-Olivan et al. 2019), agriculture (Aghelpour et al. 2020), cyber–physical systems (Yadav and Vishwakarma 2020), social networks (Bello-Orgaz et al. 2016; Camacho et al. 2020; Raychaudhuri and De 2020), and sensor networks (Zhou et al. 2020). The worldwide reaction to the COVID-19 pandemic has intensified research efforts in

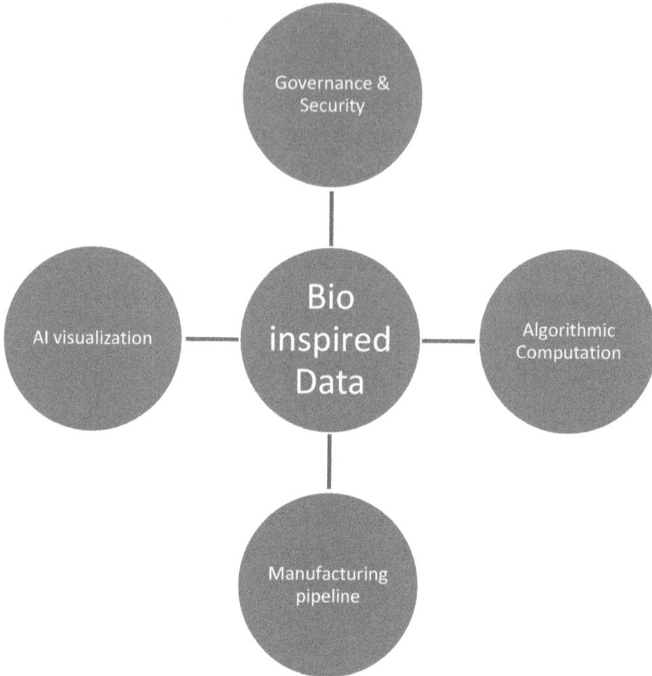

Figure 1.1 Various bioinspired future aspects.

the fields of Big Data and artificial intelligence, frequently incorporating deep neural networks for diagnoses based on computed tomography (CT) scans. However, there is a lack of substantial evidence to categorize the scales of these studies claiming to be Big Data. Our literature review reveals numerous questions that remain to be explored in hybridizing these two paradigms. Subsequently, we enumerate and discuss various research niches with respect to the previously examined literature. Figure 1.1 provides a graphical summary of our envisioned future prospects for the field.

1.5 MOTIVATION

The industrial processes can greatly benefit from the use of UAVs in Mobile Adhoc Network (MANET) and Flying Adhoc Network (FANET) networks, which are primarily powered by data-driven aircraft with novel, bioinspired paradigms. One such instance already existed in real-time printing devices, where drones are employed to collect ink when it is about to run out. For these reasons, it is imperative to investigate and concentrate on areas where data-driven technology may help aviation vehicles continue their manufacturing processes, which may be extremely challenging in particular industries, such as chemical fertilizers, where human lives are at risk.

1.6 AIRCRAFT IN MANUFACTURING APPLICATIONS

The manufacturing industry has undergone a significant transformation with the integration of UAVs, commonly referred to as drones. Originally developed for military purposes, UAVs have seamlessly transitioned into various civilian sectors, including manufacturing, due to technological advancements and the growing demand for innovative solutions. The roots of UAVs in manufacturing can be traced back to their capacity to overcome traditional limitations, offering a spectrum of advantages that enhance efficiency, safety, and overall productivity.

Initially employed for military reconnaissance and surveillance, the success of UAVs in these applications paved the way for exploring their potential in civilian industries, particularly manufacturing. The earliest applications of UAVs in manufacturing focused on aerial surveying and mapping. Equipped with high-resolution cameras, drones efficiently captured extensive manufacturing facilities, providing valuable insights into layout and infrastructure. Given the complex structures of manufacturing plants, UAVs became invaluable for inspecting equipment, pipelines, and structures, reducing the reliance on manual inspections and enhancing safety and efficiency.

Unmanned aerial vehicles exhibit swift navigation capabilities within large manufacturing facilities, enabling them to collect data and perform inspections in a fraction of the time required by human workers. This enhanced time efficiency translates to increased operational productivity. Moreover, UAVs contribute significantly to improved safety by undertaking tasks in hazardous or hard-to-reach areas, minimizing risks to human workers. This aspect is particularly crucial in manufacturing environments, which are prone to safety hazards. The most notable application in manufacturing industries is stimulated through data-driven UAVs, which play a pivotal role in various aspects. The subsequent discussion delves into the ways in which these data-driven UAVs contribute to existing manufacturing processes.

1.6.1 Data-driven UAV: a new frontier in manufacturing

1. Utilizing cameras and sensors, UAVs can conduct routine inspections of manufacturing equipment and infrastructure. Data-driven analytics play a crucial role in identifying potential issues, predicting maintenance requirements, and minimizing downtime.
2. UAVs have the capability to capture high-resolution images and videos of products on the production line. Leveraging data-driven image analysis and machine learning algorithms enables the assessment of

product quality, detection of defects, and assurance of compliance with quality standards.

3. UAVs are valuable in monitoring and managing inventory levels within expansive manufacturing facilities. Data-driven systems provide real-time insights into stock levels, track the movement of raw materials and finished products, and optimize overall supply chain processes.

4. Equipped with environmental sensors, UAVs can monitor air quality, temperature, and other environmental factors in and around manufacturing facilities. Data-driven analysis ensures compliance with environmental regulations and offers insights for implementing sustainable practices.

5. Data-driven insights derived from UAVs contribute to the optimization of manufacturing workflows. Analyzing the movement of materials and personnel can lead to more efficient layout designs, reducing bottlenecks and enhancing overall productivity.

6. In emergency situations like fires or chemical spills, UAVs can swiftly survey the situation from the air. Data-driven analysis provides essential information for emergency responders, aiding them in making informed decisions to mitigate risks.

7. UAVs equipped with sensors are effective in monitoring energy consumption in manufacturing facilities. Data-driven analysis identifies areas for energy efficiency improvements, contributing to cost savings and aligning with sustainability goals.

1.6.2 Major consideration during data drive in manufacturing system

In manufacturing systems, UAVs can use a variety of data gathering methods to obtain important data for analysis and decision-making. Here are a few typical methods for gathering data with UAVs in an industrial setting:

1. It is used in photography and videography to record visual information for examination, observation, and analysis—primarily in evaluating the state of the machinery, examining the production lines, and keeping an eye on the entire industrial plant. High-resolution cameras, red–green–blue (RGB) cameras, or even specialized cameras like thermal cameras are embedded in UAV which drive the data.

2. In thermal sectors where anomaly detection and equipment health monitoring are necessary, high-temperature fluctuations must be recorded recognizing parts of machinery that are overheating and spotting electrical problems along with ensuring that insulation is adequate. These drones are designed to be competent enough to handle this data.

3. A manufacturing facility tracking the movement of items and guaranteeing precise location during inspections utilizes precisely geospatial data for mapping and navigation.
4. UAV with sound detection sensors: To discover equipment defects, ensure optimal performance, and undertake predictive maintenance, machinery status can be monitored by listening for vibrations and acoustic signals.
5. Monitoring temperature, humidity, and other environmental parameters is crucial for quality assurance and regulatory compliance in the manufacturing units. These variables must be controlled regularly.

Selecting the right sensors and payloads for UAV data gathering in manufacturing systems is crucial, and it should be done in accordance with the particular objectives and specifications of the application. Manufacturing process monitoring, analysis, and optimization are made possible by the integration of these technologies.

1.6.3 Nondestructive testing with UAVs

Unmanned aerial vehicles with nondestructive testing (NDT) have become popular as a cost-effective and useful way to evaluate assets and structures without causing harm. Here are some examples of how UAVs are used in NDT:

1. Using UAVs for Visual Inspection: UAVs are being used to conduct visual inspections of assets and structures, looking for observable anomalies such as corrosion or surface defects throughout the manufacturing process.
2. Utilize an eddy current sensor to examine conductive materials for flaws without making direct touch to find surface fractures, corrosion, or alterations in the material's characteristics
3. Surface and subsurface defects in ferromagnetic materials can be easily recognized for magnetic particle testing (MPT) and magnetic flux leakage (MFL),typically inspecting storage tanks, pipelines, and other ferrous material-built facilities
4. A few vibration dangers might arise from improperly bolted engines or devices, which can now easily be detected by UAVs. They can monitor the vibrations in machinery and spot the malfunctions or structural vulnerability by using predictive maintenance approaches.

There are a number of benefits in using UAVs for NDT, such as enhanced efficiency, lowered safety hazards for human inspectors, and access to difficult-to-reach locations. However, when using UAV-based NDT techniques, it is crucial to adhere to industry best practices, maintain data accuracy, and comply with regulatory standards.

1.7 SUCCESSFUL IMPLEMENTATIONS OF DATA-DRIVEN UAVS IN MANUFACTURING

To the best of my knowledge through 2022, numerous industries have witnessed the successful application of data-driven UAVs in production. However, there might have been improvements and new uses since then. Here are a few instances of applications that were successful:

1. Asset inspection and maintenance is done in the energy sector (oil and gas, power plants), where industrial machinery, pipelines, and power lines are inspected and monitored by UAVs fitted with cameras and sensors. Preventive maintenance may be scheduled and possible problems can be identified with the aid of data-driven analysis, which also minimizes risks and downtime.

2. UAVs with high-resolution cameras used in the automotive manufacturing industry are able to take detailed pictures of cars while they are manufactured. Machine learning techniques and data-driven picture analysis are used to find flaws, guaranteeing premium manufacturing standards.

3. A further useful tool for logistics and distribution is the use of UAVs for aerial surveys of sizable warehouses and distribution centers. Data-driven technologies track stock movement, optimize warehouse layouts for more efficiency, and offer real-time insights into inventory levels.

4. The most crucial situation is in the manufacturing processes in heavy industry environments, where UAVs fitted with sensors keep an eye on environmental factors, air quality, and emissions near manufacturing plants. Therefore, sustainable practices are made easier and environmental standards are adhered to with the help of data-driven research.

5. Data collected by UAVs are extremely important, quick, and mobile, which is particularly useful for companies that produce chemicals, oil, or gas where disaster management and emergency response are needed at all times. UAVs that are fitted with cameras and sensors are utilized to promptly examine situations like chemical spills or industrial mishaps. Data-driven insights help emergency responders make well-informed decisions on mitigation and containment.

These illustrations outline the variety of uses for data-driven UAVs in manufacturing, from asset inspection and quality control to environmental monitoring and emergency response. It is crucial to remember that effective data-driven UAV deployment frequently requires cooperation between UAV operators, data scientists, and industry specialists to customize solutions to particular manufacturing difficulties.

1.8 INTEGRATION OF DATA-DRIVEN TECHNOLOGY WITH BIOINSPIRATIONS

Given the well-established fact that industries grapple with managing vast volumes of data efficiently and swiftly, it becomes imperative to explore the demands of bioinspired data-driven solutions and various prototypes. This exploration is particularly relevant for efficiently managing the large data-carrying UAVs, especially in manufacturing industries, which are numerous in any country.

The global gross domestic product (GDP) heavily relies on the manufacturing sector, and neglecting its advancements would result in significant losses. Therefore, this chapter focuses extensively on examining the potential of combining bioinspired prototypes with data-driven vehicles in the realm of modern UAV technology.

Biologically inspired design (BID) relies mostly on the identification of the various suitable designs based on the bioinspired prototypes and re-engineered the problems with the similar functions. Recent advancements in data science and artificial intelligence have opened avenues for developing data-driven technique to ease claim indoctrination, recovery, plotting, and estimate in intend by correlation (Wang, Ng, et al. 2020). Through the data-driven approaches new innovative designs can be created in Industry 4.0.

The suggested strategy presents an alternative method for representing biological knowledge that is based on a five-dimensional framework intended for digital twins (DT) as well as the function-based idea (Chintapalli et al. 2016). By means of relations across several dimensions, links are made between biological prototypes and their inspired bionic products using this representation approach. These relationships allow correlations between biological prototypes and engineering design criteria to be estimated, as well as biological prototypes' inspired bionic solutions. High-scoring biological prototypes can then be suggested to engineering designers as appropriate BID.

Consequently, the suggested methodology amplifies BID's flexibility in product innovations in the context of Industry 4.0 (Karimov et al. 2018). This is accomplished by saving engineering designers' labor in finding appropriate biological prototypes for comparisons and by offering illustrations of how to creatively adapt biological prototypes to meet a variety of design specifications as shown in Figure 1.2.

Understanding and mastering BID will make it simple to use the model in any frame. The suggested BID knowledge representation comprises five components, which are based on the five-dimensional model seen

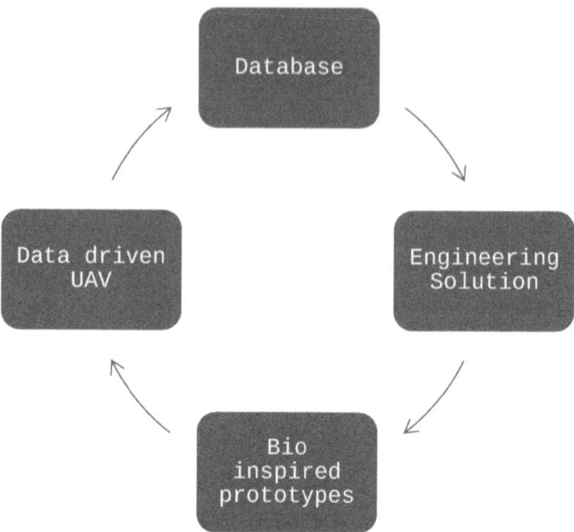

Figure 1.2 Continuous cycle of engineering solution from the rapid bioinspired paradigms.

in DT (Tao et al. 2019). As a result, the biological prototype—that may serve as sources of inspiration for engineering design—replaces the physical entity, and the engineering solutions serve as virtual entities that hold data regarding engineering design schemes that draw inspiration from biological prototypes. The BID database, which replaces the digital twin data design (Wang et al. 2014), offers comprehensive details about the function structures and methods by which biological prototypes and technological solutions fulfill their roles.

The function set takes the place of the previous service section, which included primary, secondary, and auxiliary functions that were taken from the engineering solutions and the biological prototypes. In the suggested design, the connection section is retained to show relationships between two other parts that include data exchange and mapping techniques. When compared to current modeling techniques, the suggested BID knowledge representation has a number of clear advantages (Sun et al.2020).

Firstly, biological prototypes and bionic engineering design schemes aid in the long-range translation of biological model into manufacturing elucidation by engineering designers.

Secondly, deep knowledge loom offers an efficient means of acquiring BID knowledge. As a result, engineering designers are able to locate a variety of bioinspired engineering solutions to the specific design issues that serve as a valuable source of inspiration for new ideas. Thirdly, vital information regarding the realization of biological function in engineering applications can also be found in bionic engineering solutions (Jiang et al. 2022).

The suggested heuristics potentially selects the appropriate biological prototypes for engineering designers with the two dimensions. The degree to which biological prototypes satisfy the primary design requirements of a given BID project is measured by indicators from various dimensions.

This work offers a new data-driven BID knowledge processing approach in response to the ever-increasing necessity of engineering design for smart products.

The novel hybrid BID knowledge representation, which combines the functional modeling technique with the DT, serves as the foundation for the suggested BID knowledge processing approach (Liu et al. 2019). The suggested method makes several contributions to both the theoretical and practical domains. First, to facilitate the DT with the BID, the application with the reverse engineering process, the interdisciplinary approach knowledge, and their relationships in the realization of biological strategy attempt to integrate the conceptual design. Second, algorithms that suggest the best biological prototype for a specific set of BID tasks are also included in the suggested BID knowledge processing strategy. These algorithms are created using self-updating strategies. Third, one can use the biological strategies and biomimetic designs (Del Ser et al. 2019) to process the new training and teaching using the BID.

The initial concern with the proposed BID knowledge processing approach lies in its early stage of development as a data-driven method. Consequently, there is still a reliance on manual intervention to make crucial decisions in knowledge evaluation. This reliance on manual labor may pose a challenge to the scientific robustness of the proposed approach. Hence, there is a need for more intelligent algorithms that can assist or potentially reinstate specialized assessor.

The lack of an automated online original BID data collection mechanism (Goel et al. 2022) is another noteworthy problem. This limitation restricts the planned technique to utilizing existing BID knowledge from online databases. To address this, future developments could include the creation of web crawlers-based tools. These tools would offer resources to improve the viability of the planned advance by enabling real-time data collection.

Moreover, the practical implementation of the bioinspired study in an exemplary engineering design situation is crucial to its validation. However, this stage is still in its infancy, serving as an initial step to verify the feasibility of the proposed design schema. Additional steps, including design experiments or studies into consumers' pleasure with the advanced version of the suggested approach as a new software product, are crucial to fortify the validation process. This would contribute to establishing the effectiveness and reliability of the proposed approach in practical scenarios.

1.8.1 Summary of key findings

Engaging research partners as participants in our research proved to be more challenging at times compared to researchers or research managers. Numerous practical difficulties arose in manufacturing industries, where human lives are consistently at risk, and specific data must be transmitted during critical moments, such as in spinner textile mills. In these mills, if the thread spinning does not change at a particular stage, it results in a line across the entire cloth bundle, impacting business costs and turnover.

Similar challenges were observed in chemical industries, where human errors in monitoring the boiling and flow of mixtures could lead to significant devastation. Additionally, variations in product outcomes at different temperatures created problems, with certain products not being found, posing new challenges to business operations. Consequently, these challenges prompted the exploration of solutions.

Given the crucial role of data-driven technology, especially utilized by autonomous vehicles, which gained prominence during the COVID period and is now pervasive across all sectors, the importance of efficient data collection becomes evident. Enhancing data collection through autonomous vehicles becomes even more impactful when driven by bioinspired paradigms, leveraging the liberty and flexibility offered by nature. This approach has also influenced data-driven UAVs, enabling them to contribute significantly to various manufacturing sectors, including environmental monitoring, governing supply chains, logistics, and more.

If one were to redesign this study, several changes would be necessary to create a more tailored prototype. Various studies on data-driven mechanisms and case studies from the application stage are included in this chapter in an effort to highlight the important early involvement of research partners during the original design phase. Opting for a more focused approach by researching a smaller number of applications would enable a deeper ethnographic exploration. To find out how a bigger production group feels about and has experienced public involvement, this may involve running a quick screening exercise among them.

There were difficulties with the economic evaluation, particularly getting research partners to submit resource logs. Unique data collection within a specific timeframe presented difficulties, particularly in a field like public involvement, characterized by episodic and infrequent activities. Consequently, integrating the collection of crucial peak-risk data in a more cohesive manner, particularly when analyzing ethnographic observations of manufacturing industries, could prove advantageous.

1.8.2 Recommendations for further research

Our findings highlight a number of knowledge gaps about public participation in research that might benefit from additional study, including realist evaluation to strengthen and validate the theory we have created here:

1. A thorough investigation of the ways in which the manufacturing and production sectors impact sceptics or agnostics by means of the practical difficulties brought on by time discrepancies; additional study could validate and strengthen popular acceptance.
2. More methodological research is required to fully capture the effects and results of public participation in manufacturing support, including the effects on business and the novel data-driven technological methods that are changing the world in an unfavorable way.
3. Conducting some longer-term research to assess the effect of business and safety participation on important indicators such as danger thermal sensing, gas leaking, and chemical products would be highly beneficial, despite the methodological challenges involved.
4. After data collection was complete, a last, somewhat specific but crucial question was discovered: What effect does it have on some sorts of research, especially production, when research partners go for extended stretches of time with little-to-no involvement after earlier, more intense periods?

REFERENCES

Aghelpour, P., Bahrami-Pichaghchi, H., Kisi, O. (2020).Comparison of three different bio-inspired algorithms to improve ability of neuro fuzzy approach in prediction of agricultural drought, based on three different indexes. Computers and Electronics in Agriculture, 170, 105279. https://doi.org/10.1016/j.compag.2020.105279

Andreou, A.G., Figliolia, T., Sanni, K., Murray, T.S., Tognetti, G., Mendat, D.R., Molin, J.L., Villemur, M., Pouliquen, P.O., Julian, P., Etienne-Cummings, R., Doxas, I. (2016).Bio-inspired system architecture for energy efficient, big data computing with application to wide area motion imagery. In 2016 IEEE 7th Latin American Symposium on Circuits & Systems (LASCAS), pp 1–6.

Arad, J., Housh, M., Pereiman, L., Ostfeld, A. (2013). A dynamic threshold scheme for contaminant event detection in water distribution systems. Water Research, 47(5), 1899–1908.https://doi.org/10.1016/j.watres.2013.01.017

Banerjee, S., Badr, Y. (2018). Evaluating decision analytics from Mobile Big Data using Rough Set Based Ant Colony. In Skourletopoulos, G., Mastorakis, G., Mavromoustakis, C., Dobre, C., Pallis, E., Eds., Mobile Big Data. Lecture Notes on Data Engineering and Communications Technologies, vol 10. Springer, Cham. https://doi.org/10.1007/978-3-319-67925-9_9

Bello-Orgaz, G., Jung, J.J., Camacho, D. (2016).Social big data: Recent achievements and new challenges. Information Fusion, 28, 45–59. https://doi.org/10.1016/j.inffus.2015.08.005

Bonyadi, M.R., Michalewicz, Z.(2017). Particle swarm optimization for single objective continuous space problems: A review. Evolutionary Computation, 25, 1–54. https://doi.org/10.1162/EVCO_r_00180

Camacho, D., Panizo-LLedot, A., Bello-Orgaz, G., Gonzalez-Pardo, A., Cambria, E. (2020). The four dimensions of social network analysis: An overview of research methods, applications, and software tools. Information Fusion,63, 1–33. https://doi.org/10.1016/j.inffus.2020.05.009

Cao, G., Sun, Y., Tan, R., Zhang, J., Liu, W.(2021).A function-oriented biologically analogical approach for constructing the design concept of smart product in Industry 4.0.Advanced Engineering Informatics, 49,101352. https://doi.org/10.1016/j.aei.2021.101352

Cheng, S., Zhang, Q., Qin, Q. (2016). Big data analytics with swarm intelligence. Industrial Management and Data Systems, 116(4), 646–666.https://doi.org/10.1108/IMDS-06-2015-0222

Chintapalli, S., Dagit, D., Evans, B., Farivar, R., Graves, T., Holderbaugh, M., Liu, Z., Nusbaum, K., Patil, K., Peng, B.J., Poulosky, P. (2016).Benchmarking streaming computation engines: Storm, flink and spark streaming. In 2016 IEEE International Parallel and Distributed Processing Symposium Workshops (IPDPSW), Chicago, IL, pp. 1789–1792.

Colorni, A., Dorigo, M., Maniezzo, V. (1992).An investigation of some properties of an ant algorithm. In Männer, R., Manderick, B., Eds., Proceedings of the Parallel Problem Solving from Nature Conference (PPSN 92), Brussels, Belgium, September 28–30,1992,Elsevier Publishing, Amsterdam, The Netherlands, pp. 509–520.

Del Ser,J., Osaba, E., Sanchez-Medina, J.J., Fister, I. (2019). Bioinspired computational intelligence and transportation systems: A long road ahead. *IEEE Transactions on Intelligent Transportation Systems*, 21(2), 466–495.

Diez-Olivan, A., Del Ser, J., Galar, D., Sierra, B.(2019).Data fusion and machine learning for industrial prognosis: Trends and perspectives towards industry 4.0.Information Fusion, 50, 92–111. https://doi.org/10.1016/j.inffus.2018.10.005

Goel, A., Hagopian, K., Zhang, S., Rugaber, S. (2022).Towards a virtual librarian for biologically inspired design. In Gero, J.S., Ed., Design Computing and Cognition'20, 2022, Springer, Cham, pp. 369–386. https://doi.org/10.1007/978-3-030-90625-2_21

Jiang, S., Hu, J., Wood, K.L., Luo, J. (2022). Data-driven design-by-analogy: State-of-the-art and future directions.Journal of Mechanical Design, 144(2), 020801.https://doi.org/10.1115/1.4051681

Karimov, J., Rabl, T., Katsifodimos, A., Samarev, R., Heiskanen, H., Markl, V. (2018). Benchmarking distributed stream data processing systems. In 2018 IEEE 34th International Conference on Data Engineering (ICDE), IEEE, Piscataway, NJ, pp 1507–1518.

Lin, K.C., Huang, Y.H., Hung, J.C., Lin, Y.T. (2015).Feature selection and parameter optimization of support vector machines based on modified cat swarm optimization.International Journal of Distributed Sensor Networks, 11(7). https://doi.org/10.1155/2015/365869

Liu, A., Teo, I., Chen, D.,Lu, S., Wuest, T., Zhang, Z., Tao, F. (2019). Biologically inspired design of context-aware smart products. Engineering, 5(4), 637–645. https://doi.org/10.1016/j.eng.2019.06.005

Munir, K., de Ramón-Fernández, A., Iqbal, S., Javaid, N. (2019). Neuroscience patient identification using big data and fuzzy logic—An Alzheimer's disease case study. Expert Systems with Applications, 136, 410–425. https://doi.org/10.1016/j.eswa.2019.06.049

Raff, S.,Wentzel, D., Obwegeser, N. (2020). Smart products: conceptual review, synthesis, and research directions. Journal of Product Innovation Management, 37(5), 379–404. https://doi.org/10.1111/jpim.12544

Raychaudhuri, A., De, D. (2020).Bio-inspired algorithm for multi-objective optimization in wireless sensor network. In De, D., Mukherjee, A., Kumar Das, S., Dey, N. (eds.), Nature Inspired Computing for Wireless Sensor Networks, pp. 279–301. Springer Tracts in Nature-Inspired Computing. Springer, Singapore. https://doi.org/10.1007/978-981-15-2125-6_12

Sfar, A.R., Zied, C., Challal, Y. (2017).A systemic and cognitive vision for IoT security: a case study of military live simulation and security challenges. In 2017 International Conference on Smart, Monitored and Controlled Cities (SM2C), Sfax, Tunisia, February17–19, 2017.

Shukla, A., Chaturvedi, S., Simmhan, Y. (2017). RIoTBench: An IoT benchmark for distributed stream processing systems. Concurrency and Computation: Practice and Experience, 29(21), e4257. https://doi.org/10.1002/cpe.4257

Sun, D., Gao, S., Liu, X., Li, F., Buyya, R. (2020). Performance-aware deployment of streaming applications in distributed stream computing systems. International Journal of Bio-Inspired Computation (IJBIC),15(1), 52–62. https://doi.org/10.1504/IJBIC.2020.105892

Tao, F., Sui, F., Liu, A., Qi, Q., Zhang, M., Song, B., Guo, A., Lu, S.C.Y., Nee, A.Y.C.(2019). Digital twin-driven product design framework. International Journal of Production Research, 57(12), 3935–3953. www.tandfonline.com/doi/full/10.1080/00207543.2018.1443229

Wang, T., Liu, W., Zhao, J., Guo, X., Terzija, V. (2020). A rough set-based bio-inspired fault diagnosis method for electrical substations. International Journal of Electrical Power and Energy Systems, 119, 105961. https://doi.org/10.1016/j.ijepes.2020.105961

Wang, C.J., Ng, C.Y., Brook, R.H. (2020). Response to COVID-19 in Taiwan: Big data analytics, new technology, and proactive testing. Journal of the American Medical Association, 323(14), 1341–1342. https://doi.org/10.1001/jama.2020.3151

Wang, L., Zhan, J., Luo, C., Zhu, Y., Yang, Q., He, Y., Gao, W., Jia, Z., Shi, Y., Zhang, S., Zheng, C., Lu, G., Zhan, K., Li, X., Qiu, B. (2014). Bigdatabench: A big data benchmark suite from internet services. In 2014 IEEE 20th International Symposium on High Performance Computer Architecture (HPCA), IEEE, Orlando, FL, pp. 488–499.

Xu, X., Lu, Y., Vogel-Heuser, B., Wang, L.(2021). Industry 4.0 and Industry 5.0—Inception, conception and perception. Journal of Manufacturing Systems, 61, 530–535. https://doi.org/10.1016/j.jmsy.2021.10.006

Yadav, A., Vishwakarma, D.K. (2020).A comparative study on bio-inspired algorithms for sentiment analysis. Cluster Computing, pp 1–21. Springer, Tucson, AZ.

Yan, H., Wan, J., Zhang, C., Tang, S., Hua, Q., Wang, Z. (2018). Industrial big data analytics for prediction of remaining useful life based on deep learning. IEEE Access, 6, 17190–17197. https://doi.org/10.1109/ACCESS.2018.2809681

Zhou, C., Su, F., Pei, T., Zhang, A., Du, Y., Luo, B., Cao, Z., Wang, J., Yuan, W., Zhu, Y., Song, C., Chen, J., Xu, J., Li, F., Ma, T., Jiang, L., Yan, F., Yi, J., Hu, Y., Liao, Y., Xiao, H. (2020).Covid-19: Challenges to GIS with big data. Geography and Sustainability,1(1), 77–87. https://doi.org/10.1016/j.geo sus.2020.03.005

Chapter 2

Smart data-driven sensing

New opportunities to combat environmental problems

Bhavani K and Gajendra Nagaraj

2.1 INTRODUCTION

2.1.1 Background and significance

Smart sensors are very useful in conservational approach by which one can use the digital technology to monitor, visualize, generate digital data, to control the application of resources, to improve the quality of air, and to combat the environmental challenges. Extracting knowledge or useful insights from these data can be used for smart decision-making in various applications domains. In instrumentation systems, sensors are very essential devices. At present, most of the types of sensors are smart. So, in these sensors, the sensing elements and electronics are integrated on the same chip. Therefore, the integration of electronics and sensors to make an intelligent sensor is known as a smart sensor. This sensor can make some decisions. These sensors have many benefits like higher signal to noise ratio, fast signal conditioning, auto-calibration, self-testing, high reliability, small physical size, detection and prevention of failure. Hence, this article discusses an overview of a smart sensor, its working, and its applications. A smart sensor is a device that uses a transducer to gather particular data from a physical environment to perform a predefined and programmed function on the particular type of gathered data, then it transmits the data through a networked connection [1].Smart sensors are used mainly for monitoring and control mechanisms in different environments like water level and food monitoring systems, smart grids, traffic monitoring and control, environmental monitoring, conserving energy in artificial lighting, monitoring of the remote system, fault diagnostics of equipment, transport and logistics, agriculture, telecommunications, industrial applications, animal tracking, etc. Smart sensors work by capturing data from physical environments and changing their physical properties like speed, temperature, pressure, mass, or presence of humans into calculable electrical signals. These sensors include a Digital Motion Processor (DMP). Here a DMP is one type of microprocessor that allows the sensor to perform onboard processing of the

DOI: 10.1201/9781003530077-2

smart sensor data like filtering noise, otherwise performing different kinds of signal conditioning.

2.1.2 Overview of smart data-driven sensing

Smart sensing is an emerging technology that is used in various applications like healthcare monitoring and industrial applications. Smart sensor consists of the number of sensors that collects input from the physical environment to do a specific task based on the collected inputs. The built environment offers significant opportunities and challenges for the application of such intelligent sensor systems with tangible economic, environmental, and social benefits. The emergence of dense networked embedded sensor systems in monitoring and control applications has enabled the collection of data, information processing, decision, and actuation at previously unseen temporal and spatial resolution and scale. At a larger scale, smart city infrastructures are able to implement these advances toward wider benefits in environmental monitoring, traffic management, improved utilities networks, and social services.

Data-driven technologies using smart sensors can find a solution to many glitches in environmental practices and it could improve the efficiencies of environmental degradation. Data-driven smart solutions have significant potential to improve and advance environmental sustainability in the context of smart sustainable cities, notably smart grid, advanced metering infrastructure, smart buildings, smart home appliances and tools, and smart environmental control and monitoring. There is a clear synergy between these solutions in terms of their interaction to produce combined effects greater than the sum of their separate effects with respect to the environment. Smart grid technologies provide numerous benefits associated with energy use optimization, energy management, energy conservation, cost reduction, as well as the integration of alternative energy sources in power generation, transmission, and distribution systems. They allow bidirectional energy flows and information between suppliers and consumers and provide data on real-time usage and energy pricing. This in turn provides numerous advantages, including real-time visibility, service reliability, control of cost and electricity usage, shift in peak load, capacity requirement of the grid, and energy production and sharing (prosumer). Similarly, smart building technologies have proven to be effective in curbing energy consumption. They can provide a multifunctional role in energy efficiency and greenhouse gas (GHG) emission reductions through highly advanced automatic systems for efficient and natural lighting, temperature control, window and door operation, electric appliances, and many other functions. Building Management System (BMS) allows more efficient operation, keeps the building's climate within a specified range, reduces energy consumption, reduces energy costs, and guarantees safety and security.

2.1.3 Environmental challenges in the modern world

An environment is everything that is around us, which includes both living and nonliving things such as soil, water, animals, and plants, which adapt themselves to their surroundings. It is nature's gift that helps in nourishing life on earth. The environmental challenges in the modern world encompass a range of issues that threaten the health of the planet and its inhabitants. They are complex and multifaceted, affecting ecosystems, biodiversity, human health, and the overall well-being of the planet. Some of the key environmental challenges are climate change, biodiversity loss, pollution, ocean acidification, and loss of ecosystem services. Addressing these challenges requires coordinated global efforts, policy interventions, technological innovations, and changes in human behavior to transition toward more sustainable practices. It involves a balance between economic development and environmental conservation to ensure a healthy planet for current and future generations. Environmental concerns can be defined as the negative effects of any human activity on the environment. The biological as well as the physical features of the environment are included. Some of the primary environmental challenges that are causing great worry are air pollution, water pollution, natural environment pollution, waste pollution, and so on. Some human activities that cause damage (either directly or indirectly) to the environment on a global scale include population growth, neoliberal economic policies and rapid economic growth, overconsumption, overexploitation, pollution, and deforestation. Some of the problems, including global warming and biodiversity loss, have been proposed as representing catastrophic risks to the survival of the human species [2].There is need therefore to work on these environmental problems with an aim of reducing their impacts.

2.1.3.1 Climate change

One of the most significant environmental challenges is climate change. Human activities, such as the burning of fossil fuels and deforestation, have led to an increase in GHG emissions, resulting in global warming and climate change. Climate change is one of the major concerns facing human beings globally. Climate change is a global challenge that stems from a combination of natural forces and human activities. Although we cannot control natural forces, it is within our power to mitigate the impact of human activities that contribute to adverse weather patterns. Climate change has far-reaching and multifaceted effects on the current world. These effects have manifested across various sectors, impacting ecosystems, human societies, and the global economy. The phenomenon has caused shifts in weather patterns, rising sea levels, more frequent and severe extreme weather events, and disruptions to ecosystems. It is perceived that climate change mainly entails higher temperatures. However, the temperature rise is merely the beginning of the narrative. As everything is interconnected in the ecosystem. Thus, shifts in one aspect will equally impact others. Climate change has led to increase in global temperatures, leading to more frequent and intense heatwaves, affecting human health, agriculture, and natural ecosystems. Climate change is directly contributing to humanitarian emergencies from heatwaves, wildfires, floods, tropical storms, and hurricanes, and they are increasing in scale, frequency, and intensity. From shifting weather patterns that threaten food production, to rising sea levels that increase the risk of catastrophic flooding, the impacts of climate change are global in scope and unprecedented in scale. Research shows that 3.6 billion people already live in areas highly susceptible to climate change [3]. The rising temperatures are fueling environmental degradation, natural disasters, weather extremes, food and water insecurity, economic disruption, conflict, and terrorism. Sea levels are rising, the Arctic is melting, coral reefs are dying, oceans are acidifying, and forests are burning. The world is now warming faster than at any point in recorded history. Warmer temperatures over time are changing weather patterns and disrupting the usual balance of nature. A warmer climate is expected to increase the risk of illnesses and death from extreme heat and poor air quality. Climate change will likely increase the frequency and strength of extreme events (such as floods, droughts, and storms) that threaten human health and safety.

2.1.3.2 Resource depletion

Global warming and energy resource depletion have received widespread attention over the past several decades. Understanding the need for mitigating and adapting to current changes in the globe has led to the global urgency to develop and deploy sustainable fuel alternatives. Biofuels have been recognized as a viable energy alternative. Resource depletion is the consumption of a resource faster than it can be replenished. Natural resources are commonly divided between renewable resources and non-renewable resources (see also https://mineralseducationcoalition.org/mining-minerals-information/minerals-database/). Use of either of these forms of resources beyond their rate of replacement is considered to be resource depletion. The value of a resource is a direct result of its availability in nature and the cost of extracting the resource; the more a resource is depleted, the more the value of the resource increases. They have the potential to mitigate current global warming and energy resource depletion issues. Resource depletion occurs when the renewable and non-renewable natural resources become scarce because they are consumed faster than they can recover. Environmentalists are interested in depletion accounting as a way to track the use of natural resources over time, hold governments accountable or compare their environmental conditions to those of another country. Economists want to measure resource depletion to understand how financially reliant countries or corporations are on non-renewable resources, whether this use can be sustained and the financial drawbacks of switching to renewable resources in light of the depleting resources.

2.1.3.3 Emerging issues

For more than 10 years, the United Nations Environment Programme (UNEP) has sought to identify and draw attention to emerging issues of environmental concern. The UNEP Frontiers' report continues to advance this work, signaling environmental issues and solutions for effective and timely responses [3]. Some issues may be local, relatively small-scale issues today, with a potential to become an issue of regional or global concern if not addressed early. The more important environmental problems facing the global community today include climate change, depletion of strato-spheric ozone, the worldwide spread of persistent organic pollutants, loss of biodiversity, and ocean degradation.

2.2 FUNDAMENTALS OF DATA-DRIVEN SENSING

The fundamentals of data-driven sensing refer to the foundational principles and key concepts that underlie the use of data and sensing technologies to collect, analyze, and derive insights from information. These fundamentals collectively form the basis for developing, deploying, and optimizing data-driven sensing systems across various domains, from environmental monitoring and healthcare to smart cities and industrial applications.

2.2.1 Principles of data-driven sensing

Data needs to be transported across information networks before it can be used to create new value. No amount of data will lead to accelerated impact if it is not used to inform decision-making. When an initiative is data driven, quality information is available to the right people when they need it, and they are using those data to take action. The principles of data-driven sensing to combat environmental problems involve leveraging data and advanced sensing technologies to monitor, analyze, and address environmental challenges [4]. Here is a detailed exploration of these principles:

1. **Sensing Technologies:** Sensing technologies refer to the diverse array of tools and devices designed to capture information from the physical world, converting real-world phenomena into measurable data. Some common sensing technologies include environmental sensors and Internet of Things (IoT) devices. The sensing technologies collectively contribute to the development of advanced systems for

data-driven applications [5], facilitating real-time monitoring, automation, and informed decision-making across various domains.

2. **Data Collection:** Data collection is an essential aspect of data-driven sensing, involving the acquisition of information from various sources, such as sensors, devices, or other data-generating mechanisms. It involves selecting appropriate sensors or methods based on the specific goals of the sensing application, ensuring the reliability and accuracy of the gathered information. The effectiveness of data collection directly influences the quality and relevance of insights derived from the sensing process. Emphasizing real-time data collection can provide up-to-the-minute information on environmental conditions that can enable rapid response to changes and emergencies.

3. **Data Integration:** It brings together diverse sources of information to create a unified and comprehensive dataset. The process actually involves combining data from various sensors, devices, or platforms to provide a more holistic view of the environment. Integrating data from various sensing modalities, including environmental sensors, satellite imagery, IoT devices, and other sources, provides a more nuanced understanding of the environment [6]. This multimodal integration enhances the richness of the data. In fact, integration can allow for the recognition of patterns or anomalies in the data that may not be evident when examining individual datasets. Integrated data also enables real-time decision support by providing immediate insights into the current state of the environment that is crucial for scenarios that require rapid response, such as emergency management or pollution control.

4. **Big Data Analytics:** This plays a vital role by providing the tools and techniques to process, analyze, and derive meaningful insights from vast and diverse datasets. As environments are monitored using numerous sensors, devices, and sources, big data analytics enables the extraction of valuable information, patterns, and trends. It facilitates the development of predictive models using machine learning algorithms. The models can forecast future environmental trends, enabling proactive decision-making and resource planning. Big data analytics platforms can incorporate security measures to protect sensitive environmental data from unauthorized access or manipulation.

5. **Decision Support Systems (DSS):** This provides tools and platforms that assist decision-makers in interpreting, analyzing, and utilizing the insights derived from large and complex datasets. In the realm of environmental sensing, DSS enhances the decision-making process by integrating data from various sources, offering visualization tools, and supporting scenario analysis. They can enable real-time

monitoring of environmental conditions, providing decision-makers with immediate insights into changes or critical events. DSS also assists in identifying environmental risks by analyzing data related to factors such as pollution levels, climate patterns, and natural disasters. This information enables proactive risk assessment.

6. **Citizen Science Participation:** This involves engaging the public in the collection, analysis, and interpretation of environmental data. This collaborative approach involves citizens in scientific research, contributing valuable information that complements traditional monitoring efforts. Citizens contribute observations, measurements, and data points using various tools, including mobile apps, sensors, and wearable devices. This collective effort enhances the overall dataset and aids in capturing nuances that may be missed in traditional monitoring approaches. Citizen science also allows for data collection in remote or inaccessible areas where traditional monitoring infrastructure may be limited. Participants can contribute data from their local environments, including rural or wilderness areas, contributing to a more inclusive and diverse dataset. Citizens can bring local knowledge and context to data collection efforts providing local expertise for interpreting data and understanding the specific nuances of environmental conditions in a given region. Participants can quickly share observations and data, providing a timely and responsive approach to monitoring events such as pollution incidents, wildlife sightings, or weather anomalies. It promotes a sense of community, scientific literacy, and active engagement in environmental stewardship.

7. **Open Data Sharing:** It promotes transparency, collaboration, and the widespread availability of environmental data. It involves making datasets accessible to the public, researchers, policymakers, and various stakeholders, fostering a collaborative and informed approach to addressing environmental challenges. The governments, environmental agencies, and institutions sharing their data openly demonstrate a commitment to transparency in their environmental monitoring efforts. This transparency builds trust among the public and stakeholders. Students, educators, and researchers can in turn use environmental datasets for learning, research projects, and enhancing environmental education. Shared data enables coordinated efforts in addressing global challenges such as climate change, biodiversity loss, and pollution. Access to comprehensive environmental datasets helps policymakers make evidence-based decisions on issues such as climate change mitigation, conservation, and pollution control. It supports the development of international strategies and agreements based on a shared understanding of environmental conditions. Shared environmental data also supports resilience planning by providing insights into vulnerabilities and potential risks. Communities

can use this information to develop strategies for adapting to changing environmental conditions. Open data sharing facilitates collaboration between the public and private sectors. Industries can contribute data, and public entities can share information, fostering partnerships to address environmental challenges collectively.

8. **Ethical Considerations:** These involve protecting the privacy of individuals by employing techniques such as data anonymization. Personal identifiers should be removed or masked to prevent the identification of individuals contributing data. Efforts should be made to ensure fairness and equity in the representation of diverse populations. They should ensure that the collection, use, and sharing of environmental data align with principles of fairness, transparency, accountability, and respect for individuals and communities. As technology advances and data collection becomes more prevalent, ethical considerations become paramount in safeguarding privacy, promoting equity, and avoiding potential harms.

9. **Continuous Innovation:** This drives the development and improvement of technologies, methodologies, and approaches used in collecting, analyzing, and interpreting environmental data. Continuous innovation leads to the development of more advanced and accurate sensor technologies that allows for better precision in data collection, enabling more reliable and detailed environmental monitoring. Embracing ongoing technological advancements to enhance the capabilities of environmental sensing systems thus includes adopting new sensor technologies, improving data analysis algorithms, and exploring innovative data visualization techniques. Moreover, investing in research and development addresses emerging environmental challenges and improves the accuracy and efficiency of data-driven sensing technologies.

By adhering to these principles, data-driven sensing becomes a powerful tool for understanding, managing, and mitigating environmental problems, contributing to sustainable and informed decision-making.

2.2.2 Sensors and sensor technologies

Sensors are devices or instruments that detect and measure physical properties or changes in the environment and convert this information into signals or data that can be interpreted, displayed, or used for control purposes. There are various sensor technologies designed to detect different types of physical phenomena. Sensors and sensor technologies capture, measure, and transmit information about various environmental parameters [7]. These technologies enable the collection of real-time data, which is fundamental for understanding environmental conditions, monitoring changes, and making informed decisions. In the context of data-driven environment sensing, sensors are employed to capture information related to factors such as air quality, water quality, temperature, humidity, radiation, and more. Various sensors used in this context are explained below:

1. **Environmental Sensors:** These sensors are designed to measure various environmental parameters such as temperature, humidity, pressure, and atmospheric gases. They are fundamental for understanding weather patterns and climate conditions.
2. **Air Quality Sensors:** These are used to measure concentrations of pollutants such as particulate matter (PM), carbon dioxide (CO_2), ozone (O_3), sulfur dioxide (SO_2), and nitrogen dioxide (NO_2) in the air. Air quality sensors are essential for assessing the impact of pollution on human health.
3. **Water Quality Sensors:** These are designed to monitor parameters like pH, turbidity, dissolved oxygen, and various chemical contaminants in water bodies. Water quality sensors are critical for ensuring the safety of drinking water and assessing the health of aquatic ecosystems.
4. **Biological Sensors:** These sensors detect the presence of biological entities, such as bacteria, viruses, or specific organisms. They are used in applications like monitoring waterborne pathogens or assessing the health of ecosystems.
5. **Radiation Sensors:** These are used to measure radiation levels, including ionizing and non-ionizing radiation. They are essential for monitoring radiation exposure, assessing nuclear safety, and understanding the impact of radiation on the environment.
6. **Noise Sensors:** These measure the intensity of sound in a given environment. Noise sensors are used in urban areas to monitor noise pollution and its potential impact on human well-being.
7. **Light Sensors:** These detect the intensity of light in the environment. They find applications in monitoring natural light levels, assessing the effectiveness of artificial lighting, and studying the impact of light pollution.
8. **Motion Sensors:** These sensors detect movement or changes in position. In the context of environment sensing, motion sensors can be used to monitor wildlife activity, track the movement of objects, or assess changes in the landscape.

Advancements in sensor technologies continue to occur, with ongoing developments in areas such as miniaturization, energy efficiency, and integration with wireless communication technologies, leading to the creation of more sophisticated and versatile sensing devices. These sensors find applications in various industries, including healthcare, automotive, aerospace, environmental monitoring, and consumer electronics. When sensors are used, ensuring the accuracy of sensor measurements becomes crucial for reliable environmental monitoring. Regular calibration will also be needed to maintain precision. Many sensors, when deployed in remote or off-grid locations, make power consumption a significant consideration. As sensors become more connected, ensuring the security of the transmitted data will also be crucial. Thus, the sensing technologies collectively contribute to the development of advanced systems for data-driven applications, facilitating real-time monitoring, automation, and informed decision-making across various domains. The key technologies that play a critical role in this regard are:

1. IoT: Many sensors are part of IoT systems, allowing them to connect to the internet and share data in real time. IoT enables the creation of interconnected sensor networks for comprehensive environmental monitoring.
2. **Wireless Sensor Networks (WSNs):** WSNs consist of spatially distributed sensors that communicate wirelessly. These networks are commonly used for monitoring large areas, enabling data collection from remote or hard-to-reach locations.
3. **Nanotechnology:** Nanoscale sensors are designed to operate at the molecular or atomic level, allowing for highly sensitive measurements. They are used for detecting trace amounts of pollutants and contaminants.
4. **Remote Sensing:** Technologies such as satellite-based sensors, aerial surveys, and drones provide a broader perspective on environmental conditions. Remote sensing allows for large-scale and continuous monitoring of landscapes, oceans, and atmospheric conditions.
5. **Machine Learning and Sensor Fusion:** Advanced analytics, including machine learning algorithms, are applied to sensor data for improved analysis and decision-making. Sensor fusion combines data from multiple sensors to enhance the accuracy and reliability of measurements.
6. **Energy Harvesting:** To address power supply challenges in remote locations, some sensors use energy harvesting technologies to generate power from ambient sources such as sunlight, vibrations, or temperature differentials.

Encryption and secure communication protocols need to be implemented to protect against unauthorized access. Handling large volumes of data generated by sensors requires integration with databases, cloud platforms, and analytics tools for effective data utilization. In conclusion, sensors and sensor technologies are foundational elements in the realm of data-driven environment sensing. Their continuous innovation and integration into sophisticated monitoring systems contribute to a deeper understanding of environmental dynamics, enabling more effective decision-making for sustainable and resilient ecosystems.

2.2.3 Data collection and processing techniques

Data collection and processing techniques are pivotal components in the realm of data-driven environment sensing. These techniques involve the systematic gathering of environmental data and the application of various methods to transform raw data into meaningful information. The integration of advanced technologies and methodologies enables a comprehensive understanding of environmental conditions. Here is an elaboration on data collection and processing techniques in the context of data-driven environment sensing:

- Deploying networks of sensors across an area to capture real-time data on environmental parameters. WSNs and IoT devices contribute to large-scale and remote monitoring.
- Utilizing satellites to capture high-resolution images of the Earth's surface, atmosphere, and oceans. Remote sensing from space provides a broad and continuous view of environmental conditions.
- Using drones or aircraft to capture detailed images and data for specific regions. Aerial surveys are particularly useful for localized and high-resolution monitoring.
- Engaging the public in data collection efforts, allowing citizens to contribute observations, measurements, and photographs. This approach enhances data density and community involvement.
- Utilizing mobile applications that enable individuals to report environmental observations. These apps often include features for geotagging, allowing for spatial mapping of data.
- Implementing IoT sensors and devices in urban environments to monitor air quality, traffic, waste management, and other parameters. This contributes to sustainable urban planning.
- Employing IoT devices and sensors in precision agriculture to monitor soil conditions, crop health, and irrigation needs, optimizing resource use [7].
- Utilizing Global Positioning System (GPS) and Global Navigation Satellite System (GNSS) data for tracking and mapping environmental parameters. This is crucial for understanding movement patterns and changes over time.

Data Processing Techniques involve identifying and rectifying errors, outliers, or missing values in the raw data. Quality control ensures the reliability and accuracy of the dataset. Analyzing data in relation to geographic location. Geographic Information System (GIS) techniques are used to understand spatial patterns and relationships.

1. **Temporal Analysis:** The studying data trends and changes over time. Time-series analysis helps identify patterns, anomalies, and seasonality in environmental parameters.
2. **Machine Learning and AI:**
   - **Classification and Regression:** Using machine learning algorithms to classify environmental data into categories or predict numerical values. This is applied, for instance, in land cover classification or predicting pollutant levels.
   - **Clustering and Anomaly Detection:** Employing clustering algorithms to group similar environmental data points. Anomaly detection identifies irregularities that may indicate environmental disturbances.
3. **Statistical Analysis:**
   - **Descriptive Statistics:** Calculating measures such as mean, median, and standard deviation to summarize and describe the central tendencies of data. This provides an overview of data distribution.
   - **Hypothesis Testing:** Applying statistical tests to validate hypotheses about environmental parameters. This is important for drawing meaningful conclusions from the data.
4. **Data Fusion:**
   - **Sensor Fusion:** Combining data from multiple sensors to improve accuracy and reliability. Sensor fusion techniques enhance the overall quality of environmental monitoring.
   - **Source Integration:** Integrating data from various sources, including satellite imagery, sensor networks, and citizen science contributions. This allows for a more comprehensive understanding of the environment.
5. **Visualization and Reporting:**
   - **Graphs, Maps, and Dashboards:** Creating visual representations of data to facilitate interpretation. Graphs, maps, and dashboards enhance communication and decision-making based on environmental insights.
   - **Environmental Impact Assessment (EIA):** Producing comprehensive reports that assess the potential environmental impacts of specific activities or projects. EIA involves synthesizing diverse data types for decision support.
6. **Big Data Technologies:**
   - **Data Storage and Retrieval:** Utilizing big data storage solutions to manage large volumes of environmental data. Distributed databases and cloud-based storage enhance scalability and accessibility.

- **Parallel Computing:** Employing parallel processing techniques to analyze vast datasets efficiently. This accelerates computation and supports real-time or near-real-time analysis.

7. **Ethical Considerations:**
   - **Privacy Protection:** Implementing measures to safeguard the privacy of individuals contributing data. This includes anonymization, consent mechanisms, and secure data transmission.
   - **Transparency and Accountability:** Ensuring transparent communication about data collection and processing methods. Accountability measures are in place to address any misuse or misinterpretation of data.

In conclusion, the effective integration of data collection and processing techniques is essential for deriving meaningful insights from the vast and diverse environmental datasets generated by modern sensing technologies. These techniques contribute to evidence-based decision-making, sustainable resource management, and the overall advancement of environmental science.

2.3 SMART TECHNOLOGIES IN ENVIRONMENTAL MONITORING

2.3.1 Internet of Things applications

The IoT refers to the network of interconnected devices embedded with sensors, software, and other technologies, enabling them to collect and exchange data. IoT applications span a wide range of industries, enhancing efficiency, convenience, and decision-making processes. The IoT has significantly transformed the field of environmental monitoring by providing a network of interconnected devices capable of collecting, transmitting, and analyzing real-time data. IoT applications in environmental monitoring offer a range of benefits, including increased efficiency, precision, and accessibility of environmental data. Here are several key applications of IoT in environmental monitoring:

1. **Air Quality Monitoring:**
   - **Sensor Networks:** IoT-enabled air quality monitoring systems use networks of sensors to measure concentrations of pollutants such as PM, NO_2, SO_2, and O_3.
   - **Real-Time Data:** Continuous monitoring and real-time data transmission enable quick responses to changes in air quality, allowing for timely interventions and public awareness.

2. **Water Quality Management:**
   - **Sensor Deployments:** IoT devices equipped with water quality sensors are deployed in bodies of water to measure parameters like pH, turbidity, dissolved oxygen, and chemical contaminants.
   - **Early Warning Systems:** Real-time monitoring allows for the early detection of water pollution events, helping authorities implement rapid responses and safeguard water resources.
3. **Smart Agriculture:**
   - **Precision Farming:** IoT applications in agriculture involve sensors that monitor soil moisture, temperature, and nutrient levels, optimizing irrigation and fertilization practices.
   - **Crop Health Monitoring:** IoT devices equipped with cameras and sensors monitor crop health, detect diseases, and assess the overall condition of crops.
4. **Waste Management:**
   - **Smart Bins:** IoT-connected waste bins equipped with sensors monitor fill levels. This data is used to optimize waste collection routes, reduce operational costs, and enhance efficiency.
   - **Environmental Impact Assessment:** Monitoring waste composition and its impact on the environment helps in designing sustainable waste management strategies.
5. **Climate Monitoring and Prediction:**
   - **Weather Stations:** IoT-enabled weather stations collect data on temperature, humidity, wind speed, and atmospheric pressure.
   - **Data Analytics:** Continuous monitoring and analysis of climate data contribute to improved weather predictions, helping authorities and communities prepare for extreme weather events.
6. **Wildlife Conservation:**
   - **Animal Tracking:** IoT devices, such as GPS-enabled collars, are used to track the movement of wildlife, providing valuable insights into migration patterns and habitat use.
   - **Monitoring Endangered Species:** Sensors help monitor the health and behavior of endangered species, supporting conservation efforts.
7. **Noise Pollution Monitoring:**
   - **Smart Cities Initiatives:** IoT sensors measure noise levels in urban areas, allowing for the identification of noise pollution hotspots.
   - **Public Awareness:** Real-time data on noise pollution can be shared with the public, raising awareness and promoting community engagement in noise reduction efforts.
8. **Smart Forestry:**
   - **Forest Fire Detection:** IoT devices equipped with sensors detect signs of forest fires, allowing for rapid response and containment.

- **Environmental Data Logging:** Sensors in forests collect data on temperature, humidity, and soil conditions, aiding in sustainable forestry practices.

9. **Ocean and Marine Monitoring:**
   - **Buoy Networks:** IoT-connected buoys equipped with sensors monitor ocean conditions, including temperature, salinity, and sea level.
   - **Marine Life Tracking:** IoT devices are used to track marine animals, study migration patterns, and monitor the impact of human activities on marine ecosystems.

In summary, IoT applications in environmental monitoring contribute to a more connected and data-driven approach to understanding and managing the environment. These technologies empower decision-makers, researchers, and communities with real-time information, facilitating sustainable practices and enhancing overall environmental stewardship.

2.3.2 Remote sensing and satellite technology

Remote sensing and satellite technology play a pivotal role in environmental monitoring, providing valuable tools for collecting data on Earth's land, oceans, atmosphere, and ecosystems from a distance. These technologies enable comprehensive and continuous observation of the planet, offering insights into environmental changes, natural processes, and human activities [8]. Here is an overview of the key aspects of remote sensing and satellite technology in environmental monitoring:

1. Remote Sensing Technologies :
   a. Electromagnetic Spectrum:
      - Remote sensing instruments capture data across various portions of the electromagnetic spectrum, including visible, infrared, and microwave wavelengths.
      - Different regions of the spectrum provide information about specific environmental features, such as vegetation health, land cover, and sea surface temperatures.
   b. **Passive and Active Remote Sensing:**
      - **Passive Sensors:** Detect natural radiation emitted or reflected by the Earth's surface. Examples include optical and thermal infrared sensors.
      - **Active Sensors:** Transmit energy and measure the reflected or emitted signals. Radar and light detecting and ranging (lidar) are examples of active sensors.

    c. **Optical Remote Sensing:**
- Utilizes visible and near-infrared light to capture images and information about the Earth's surface.
- Applications include land cover mapping, vegetation monitoring, and assessment of coastal and aquatic environments.

    d. **Thermal Infrared Remote Sensing:**
- Measures the heat emitted by Earth's surface.
- Useful for assessing land surface temperatures, identifying thermal anomalies, and monitoring volcanic activity.

    e. **Radar Remote Sensing:**
- Uses microwave signals to penetrate clouds and provide all-weather imaging.
- Applications include terrain mapping, monitoring ice and snow cover, and detecting changes in land surface elevation.

    f. **Lidar:**
- Employs laser beams to measure distances and create high-resolution 3D maps.
- Used for topographic mapping, forest structure analysis, and urban planning.

2. **Satellite Technology:**

    a. **Earth Observation Satellites:**
- Dedicated satellites equipped with various sensors for environmental monitoring.
- Examples include optical imaging satellites (e.g., Landsat), synthetic aperture radar (SAR) satellites, and weather satellites.

    b. **Constellations and Orbits:**
- Satellite constellations provide global coverage and frequent revisits to specific areas.
- Different orbits, such as polar, geostationary, and sun-synchronous, serve specific monitoring purposes.

    c. **High-Resolution Imaging:**
- Advanced optical and radar satellites offer high-resolution images, enabling detailed mapping and monitoring of small-scale environmental features.
- Beneficial for urban planning, agriculture, and disaster response.

    d. **Data Fusion and Multi-Sensor Integration:**
- Combining data from multiple sensors and satellites enhances the richness and accuracy of environmental information.
- Data fusion improves the understanding of complex environmental processes.

e. Satellite-Based Atmospheric Monitoring:
- Weather satellites provide continuous monitoring of atmospheric conditions, including cloud cover, precipitation, and atmospheric composition.
- These satellites contribute to weather forecasting, climate studies, and disaster management.

f. Ocean and Marine Monitoring:
- Satellite altimeters measure sea surface heights, aiding in ocean circulation and climate studies.
- Ocean color sensors assess phytoplankton concentrations and water quality.

g. Deforestation and Land Use Change:
- Satellite data helps monitor deforestation, land use changes, and urban expansion.
- Essential for conservation efforts, biodiversity assessment, and sustainable land management.

h. Natural Disaster Monitoring:
- Satellites provide rapid assessment of natural disasters, such as hurricanes, wildfires, and earthquakes.
- Enables timely response and supports disaster management and recovery efforts.

i. Climate Change Monitoring:
- Satellite observations contribute to monitoring climate variables, including temperature, sea level rise, and ice melt.
- Facilitates the assessment of long-term trends and impacts of climate change.

3. **Applications in Environmental Monitoring:**

a. **Precision Agriculture:** Satellite data assists farmers in optimizing crop management, monitoring soil conditions, and assessing crop health.

b. **Water Resource Management:** Satellite imagery helps monitor water bodies, track changes in water levels, and assess water quality.

c. **Urban Planning:** Satellite technology supports urban planners in monitoring urban growth, infrastructure development, and environmental impacts.

d. **Ecological Monitoring:** Enables the monitoring of ecosystems, biodiversity, and changes in vegetation cover over time.

e. **Disaster Response and Recovery:** Rapid satellite imaging aids in assessing the extent of natural disasters and supports emergency response and recovery efforts.

f. **Global Climate Studies:** It contributes to the understanding of global climate patterns, GHG concentrations, and the impact of climate change.

g. **Environmental Impact Assessment:** Satellite data assists in assessing the environmental impact of human activities, infrastructure projects, and industrial processes.

h. **Illegal Logging and Fisheries Monitoring:** Satellites help monitor and combat illegal logging activities in forests and illegal fishing practices in oceans.

In conclusion, remote sensing and satellite technology provide a comprehensive and powerful means of monitoring and understanding Earth's environment. These technologies contribute essential data for scientific research, policymaking, and sustainable management of natural resources and ecosystems. The continuous advancement of satellite technology enhances our ability to address environmental challenges and promote the well-being of our planet.

2.3.3 Blockchain and decentralized environmental data management

Blockchain technology, initially designed for secure and transparent financial transactions using cryptocurrencies, has found applications beyond finance, and one notable area is in environmental data management. Blockchain, a decentralized and distributed ledger, offers a secure and transparent way to record and verify transactions. When applied to environmental data management, blockchain can address challenges related to data integrity, trust, and decentralized collaboration [9]. Here is an elaboration on how blockchain and decentralized approaches are utilized in environmental data management:

1. Data Integrity and Transparency:
    - **Immutable Records:** Blockchain's key feature is immutability. Once data is recorded on the blockchain, it cannot be altered or deleted. This ensures the integrity of environmental data, preventing tampering or manipulation.
    - **Transparent Transactions:** Every transaction or data entry on the blockchain is visible to all participants in the network. This transparency builds trust among stakeholders, as they can independently verify the accuracy of the recorded data.
2. Decentralization and Trust:
    - **Decentralized Architecture:** Traditional environmental data management systems often rely on centralized databases, which can be vulnerable to hacking or manipulation. Blockchain operates

on a decentralized network of nodes, reducing the risk of a single point of failure.

- **Trustless Collaboration:** The decentralized nature of blockchain allows multiple parties, including government agencies, researchers, and businesses, to collaborate without relying on a central authority. Each participant has equal access to the data and can trust the information recorded on the blockchain.

3. **Smart Contracts for Automation:**
   - **Self-Executing Contracts:** Smart contracts, programmable scripts on the blockchain, can automate certain actions when predefined conditions are met. In environmental data management, smart contracts can automate data validation, trigger alerts for anomalies, or execute predefined actions based on environmental conditions.
   - **Efficiency and Accuracy:** Automation through smart contracts reduces manual intervention, minimizing the risk of human errors and ensuring the efficiency and accuracy of data-related processes.

4. **Data Ownership and Privacy:**
   - **User Control:** Blockchain enables individuals or organizations to have greater control over their data. Participants own cryptographic keys that control access to their data, determining who can view or update specific information.
   - **Privacy Preservation:** Although blockchain transactions are transparent, participants' identities are often pseudonymous. This pseudonymity, combined with cryptographic techniques, helps preserve the privacy of individuals or organizations contributing data.

5. **Supply Chain Traceability:**
   - **Provenance Tracking:** Blockchain facilitates the tracking of environmental data throughout the supply chain. For instance, in the food industry, it can be used to trace the origin and journey of products, ensuring transparency and authenticity.
   - **Anti-Counterfeiting Measures:** By recording key environmental indicators on the blockchain, it becomes more challenging for malicious actors to counterfeit or manipulate environmental data associated with products or supply chains.

6. **Tokenization and Incentives:**
   - **Token-Based Systems:** Some blockchain implementations introduce tokens that represent environmental assets or contributions. These tokens can be traded or redeemed for specific benefits, creating economic incentives for environmentally friendly actions.
   - **Carbon Credits and Offsets:** Blockchain can be used to create transparent and tradable carbon credits, allowing organizations

to offset their carbon footprint by investing in environmentally friendly projects.

7. **Interoperability and Standardization:**
   - **Cross-Platform Compatibility:** Blockchain's ability to facilitate interoperability enables seamless integration with existing systems and databases. This promotes standardization in environmental data formats and protocols.
   - **Collaborative Ecosystems:** Multiple stakeholders using different technologies can contribute to and access the same blockchain-based environmental data ecosystem, fostering collaboration and data exchange.

8. **Resilience to Attacks:**
   - **Security Measures:** Blockchain's cryptographic features enhance the security of environmental data. It is resistant to various cyber-attacks, ensuring the reliability and availability of data, even in the face of adversarial activities.
   - **Distributed Consensus:** Consensus mechanisms in blockchain networks, such as proof-of-work or proof-of-stake, ensure that data changes are only accepted when agreed upon by the majority of participants.

9. **Regulatory Compliance:**
   - **Auditable Records:** Blockchain provides auditable records of all transactions or data entries. This feature is valuable for regulatory compliance in environmental monitoring, where accurate and traceable records are essential.
   - **Tamper-Evident Records:** Regulatory bodies can easily verify that environmental data has not been tampered with since its original recording, enhancing trust in compliance reporting.

10. **Global Collaboration for Climate Action:**
    - **International Cooperation:** Blockchain can facilitate international collaboration on climate-related initiatives by providing a secure and transparent platform for sharing environmental data among countries.
    - **Enabling Global Agreements:** Blockchain's features can support the implementation and verification of commitments made in global environmental agreements, such as those related to climate change mitigation and biodiversity conservation.

In summary, blockchain and decentralized environmental data management offer a robust framework for ensuring data integrity, transparency, and collaboration in monitoring and addressing environmental challenges. These technologies provide a foundation for building trust among diverse stakeholders and promoting sustainable practices in the management of environmental data.

2.4 CASE STUDIES: SUCCESSFUL IMPLEMENTATIONS

2.4.1 Air quality monitoring in urban areas

Case Study: "Clean Air India: Real-Time Air Quality Monitoring for Sustainable Urban Living"

Rapid urbanization and industrial growth in India have led to escalating concerns about air quality and its impact on public health. The "Clean Air India" project was initiated in a major urban center to implement a real-time air quality monitoring system using advanced technologies [10]. At the core of Clean Air India was the strategic deployment of cutting-edge air quality sensors in key urban locations. These sensors measured various pollutants such as PM, NO_2, SO_2, CO, and O_3. Positioned strategically in areas prone to pollution, the sensors continuously collected real-time data, creating a comprehensive and dynamic picture of the air quality in urban environments.

Clean Air India placed a strong emphasis on public awareness and engagement. The real-time air quality data collected was made accessible to the public through online platforms, mobile applications, and public displayed in urban centers. This transparency not only informed residents about the current air quality but also raised awareness about the sources and impact of air pollution. Citizens were encouraged to actively engage with the data, fostering a sense of collective responsibility for the air they breathe.

Clean Air India actively collaborated with policymakers, urban planners, and environmental agencies to influence air quality regulations and urban planning strategies. The real-time data generated by the initiative served as a crucial resource for evidence-based policymaking. By identifying pollution hotspots and contributing to the formulation of air quality standards, Clean Air India targeted to shape sustainable urban development practices.

While Clean Air India has achieved notable success, challenges remain, including the need for widespread sensor deployment, addressing issues of data accuracy and ensuring inclusivity in data accessibility. The initiative is poised for future advancements, exploring innovations such as satellite-based monitoring and expanding its reach to cover more urban areas. Ongoing efforts are directed toward enhancing the reliability and scalability of the sensing infrastructure.

Clean Air India stands as a beacon of progress in the pursuit of sustainable urban living through real-time air quality monitoring. By combining advanced sensing technologies, public engagement, health considerations, and policy advocacy, this initiative contributed to the creation of healthier and more resilient cities in India. Clean Air India exemplified the transformative power of real-time data in addressing pressing environmental challenges and underscores the importance of collective action for the well-being of urban communities.

2.4.2 Water pollution detection and prevention

Case Study: "Clean Water India: Real-Time Water Pollution Detection and Prevention for Sustainable Aquatic Ecosystem"

Water pollution poses significant challenges to India's aquatic ecosystems, affecting both environmental health and public well-being. The "Clean Water India" project was initiated to implement a real-time water pollution detection and prevention system [11]. This case study explores the objectives, implementation strategies, outcomes, and challenges encountered during the Clean Water India initiative.

At the heart of Clean Water India is the deployment of state-of-the-art water quality sensors across key water bodies. These sensors are equipped to measure various parameters, including chemical pollutants, dissolved oxygen levels, pH, and temperature. Placed strategically at critical points, these sensors continuously collect real-time data, providing a comprehensive understanding of the dynamic water quality conditions.

The real-time data generated by the water quality sensors enables early detection of pollution incidents. Machine learning algorithms analyze the data patterns and trigger alerts in case of abnormal variations, allowing authorities and environmental agencies to respond rapidly to potential pollution events. This proactive approach is instrumental in minimizing the impact of pollution on aquatic ecosystems and human communities dependent on these water sources.

Clean Water India places a strong emphasis on community involvement and awareness. Local communities living near water bodies are actively engaged in the project, from the installation and maintenance of sensors to the interpretation of data. Awareness campaigns and educational programs are conducted to inform residents about the importance of water quality, the potential sources of pollution, and the collective role they play in preserving their water resources.

The initiative integrates real-time water quality data with GIS, providing a spatial context to pollution monitoring. GIS mapping allows for the visualization of pollution hotspots, helping authorities target interventions more effectively. The integration of GIS enhances the overall effectiveness of pollution detection and prevention strategies.

Clean Water India real-time data and insights contribute to evidence-based policymaking in the realm of water resource management. The project collaborates with government agencies and policymakers, offering valuable inputs for the formulation of water quality standards and regulations. The data generated by the initiative has the potential to influence regional water policies, emphasizing the importance of preventive measures for sustainable aquatic ecosystems.

Although Clean Water India has demonstrated significant success, challenges persist. Ensuring the scalability and affordability of water quality sensors for widespread deployment remains a priority. Additionally, addressing issues related to data privacy and community participation requires ongoing efforts. The initiative looks forward to refining its approach, exploring emerging technologies, and expanding its geographical reach to further protect India's water resources.

Clean Water India stands as a beacon of innovation in the domain of real-time water pollution detection and prevention. By integrating advanced sensing technologies, community engagement, and policy advocacy, this initiative not only safeguards the health of aquatic ecosystems but also fosters a collective commitment to sustainable water management in India. Clean Water India serves as a model for leveraging technology for environmental conservation and underscores the critical role of real-time data in ensuring the sustainability of vital water resources.

2.4.3 Soil and land management

Case Study: Green Fields India: Real-Time Soil and Land Management for Sustainable Agriculture

The Green Fields India initiative showcased the transformative potential of real-time soil and land management for sustainable agriculture. By combining technology, community engagement, and precision farming techniques, the initiative not only enhanced agricultural productivity but also contributed to the long-term health of the soil and surrounding ecosystems. The case study provides valuable insights for scaling similar initiatives across India and other regions facing agricultural and environmental challenges [12].

At the core of Green Fields India is the strategic deployment of advanced sensing technologies across agricultural landscapes. These technologies include a network of soil sensors equipped with probes to measure critical parameters such as moisture content, nutrient levels, and pH. These sensors, placed at key locations within farmlands, operate in real time, providing a continuous stream of data that offers unparalleled insights into the dynamic nature of soil conditions.

The real-time data generated by the soil sensors forms the foundation for implementing precision agriculture practices. Integrated with satellite imagery and weather forecasts, this data allows farmers to make informed decisions tailored to the specific needs of their fields. The precision agriculture techniques, including optimized irrigation schedules, targeted fertilization, and crop management strategies, contribute to resource efficiency, reduced environmental impact, and improved overall farm productivity.

Green Fields India incorporates sophisticated DSS designed to empower farmers with actionable insights. These systems, driven by data analytics and machine learning algorithms, analyze the real-time sensor data alongside historical patterns. The result is a personalized set of recommendations for farmers, guiding them on optimal farming practices. For instance, the system may advise on the precise timing and quantity of irrigation, fostering water conservation efforts while maximizing crop yields.

While Green Fields India has demonstrated significant success, challenges persist. Ensuring the affordability and accessibility of sensing technologies for all farmers, especially those with smaller landholdings, remains a priority. Addressing concerns related to data privacy and ownership is an ongoing consideration. The initiative looks toward the future with a commitment to refining its approach, integrating emerging technologies, and expanding its impact to further regions.

In conclusion, Green Fields India stands as a beacon of innovation in the domain of real-time soil and land management for sustainable agriculture. By seamlessly integrating advanced sensing technologies, precision agriculture practices, and community engagement, this initiative not only enhances farm productivity but also lays the foundation for a more resilient and environmentally conscious agricultural landscape in India.

2.5. CHALLENGES AND LIMITATIONS

Data-driven environmental sensing offers immense potential for monitoring and managing the environment, but it comes with several challenges and limitations. Addressing the challenges requires collaborative efforts from researchers, policymakers, technologists, and the public to develop innovative solutions and frameworks that promote the responsible use of data-driven environmental sensing technologies [13]. Data-driven environmental sensing faces several technological barriers that can impact the successful implementation and effectiveness of monitoring systems.

2.5.1 Data security and privacy concerns

The proliferation of data-driven environmental sensing technologies brings forth significant concerns related to data security and privacy. As environmental sensing systems collect and transmit vast amounts of sensitive information, ensuring the protection of this data from unauthorized access and potential misuse becomes a paramount challenge.

One of the primary concerns is the safeguarding of environmental data from cyber threats. As data-driven sensing relies on interconnected networks and systems, the risk of unauthorized access, data breaches, and cyber-attacks increases. Ensuring the integrity and confidentiality of the collected data is critical, especially when dealing with information that may have implications for public health, environmental regulations, or strategic decision-making.

Privacy issues further complicate the landscape of data-driven environmental sensing. The granularity of data collected, including location-specific information and personal identifiers, raises ethical questions regarding the privacy of individuals. Balancing the need for detailed environmental insights with the protection of individual privacy becomes a delicate task, requiring the implementation of robust privacy-preserving measures.

In many instances, environmental sensing initiatives involve citizen participation and the collection of data from decentralized sources. While citizen engagement is valuable for expanding data coverage, it introduces challenges related to data privacy. Individuals contributing data may inadvertently disclose personal information, necessitating careful consideration of anonymization and aggregation techniques to protect their identities.

The advent of stringent data protection regulations, such as the General Data Protection Regulation (GDPR) in the European Union, has further emphasized the importance of addressing privacy concerns. Compliance with these regulations requires organizations to implement transparent data practices, secure storage and transmission protocols, and employ mechanisms for obtaining informed consent from individuals contributing data.

To mitigate data security and privacy concerns, environmental sensing initiatives must adopt robust encryption methods, authentication protocols, and access control mechanisms. Implementing secure data storage practices and employing anonymization techniques are also essential components of a comprehensive security strategy. Moreover, fostering public awareness and understanding of how data is collected, used, and protected can enhance transparency and build trust among individuals contributing to environmental sensing efforts.

In summary, as data-driven environmental sensing continues to evolve, addressing data security and privacy concerns is imperative for the responsible and ethical deployment of sensing technologies. By implementing rigorous security measures, adhering to privacy regulations, and fostering transparency, organizations can strike a balance between harnessing the benefits of environmental data and safeguarding the rights and privacy of individuals contributing to these initiatives.

2.5.2 Technological barriers

The effective implementation of data-driven environmental sensing initiatives encounters various technological barriers that impact the quality and efficiency of sensing systems. One primary challenge lies in the limitations of environmental sensors. These sensors may face issues related to resolution and accuracy, affecting the precision of the collected data. Additionally, ensuring compatibility and interoperability among diverse sensor types proves challenging, hindering seamless integration within sensing networks.

The sheer volume and velocity of data generated by environmental sensors present another significant technological barrier. Dealing with the vast amount of data, often in real-time applications, poses challenges in terms of storage, processing, and analysis. This big data challenge requires robust computational capabilities and efficient data transmission infrastructure, especially in remote or widespread sensing networks.

The complexity of data integration is a key technological hurdle. Environmental data comes from heterogeneous sources with varying formats, making integration a complex task. Moreover, the interdisciplinary nature of environmental sciences, combining data from fields such as meteorology, biology, and chemistry, adds an additional layer of complexity to the integration process.

Computational resources represent a critical bottleneck. Real-time analysis and complex modeling demand significant processing power, which may be limited in certain environments. The associated infrastructure costs, including investments in high-performance computing, can pose financial challenges for organizations with budget constraints.

Addressing technological barriers in data-driven environmental sensing also involves grappling with cybersecurity concerns. Protecting sensitive environmental data from unauthorized access and cyber threats is crucial. Striking a balance between harnessing data-driven insights and safeguarding individual privacy, particularly with personal or location-specific data, is an ongoing challenge.

Energy efficiency emerges as a technological barrier, particularly in remote or off-grid locations. Many sensors rely on a continuous power source, and optimizing energy consumption is essential. Integrating sustainable energy sources to power sensing infrastructure becomes a complex task, necessitating innovative solutions.

Remote sensing introduces additional challenges, including communication delays in inaccessible areas. Maintaining and repairing sensing equipment in remote locations pose logistical challenges that need to be addressed for sustained functionality. Overall, overcoming these technological barriers requires a multidimensional approach involving research and development, strategic investments, collaboration among stakeholders, and adherence to industry standards. Successfully navigating these challenges is essential for unlocking the full potential of data-driven environmental sensing in providing actionable insights for environmental management.

2.5.3 Ethical and regulatory challenges

The rapid advancement of data-driven environmental sensing technologies raises a spectrum of ethical challenges that necessitate careful consideration. One prominent concern revolves around the ethical implications of collecting and utilizing vast amounts of data, often including sensitive information [14]. Striking a balance between the benefits of enhanced environmental insights and protecting individual privacy is a complex ethical challenge. As sensing systems become more sophisticated and interconnected, ensuring that data collection and usage align with ethical standards becomes imperative.

Incorporating citizen participation in environmental sensing introduces ethical considerations related to informed consent and community engagement. Ethical dilemmas may arise when individuals contribute personal or location-specific data without a full understanding of how the data will be used. Implementing transparent communication strategies, obtaining informed consent, and providing clear avenues for individuals to opt-out are crucial components in addressing these ethical concerns and fostering a sense of trust among participants.

Regulatory frameworks also play a pivotal role in shaping the ethical landscape of data-driven environmental sensing. The absence of standardized regulations specific to environmental data can pose challenges in ensuring consistent ethical practices across different initiatives. As a result, navigating the ethical terrain requires adherence to broader data protection and privacy regulations, such as the GDPR in the European Union, and adapting these principles to the unique challenges presented by environmental sensing [15].

The sharing and collaboration of data among diverse stakeholders, including government agencies, research institutions, and industry partners, introduce regulatory challenges. Establishing effective data-sharing agreements that balance the need for collaboration with individual data rights is critical. Regulatory frameworks must evolve to facilitate responsible data sharing while safeguarding against potential misuse or unintended consequences.

In summary, addressing the ethical and regulatory challenges in data-driven environmental sensing requires a holistic approach. Ethical considerations should be integrated into the design, implementation, and communication of sensing initiatives. Regulatory frameworks need to adapt to the dynamic nature of environmental sensing, providing clear guidelines and standards. Striking the right balance between technological advancement, data utilization, and ethical principles is crucial to harnessing the full potential of data-driven environmental sensing responsibly and sustainably.

2.6. RECOMMENDATIONS AND BEST PRACTICES

2.6.1 Strategies for implementing data-driven sensing solutions

Implementing effective data-driven sensing solutions in environmental monitoring requires a comprehensive approach that encompasses technology, collaboration, and strategic planning. One key strategy involves leveraging advanced sensor technologies capable of collecting high-quality, real-time data. Investing in state-of-the-art sensors ensures the accuracy and reliability of environmental measurements, laying a solid foundation for data-driven insights.

Interdisciplinary collaboration stands as a critical strategy in the implementation of data-driven sensing solutions. Establishing partnerships between government agencies, research institutions, industry stakeholders, and the public fosters a collaborative ecosystem. This collaboration enables the integration of diverse data sources, combining insights from fields such as meteorology, biology, and chemistry to create a more holistic understanding of environmental conditions.

The development and adoption of standardized data formats and protocols play a crucial role in facilitating interoperability and seamless integration. Establishing industry-wide standards ensures that data from different sensors and sources can be easily harmonized, enhancing the efficiency of data-driven environmental sensing initiatives. These standards also contribute to data sharing and collaboration among diverse stakeholders.

Investing in robust computational infrastructure is essential for processing and analyzing large datasets in real time. High-performance computing resources enable the application of advanced analytics and modeling techniques, providing actionable insights for decision-making. Strategic planning for the scalability of computational resources ensures that the system can handle the increasing volume of data generated over time.

Cybersecurity measures form a fundamental strategy in safeguarding sensitive environmental data. Implementing encryption, authentication, and access control mechanisms helps protect against unauthorized access and potential cyber threats. Regular audits and updates to cybersecurity protocols ensure that data remains secure throughout the sensing infrastructure.

Ethical considerations should be embedded into the strategy for data-driven sensing solutions. Developing clear and transparent communication strategies, obtaining informed consent from participants, and implementing privacy-preserving measures contribute to the ethical conduct of environmental sensing initiatives. Engaging with the public through awareness campaigns and educational programs builds trust and encourages responsible participation.

Continuous monitoring and evaluation of sensing systems are integral to their success. Implementing feedback loops for system performance, data quality, and user satisfaction allows for adaptive improvements. Regular assessments ensure that the sensing infrastructure remains effective, efficient, and aligned with the evolving needs of environmental monitoring.

In conclusion, the successful implementation of data-driven sensing solutions in environmental monitoring requires a multifaceted strategy that integrates cutting-edge technology, collaboration, standardization, computational infrastructure, cybersecurity, ethical considerations, and continuous improvement. By adopting these strategies, organizations can harness the full potential of data-driven environmental sensing to address challenges, make informed decisions, and contribute to sustainable environmental management.

2.6.2 Collaboration between government, industry, and academia

Public–private–academic partnerships are instrumental in overcoming challenges that may be beyond the scope of any single sector. Government agencies provide a regulatory framework that ensures responsible data collection and usage. Industry partners offer the technological infrastructure required for large-scale implementation, while academia contributes scientific insights that underpin the reliability and accuracy of environmental data.

Data sharing is a key outcome of collaboration between government, industry, and academia. By pooling resources and expertise, these stakeholders create a more comprehensive and diverse dataset, enriching the quality of environmental information. Shared data promotes transparency and facilitates a collective understanding of environmental conditions, enabling more informed decision-making.

Challenges such as data interoperability, standardization, and ethical considerations are effectively addressed through collaboration. Joint efforts enable the development of standardized protocols for data formats and sharing agreements. Ethical frameworks are established, ensuring that data collection and usage adhere to privacy regulations and ethical principles.

In conclusion, the collaboration between government, industry, and academia is essential for the success of data-driven environmental sensing. This collaborative approach harnesses the strengths of each sector, fostering innovation, ensuring regulatory compliance, and advancing scientific understanding. By working together, these stakeholders create a synergistic ecosystem that maximizes the potential of data-driven sensing to address complex environmental challenges and contribute to sustainable environmental management.

2.6.3 Community engagement and participation

Community engagement and participation are integral components of successful data-driven environmental sensing initiatives, fostering a sense of shared responsibility and inclusivity. Engaging the community in the sensing process transforms citizens from passive observers to active contributors, amplifying the reach and impact of environmental monitoring efforts. One key aspect of community engagement is raising awareness about the goals and benefits of data-driven environmental sensing. Outreach campaigns, workshops, and educational programs inform the community about the purpose of sensing initiatives, the types of data collected, and how their participation contributes to a better understanding of local environmental conditions. This awareness-building establishes a foundation for trust and collaboration.

Participatory sensing, where community members actively collect and contribute environmental data, is a powerful mechanism for engagement. Equipping individuals with user-friendly sensors and mobile applications empowers them to become citizen scientists, capturing real-time data in their surroundings. This grassroots involvement not only expands data coverage but also promotes a sense of ownership and connection to the environmental issues at hand.

Collaborative decision-making is facilitated through community engagement, where local knowledge and perspectives are incorporated into the design and implementation of sensing projects. Soliciting input from community members helps identify specific environmental concerns, potential data collection points, and the most relevant parameters to monitor. This co-creation of projects ensures that sensing initiatives align with community needs and priorities. Transparency and open communication are vital in maintaining community trust. Providing accessible information about how data is collected, processed, and used, as well as addressing concerns related to privacy and data security, builds a foundation of transparency. Regular communication channels, including community meetings and online platforms, foster ongoing dialogue and encourage continuous community involvement.

In summary, community engagement is essential not only for accurate data collection but also for creating a collaborative, inclusive, and sustainable approach to addressing environmental challenges. By involving the community, data-driven environmental sensing initiatives can benefit from local knowledge, increased data coverage, improved data quality, and enhanced community awareness and empowerment.

REFERENCES

1. Ted Mortenson. *Smart Sensor Explained, Industrial Sensors.*www.realpars.com/blog/smart-sensor, August 31, 2020.
2. Piskulova, Natalia. "Challenges for the environmental restructuring of the global economy." *Global Governance in the New Era: Concepts and Approaches.* Singapore: Springer Nature Singapore, 2023. 185–192.
3. Poole, Carol Jean, et al. "Taking state of biodiversity reporting into the information age—a South African perspective." *Frontiers in Ecology and Evolution* 11 (2023): 224.
4. Hassani, Sahar, and UlrikeDackermann. "A systematic review of advanced sensor technologies for non-destructive testing and structural health monitoring."*Sensors* 23.4 (2023): 2204.
5. Deroco, PatriciaBatista, Dagwin WachholzJunior, and Lauro TatsuoKubota. "Paper-based wearable electrochemical sensors: A new generation of analytical devices."*Electroanalysis* 35.1 (2023): e202200177.

6. Cacciuttolo, Carlos, et al. "Low-cost sensors technologies for monitoring sustainability and safety issues in mining activities: advances, gaps, and future directions in the digitalization for smart mining."*Sensors* 23.15 (2023): 6846.

7. Ding, Suiting, Arnold Tukker, and Hauke Ward. "Opportunities and risks of internet of things (IoT) technologies for circular business models: A literature review." *Journal of Environmental Management* 336 (2023): 117662.

8. Gagliardi, Valerio, et al. "Satellite remote sensing and non-destructive testing methods for transport infrastructure monitoring: Advances, challenges and perspectives." *Remote Sensing* 15.2 (2023): 418.

9. Kumar, Avinash, et al. "Blockchain-based decentralized management of IoT devices for preserving data integrity."*Blockchain Technology Solutions for the Security of Iot-Based Healthcare Systems.* Academic Press, London Wall, UK, 2023. 263–286.

10. Jain, Prem Chand, et al. "IoT based real time monitoring of Delhi-NCR air pollution using low power wide area network." *IFIP International Internet of Things Conference.* Cham: Springer Nature Switzerland, 2023.

11. Sahoo, Sushil Kumar, and Shankha Shubhra Goswami. "Theoretical framework for assessing the economic and environmental impact of water pollution: A detailed study on sustainable development of India." *Journal of Future Sustainability* 4.1 (2024): 23–34.

12. AlZubi, Ahmad Ali, and Kalda Galyna. "Artificial Intelligence and Internet of Things for Sustainable Farming and Smart Agriculture." *IEEE Access* 11(2023): 78686–78692.

13. Briard, Tristan, et al. "Challenges for data-driven design in early physical product design: A scientific and industrial perspective." *Computers in Industry* 145 (2023): 103814.

14. Lemke, Claudia, Dagmar Monett, and Manuel Mikoleit. "Digital ethics in data-driven organizations and AI ethics as application example." *Apply Data Science: Introduction, Applications and Projects.* Wiesbaden: Springer Fachmedien Wiesbaden, 2023. 31–48.

15. Ünal, Perin, Özlem Albayrak, Moez Jomâa, Arne J. Berre. "Data-driven artificial intelligence and predictive analytics for the maintenance of industrial machinery with hybrid and cognitive digital twins." In: Curry, E., Auer, S., Berre, A.J., Metzger, A., Perez, M.S., Zillner, S. (eds.), *Technologies and Applications for Big Data Value.* Cham: Springer, 2022.

Chapter 3

IoT-based smart car parking using particle Swarm optimization techniques

Dipra Mitra, Shika Gupta, Gurpreet Singh Panesar, and Ganesh Gupta

3.1 INTRODUCTION

The present transportation infrastructure and parking facilities are regarded insufficient to support the surge of automobiles on the road due to the significant numbers of newly registered vehicles that have been recorded in recent years compared to prior years. Consequently, issues such as heavy traffic and a lack of parking spaces ultimately arise. The traffic issues have been addressed using a variety of techniques. Although there are other ways to solve the issue, the proposal primarily focuses on the parking structure. In this approach, problems like traffic tend to be Internet of Things (IoT)-related. Therefore, a number of parking management systems have been set up to reduce such traffic problems and enhance the convenience of automobile use.

The deployed smart parking system was created by using cutting-edge technology and research. Provided, the formatter will need to create these components, incorporating the applicable criteria that follow from a variety of academic fields. It is intended that by deploying it in the parking lot, it would be able to address the aforementioned issues of the customers there. The smart parking system is thought to be advantageous for customers, parking operators, as well as environmental preservation. The information acquired via the use of the smart parking system can be used by car park managers to forecast upcoming parking patterns.

Altering pricing strategies in the light of collected data is another way to increase the company's profit. One way to help the environment is to cut down on air pollution, which is caused by cars and trucks. This is because fewer people are driving. Additionally, fuel efficiency will improve since it is inversely related to vehicle mileage. Smart parking systems benefit customers by allowing them to fully utilize parking spaces and create more efficient systems. The data collected by the smart parking system significantly cuts down on the amount of time it takes for vehicles to travel and for the system to find a parking spot, making it more efficient overall. Drivers can utilize the provided information to readily find empty parking spots elsewhere and avoid fully booked garages [1].

DOI: 10.1201/9781003530077-3

There will be less need for people to park their cars illegally on the side of the road, which helps alleviate traffic congestion, as more cars are absorbed into parking lots. First and foremost, transportation congestion may be reduced. In the long run, this would all make things easier for shoppers. A decrease in the number of staff in the garage and the predominance of car owners is the project's goal, as indicated by the countdown in the 7-meter clock at the entrance of the automobile. As the owner is utilized to determine the amount of automobiles absorbed into the garage, the exit is tallied upward. At its peak, the meter will trigger the door, unless one of the vehicles has departed; the garage door will not open due to the electronic lock. It is possible to enhance this project by including several methods that save time and reduce obstructions, such as global positioning system (GPS) depletion [2] in congestion caused by vehicles is a major problem that has been becoming worse over the past few decades. The parking problem has been and is still a major issue in urban areas because of the increasing size of luxury vehicles and the restricted number of spaces available for parking. People in urban areas all around the globe frequently partake in the (at times annoying) activity of searching for parking spots. For this mission, one million barrels of gasoline are required. It is common practice for smart parking systems to collect information about nearby parking spots and then use real-time analysis to assign vehicles to those locations. It makes use of cheap sensors, data collected in real time, and automated payment systems that can be accessed through mobile devices.

3.2 RELATED LITERATURE

Some demand response (DR) systems for a smart grid were investigated in references [2–7] to handle the possible high peak demand of electric vehicle (EV) charging in home distribution zones. To address the optimization operation challenge in a distribution system operator (DSO), references [8,9] offered a hierarchical coordinated charging architecture and a vehicle-to-grid (V2G) scenario. Taking into account the pros and cons of renewable energy sources (RES), namely its intermittent and uncontrollable nature, a number of optimization strategies for controlling a charging station's many energy sources are presented in references [10–13]. In addition, to lessen the effect of uncertainty and erroneous prediction, an energy storage (ES) unit was employed in the calculations shown in references [13,14]. Finding out where parking lots (PLs) are and how much capacity they have is another strategy discussed in references [15–17]. In reference [18], a system for real-time station recommendations for electric taxis was suggested, beginning with a few potential charging stations. To optimize earnings, the authors of reference [19] suggested coordinating the bidding of V2G's supplementary services. Additional research was conducted to identify the optimal charging price in references [13,20]. Reference [21] offered a new approach that would enable vehicle-to-vehicle (V2V) charging. The goal was to reduce the overall expense, which encompassed not only the cost of electricity but also the harm caused to EV batteries by excessive charging cycles. Binary variables representing the EVs' charging and discharging states were used to make the decision. Evaluating performance with such a tiny number of EVs required a tremendous amount of computer power. In reference [22], a real-time charging scheme was suggested as a means to coordinate the charging decisions of EVs and to meet demand in response to changes in electricity prices and requests for demand curtailment from utility companies. Our goal was to maximize the number of EVs that could be charged during each schedule period while minimizing the electricity bill. To ascertain how to turn each charging pole on and off, the schedule used a binary decision method. To find the best strategy for charging EVs during the day, reference [23] suggested a PL management system, which differs from the binary choice technique. Keeping in mind the cost of electricity and a penalty for unmet EV demand, the goal was to minimize the charging station's cost. The authors also used a game-theoretic approach to determine the best time to charge EVs in reference [24]. Candidate solutions were easy to find, though, because the vehicle profiles they employed overlapping for 30 time intervals and the charging demands were quite modest. For the purpose of calculating the charging schedule, certain optimization intelligent-based techniques were suggested in references [25,26], and new accelerated particle swarm optimization (APSO) algorithms were developed to address this issue in reference [27]. Making sure that all EVs had the best average state of charge at the following time step was the main goal. The optimization model, however, solely considered the upcoming time step's charging of EVs while making its conclusion. Therefore, without taking the entire EV parking period into account, this may not give an optimal schedule. The handling of EVs with and without appointments was suggested in references [28–30] using separate layers, scenarios, and schemes. Taking into account the aggregator's income, client requests, and costs, two scenarios were suggested in reference [28] to establish a billing schedule. Consumers who were already familiar with the aggregator were charged in the static situation, but consumers who dropped in and out at will were charged in the dynamic scenario. In the static timetable, the decision was set for these EVs. Even if a better alternative is available, the determined schedule cannot be adjusted if a new customer comes without early notice. According to reference [29], to maximize PL revenue or the total number of EVs that meet their needs, a PL management system for a centralized EVs charging system was suggested. The two types of EVs were handled by defining a two-layer scheduling mechanism. They handled the atypical EVs using the first-in, first-serve (FIFS) and earliest-deadline-first (EDF) procedures, and they utilized an optimization approach to establish the routine layer's timetable. Their optimization technique is limited to regular EVs; however, irregular

EVs are handled by FIFS and EDF, which could not result in an optimal decision. For EV charging and discharging, a globally optimal scheduling scheme and a locally optimal scheduling method were suggested in reference [30]. With each time slot's charging and discharging power taken into account, the goal was to reduce the overall cost of all EVs. The designated EVs were handled by a global scheduling system, while the non-appointed EVs were handled by a local one. It is possible for the power load to surpass the transformer limit in their local optimization model.

3.2.1 Existing system

We have switched to a more traditional method of parking our cars. A standard parking system is present in most homes, businesses, and recreational centers. There will be less congestion and better organization because of the designated parking spaces. Their energy consumption is reduced, as the sole energy requirement is for illumination [4]. A power outage will not impact the reliability of traditional garages. There is no set time when traditional garages are open. Finding an empty spot to park is the first order of business in certain conventional parking lots due to space limits. With an easy-to-understand layout, the starting cost is reasonable. Another way to give car protection tactics is to assign certain people to specific lots. The amount of land accessible for urban development is decreasing, and keeping the specifications intact is essential.

As the number of cars on the road continues to rise, a more permanent and alternative solution is required. Everybody knows that the number of cars on Indian roads is going up corresponding to the country's growing population every day. Consequently, parking is an issue for both cars and the areas they occupy. The current parking system is the backbone of our country, and we must accept the fact that it needs improvement to keep up with the times. Finding a job takes up a lot of their time. When drivers are in a lousy parking spot, their tension and irritation levels go up. In densely populated parking lots, traffic builds up as most cars are taking their time looking for a spot. The vast majority of shoppers prefer to park near frequently used amenities, such as escalators, elevators, and entrances to shopping centers. Therefore, while some parking spots would go unused, others will be jam-packed with cars.

Building or installing something can be quite expensive. There is always a hefty price tag associated with a dependable and effective parking management system. This is because the system is comprised of many different parts. More and more individuals are finding it difficult to park and drive as a result of our current system and the proliferation of cars. Our current system uses the template to style the text and format the paper when a customer enters the parking lot. Please do not change the specified margins, column widths, line spacing, or text typefaces. However, you might find some oddities. One notable difference is the significantly larger head margin compared to the norm in this design. The purpose of this metric and others like it is to ensure that your paper is a cohesive component of the proceedings rather than a stand-alone piece of writing. Kindly refrain from making any changes to the existing names. Many factors, including the statistical feature, automated ticketing, statistical reporting, and others, contribute to the high price tag. It may be out of reach for certain organizations. The organization still has to do a lot of regular maintenance on the system, even if it is automated. The purpose of this is to verify that there are no problems and that everything is functioning properly. The maintenance could be done on a monthly basis. A lot of people do not know how to use the parking control system. This might make it much more difficult for them to use, which would make parking an even bigger hassle. Being a machine, the system is bound to experience malfunctions. Vehicles parked within buildings may become immobile and unable to enter when this occurs [5]. It could malfunction and lead to cars being parked in the wrong spots. One of the most common problems in modern times is the lack of available parking spots. Roads are becoming more congested due to the greater number of vehicles compared to the available parking spaces.

Overcrowding, distorted vehicles caused by lack of space, and exorbitant parking costs are some of the potential outcomes of this situation. Another problem is that in Indian cities, on-street parking is more popular than off-street parking. Particularly on heavily traveled roadways, delays caused by on-street parking issues are common. A fair tariff is necessary for these kinds of places to be used effectively. In Indian cities, parking spaces are almost typically built for residential units as people think they will not sell if there are not any. Vacant parking spaces contribute to a general increase in construction costs [6]. As parking construction costs are constant regardless of the type of building, lower-income residents, who are less likely to own cars, wind up paying a disproportionate share of parking rates. Other people's parking costs are essentially covered by this. Overcrowding of residents' cars is another typical issue caused by businesses' and individuals' excessive car ownership. Though the vehicle parking still appears to be a challenging problem in large cities. With less room to maneuver, drivers waste more time searching for parking spots, which in turn worsens traffic and exacerbates their frustration. This issue could be addressed by utilizing sensors connected to the IoT placed throughout the city to detect when parking spots are empty[7].

3.2.2 Proposed system

To decrease the number of garage staff is the project's objective. People who own cars, those who use parking lots, and those concerned about the environment are all believed to benefit from the smart parking system. Vehicle parking providers can utilize the data obtained from the smart parking system to predict how customers will park in the future. Drivers can avoid overcrowded parking lots and quickly find empty spots elsewhere with the help of the provided information.

Customers also gain from fully utilized parking spots, optimized systems, and new, improved mechanisms. The smart parking system improves efficiency by reducing vehicle travel and search times with the help of the data it provides.

The sensor does more than just identify the car; it also provides information about the vehicle's condition and the amount of time it has been parked. The control units receive data, process it, and then relay the findings to a centralized monitoring system. Using user datagram protocol (UDP), the controller communicates parking spot data to the centralized monitoring system [8].

Afterwards, the user's mobile phone receives notifications regarding their assigned parking slot, the total time spent in the slot, their billing information, and directions. Each day, the smart car parking system records the number of parked and reserved vehicles. An elite car's membership can be confirmed with the use of a barcode scanner or license plate reader as soon as the vehicle enters a parking spot.

The gate will open, the specified spot will be shown, and the car will pass through if it is an elite member. If the customer is a regular, you can tell which spot is available and which one the car should use. On the app or website, there will be four signals, with one reserved for the elite. There should be three people on the team: One for confirmed reservations, one for available spots, and one for parked vehicles. Updates to the data will be made often.

Smart parking could make these a reality: Providing precise real-time predictions and vehicle/spot occupancy sensors points both residents and visitors in the direction of available parking spots, and maximizes the utilization of parking spots makes parking easier for everyone involved, from drivers to business owners, and simplifies operations overall. The city's traffic flow can be improved by utilizing IoT technologies, which makes it possible to make decisions based on data using tools like real-time status via apps and reports from previous analytics. Smart parking greatly improves city environments by reducing emissions of carbon dioxide and other pollutants. A substantial boost to revenue is made feasible by smart parking, which allows for better real-time monitoring and control of parking spot availability and provides tools to enhance human resource (HR) operations.

A frequent method of allocating parking spaces to vehicles is the automated car parking system, which allows for the automatic parking and retrieval of vehicles. By removing the necessity for roadways and ramps, the technology enables the parking station to make better use of its floor space volume. Automated parking systems are space-efficient, taking up only a third as much room as conventional parking methods, and they can be constructed either above or below ground, economical from multiple angles.

Automated car parking system maximizes the use of available space, whether it is above or below ground. With its reduced environmental impact, time impact, opportunity for theft and vandalism, and real cost benefits, the automated car parking system is the new standard for parking. This is according to urban planning guiding principles. The increasing volume of traffic is making these technologies more important. These technologies can be integrated with consumer electronics.

In order for this structure to function, it is essential to install each section internally. The proposed fix counts the number of cars entering the park and then tells the entry to close to all vehicles once the maximum number of cars has been achieved. To top it all off, the same counter that was made for the entry gate is also used to count the cars exiting the park, but in the opposite order. Here, a small push button serves as the basis for the park's operation, which is dependent on a piezoelectric sensor that measures the weight of the automobile. A single press of the button causes the counter to go up or down by one. Regulating is made easier with the proposed system.

Massive traffic jams in popular tourist areas and empty streets everywhere else can be avoided with effective traffic control in cities. Although this can be managed with smart traffic lights, it is also affected by the design and layout of the roads. For instance, green lights could automatically change their duration depending on the amount of traffic, staying longer in more congested regions and shorter in less congested ones. In addition, roads and bridges can be equipped with sensors to track their condition and receive repairs when they show signs of severe damage [9].

Air pollution is a major problem in many big cities, because the air is so thin and full of particle matter that it can damage people's lungs over time. But there are ways to use the IoT and artificial intelligence (AI) to reduce air pollution. IoT makes this possible by collecting data on urban pollution from a wide range of sources (e.g., pollen levels, airflow direction, weather, traffic levels, etc.), and then making predictions about pollution to spot patterns and manage them[10].

3.3 SYSTEM ARCHITECTURE

To decrease the number of garage staff is the project's objective. People who own cars, those who use parking lots, and those concerned about the environment are all believed to benefit from the smart parking system. Vehicle parking providers can utilize the data obtained from the smart parking system to predict how customers will park in the future.

Drivers can simply find available parking spots elsewhere and avoid fully used parking lots with the help of the provided information. Customers also gain from fully utilized parking spots, optimized systems, and new, improved mechanisms.

The smart parking system informs a more efficient system, which in turn reduces vehicle travel and search times. There will be less need for people to park their cars illegally on the side of the road, which helps alleviate traffic congestion, as more cars are absorbed into parking lots. First and foremost, transportation congestion may be reduced. One goal of smart parking is to make finding a parking spot easier and faster. Environmental benefits of precisely guiding a vehicle to an open area include reduced noise, CO_2 emissions, and other pollutants. When combined with smart settings, smart parking allows for the measurement of air quality and the availability of parking spots as given in Figure 3.1.

Reducing the garage's workforce is the project's main purpose. Everyone from drivers to other parking lot users to ecologists think the smart parking system is a win–win. Parking lots might use the smart parking system data to foretell customers' parking patterns. Drivers can easily locate empty parking spots and avoid overcrowded parking lots by using the given information.

Figure 3.1 System architecture.

It is in the best interest of users to optimize parking spots and create more efficient procedures so that all available spaces are completely utilized. The data given by the smart parking system can drastically reduce vehicle travel and search times. With more automobiles parked in lots, there will be less need for people to park illegally by the road, which helps alleviate traffic congestion. As reducing traffic delays is a top priority.

The sensor not only recognizes the vehicle but also gives other details like the vehicle's condition and the amount of time it has been parked for. The data is processed by the control units, which then report their findings to a centralized supervisory system. The controller communicates with the central supervisory system via UDP to transmit parking spot information.

The next step is for the user's mobile phone to receive details such as their allocated time slot, the total amount of time parked, billing information, and directions. The smart automobile parking system keeps a daily log of all parked cars, including those with reservations. A license plate reader or barcode scanner can detect if a vehicle is owned by the elite simply by scanning its number as it pulls into a parking area.

The parking lot serves as the organization's hub, which is why this concept was developed. It has elements like "Address" to indicate its location and "Name" to distinguish it from other parking lots. The parking lot will have multiple parking floors.

Parking places: Every parking floor has a large number of parking places. We have a system that can handle different parking spots; among them are electric, motorbike, huge, tiny, and disabled.

Account: There will be two distinct types of accounts in the system: An administrator's account and a parking attendant's account.

Parking ticket: The topic of discussion in this session will be parking tickets. Customers will get a ticket when they enter the parking lot. Vehicle automobiles will be parked in the designated spaces. Our technology will be compatible with a variety of automobiles. Car, truck, EV, van, and motorcycle round out the list.

The entrance panel and exit panel: The entrance panel prints tickets, and the exit panel makes it easier to pay the ticket price.

3.4 RESULT AND DISCUSSION

Figure 3.2 shows how the parking area is used with time-out time in the mobile number of each and every passenger in the parking lot. If park owners want to know about the vehicle parked in the parking lot, Figure 3.3 shows the management panel fetching the details of the vehicle if it is doing some unethical work. At night, Figure 3.4 shows what cars look like while parked at night. There will be a display board on every parking level that lists the available places for each type of place. Keeping clients updated on

Check-In-List

Carnumber	unique-id	check in	Entry cam id	mobile no
AMSAF4334	5653387111	5/5/2023-6/5/2023	11125566	9874766479
PMSAF4384	8953387111	6/5/2023-6/7/2023	11125567	8874766479
WBSAF4884	8973387111	7/5/2023-8/5/2023	11125568	8774766478
APSAF4884	7004338711	7/5/2023-8/5/2023	11125569	7004766478
WB089564	7074338711	7/6/2023-8/6/2023	11125570	7004766478
UK089564	8074338711	7/6/2023-8/6/2023	11125571	7004766479

Figure 3.2 Verify parking area.

the latest free parking availability is the responsibility of this group. The training will cover every possible task that an attendant may have, such as taking payments and scanning tickets. Parking ticket payments are handled through the customer information portal, which is taught in this session. Payment will be reflected in the ticket once it has been accepted by the information portal. Customer EV charging and payment will be facilitated by the electric panels.

3.5 CONCLUSION AND FUTURE SCOPE

Automation is a step in the right direction for the transportation sector's future prosperity. This design offers a workable solution to the common

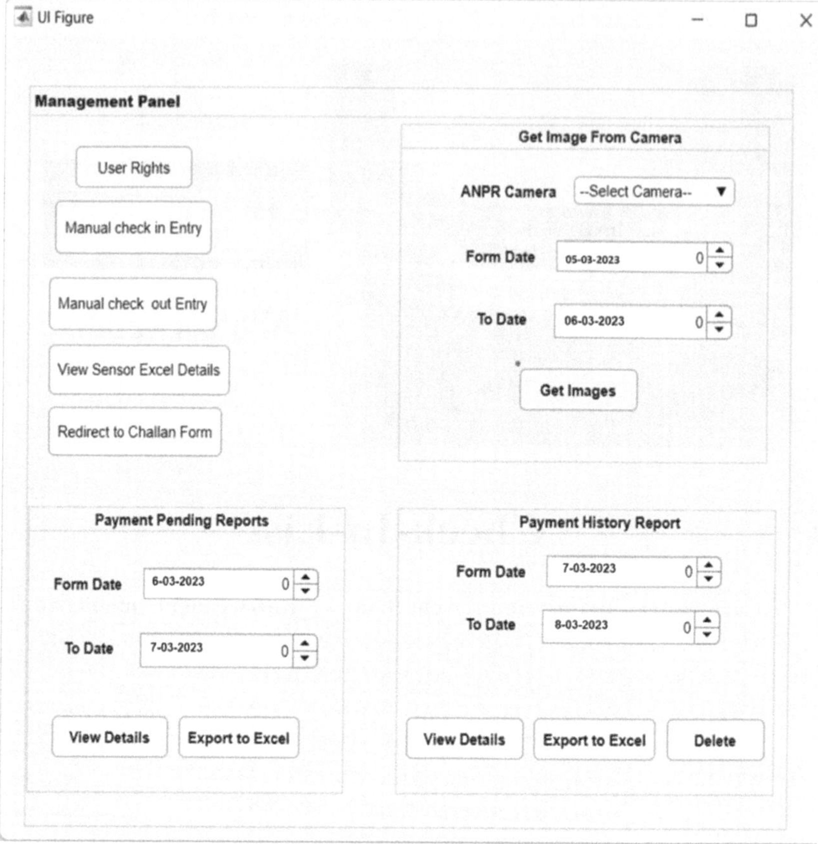

Figure 3.3 Management panel this panel will help user to get the detail reports on parking.

problem it raises. Accurate results were obtained from the designed, manufactured, and tested smart vehicle parking system when the threshold distance was calibrated and the barrier was identified. The light-emitting diodes (LEDs) changed between the two states instantaneously based on whether a car was in the parking lot or not. The layout is flexible and can be altered to better utilize the available area. It can even be positioned in a confined space. Based on the quantity of yellow LEDs discovered, a well-known information board is displayed, showing the number of parking spaces that are available. One may argue that with accuracy, it is possible to construct an automated smart car parking system that minimizes unnecessary driving and wasting of time and gasoline, and significantly

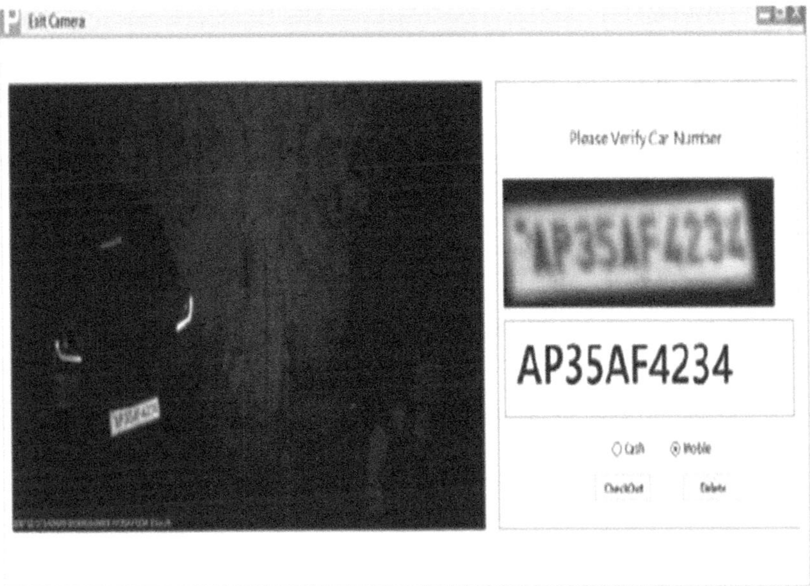

car number	unique ID	Exit camera entry	Entry camera entry	Amount	Check out time	Check in time
DL542689	121	1456	1456	40	9.00AM	6.30AM
WB89653	122	1458	1458	50	9.30AM	5.30AM
UP546231	123	1457	1457	43	11.00PM	10.00AM
JH789654	124	1459	1459	45	10.00AM	1.30PM

Figure 3.4 Finding parking area at night.

streamlines the parking process by connecting a few simple electrical components.

This chapter discusses the various types of smart parking systems. The various applications of the smart parking system that are being offered demonstrate how effective it is at reducing traffic, especially in metropolitan areas where there is a shortage of parking spaces and traffic congestion.

It accomplishes this by directing patrons and optimizing parking spots. The examination of all the sensor technologies used to detect cars is one of the most crucial parts of the smart parking system.

It is possible to assess the benefits and drawbacks of each sensor technology. As mentioned before, there are certain disadvantages to employing a visual-based approach for vehicle detection, but they are greatly exceeded by the advantages.

REFERENCES

1. A. Dalvi, EV sales in India charge past 1.5 million units in CY2023, stellar growth in all segments. Available Dec 2023. www.autocarpro.in/analysis-sales/ev-sales-in-india-charge-past-15-million-units-in-cy2023-stellar-gro wth-in-all-segments-118410
2. S. Shao, M. Pipattanasomporn, S. Rahman, Grid integration of electric vehiclesand demand response with customer choice, IEEE Trans. Smart Grid 3 (March(1)) (2012) 543–550.
3. M. Pipattanasomporn, M. Kuzlu, S. Rahman, An algorithm for intelligent home energy management and demand response analysis, IEEE Trans. Smart Grid 3(December (4)) (2012) 2166–2173.
4. Z. Tan, P. Yang, A. Nehorai, An optimal and distributed demand response strategy with electric vehicles in the smart grid, IEEE Trans. Smart Grid 5(March (2)) (2014) 861–869.
5. F. Rassaei, W.S. Soh, K.C. Chua, Demand response for residential electric vehicles with random usage patterns in smart grids, IEEE Trans. Sustain. Energy 6 (October (4)) (2015) 1367–1376.
6. M. Shafie-khah, E. Heydarian-Forushani, G.J. Osorio, F.A.S. Gil, J. Aghaei, M. Barani, J.P.S. Catalao, Optimal behavior of electric vehicle parking lots as demand response aggregation agents, IEEE Trans. Smart Grid 7 (November(4)) (2016) 2654–2665.
7. W. Lausenhammer, D. Engel, R. Green, Utilizing capabilities of plug in electric vehicles with a new demand response optimization software framework: Okeanos, Int. J. Elect. Power Energy Syst. 75 (February) (2016) 1–7.
8. Z. Xu, Z. Hu, Y. Song, W. Zhao, Y. Zhang, Coordination of PEVs charging across multiple aggregators, Appl. Energy 136 (December) (2014) 582–589.
9. L. Jian, X. Zhu, Z. Shao, S. Niu, C.C. Chan, A scenario of vehicle to grid implementation and its double-layer optimal charging strategy for minimizing load variance within regional smart grids, Energy Convers.Manag. 78 (February) (2014) 508–517.
10. B. Skugor, J. Deur, Dynamic programming-based optimisation of charging an electric vehicle fleet system represented by an aggregate battery model, Energy 92 (December) (2015) 456–465.
11. M. Yazdani-Damavandi, M.P. Moghaddam, M.R. Haghifam, M. Shafie-Khah, J.P.S. Catalão, Modeling operational behavior of plug-in electric vehicles' parking lot in multienergy systems, IEEE Trans. Smart Grid 7 (January (1))(2016) 124–135.
12. M. Shafie-Khah, E. Heydarian-Forushani, M.E.H. Golshan, P.Siano, M.P. Moghaddam, M.K. Sheikh-El-Eslami, J.P.S. Catalao, Optimal trading of plug-in electric vehicle aggregation agents in a market environment for sustainability, Appl. Energy 162 (January) (2016) 601–612.

13. A.S.A. Awad, M.F.S. Awad, T.H.M. EL-Fouly, E.F. El-Saadany, M.M.A. Salama, Optimal resource allocation and charging prices for benefit maximization in smart PEV-parking lots, IEEE Trans. Sustain. Energy 8 (July (3)) (2017) 906–915.

14. C. Jin, J. Tang, P. Ghosh, Optimizing electric vehicle charging with energy storage in the electricity market, IEEE Trans. Smart Grid 4 (March (1)) (2013) 311–320.

15. A.Y.S. Lam, Y.W. Leung, X. Chu, Electric vehicle charging station placement: formulation, complexity, and solutions, IEEE Trans. Smart Grid 5 (November(6)) (2014) 2846–2856.

16. M.J. Mirzaei, A. Kazemi, O. Homaee, A probabilistic approach to determine optimal capacity and location of electric vehicles parking lots in distribution networks, IEEE Trans. Ind. Inf. 12 (October (5)) (2016) 1963–1972.

17. X. Dong, Y. Mu, H. Jia, J. Wu, X. Yu, Planning of fast EV charging stations on around freeway, IEEE Trans. Sustain. Energy 7 (October (4)) (2016) 1452–1461

18. Z. Tian, T. Jung, Y. Wang, F. Zhang, L. Tu, C. Xu, C. Tian, X.Y. Li, Real-time charging station recommendation system for electric-vehicle taxis, IEEETrans. Intell. Trans. Syst. 17 (November (11)) (2016) 3098–3109.

19. M. Ansari, A.T. Al-Awami, E. Sortomme, M.A. Abido, Coordinated bidding of ancillary services for vehicle-to-grid using fuzzy optimization, IEEE Trans Smart Grid, 6, 1, 261–270. doi: 10.1109/TSG.2014.2341625.

20. P.F. Felzenszwalb, R.B. Girshick, D. McAllester, D. Ramanan, Object detection with discriminatively trained part-based models, IEEE Trans. Pattern Anal. Mach. Intell. 32 (September (9)), (2010) 1627–1645.

21. G.A.Bertone, Z.H.Meiksin,N.L.Carroll, Investigation of a capacitance-based displacement transducer, IEEE Trans. Instrum. Meas. 39 (April) (1990) 424–428..

22. R.C. Luo, Z. Chen, An innovative micro proximity sensor, Proc. 1993 JSME Int. Conf. Advanced Mechatronics(August) (1993) 621–625.

23. D. Mitra, S. Goswami, D. Hati, S. Roy, Comparative study of IOT protocols, PalArch's J. Archaeol. Egypt/Egyptol. 17(7) (2020)12527–12537.

24. D. Mitra, S. Gupta, Data security in IOT using Trust Management Technique, In 2021 2nd International Conference on Computational Methods in Science and Technology (ICCMST), IEEE (2021, December)14–19.

25. D. Mitra, S. Sarkar, D. Hati, A comparative study of routing protocols, Eng. Sci. 2(1) (2016)46–50.

26. S. Sani, A.Bera, D. Mitra, K.M. Das, COVID-19 detection using chest X-ray images based on deep learning, Int. J. Softw. Sci. Comp. Intell. 14(1) (2022)1–12.

27. D. Mitra, S. Gupta, Plant disease identification and its solution using machine learning, In 2022 3rd International Conference on Intelligent Engineering and Management (ICIEM), IEEE (2022, April) 152–157.

28. N. Kumar, A.K. Dahiya, K. Kumar,S.Tanwar, Application of IoT in agriculture, In 2021 9th International Conference on Reliability, Infocom Technologies and Optimization (Trends and Future Directions)(ICRITO), IEEE (2021, September) 1–4.

29. N. Kumar, K. Kumar, A. Kumar, Application of internet of things in image processing, In 2022 IEEE Delhi Section Conference (DELCON),IEEE (2022, February) 1–5.
30. D. Mitra, S. Gupta, P. Kaur, An algorithmic approach to machine learning techniques for fraud detection: A comparative analysis, In 2021 International Conference on Intelligent Technology, System and Service for Internet of Everything (ITSS-IoE), IEEE(2021, November) 1–4.

Chapter 4

Impact of emerging robots using the bioinspired processed data drive mechanisms

Agha Imran Husain and Anu Priya Sharma

4.1 INTRODUCTION

Over the past few years, robotics has evolved from inflexible, preprogrammed machines to adaptable, biologically inspired systems. Data-processing technologies, especially those that mimic biological systems, have hastened this progress. Bioinspired data-driven mechanisms have revolutionized robotics by allowing robots to observe, learn from, and adapt to dynamic settings with unprecedented efficiency and autonomy (Li et al., 2021). This chapter explores new robots with bioinspired processed data-driven systems that are having huge impacts on several fields. Using biological systems as inspiration, researchers have developed neural networks and evolutionary algorithms to imitate brain functions. These bioinspired data-processing methods have revolutionized robotics by allowing machines to learn from mistakes, optimize their behavior, and adapt to changing environments. Bioinspired processed data-driven processes in robots have opened up various possibilities, spurring innovation in many domains. These medical robots assist with treatments, monitor vital signs, and provide companionship. By analyzing vast amounts of data in real-time, they help doctors make better decisions and provide personalized care.

This new generation of bioinspired, data-driven robots impacts several industries. These robots are redefining what is possible and solving some of humanity's biggest issues, from healthcare and agriculture to environmental monitoring and more.

4.2 RELATED LITERATURE

There has been a lot of attention paid to bioinspired intelligence. The paper by Li et al. (2021) provides a comprehensive survey of bioinspired intelligence, with a focus on neuroscience approaches to various robotic applications. The bioinspired shunting model and its variants (additive model and gated dipole model) are introduced, and their main characteristics are given in detail. Two main applications to real-time path planning and control of

DOI: 10.1201/9781003530077-4

robotic systems are reviewed. A bioinspired neural network framework, in which neurons are characterized by the neurodynamics models, is discussed for mobile robots, cleaning robots, and underwater robots. The bioinspired neural network has been widely used in real-time collision-free navigation and cooperation without any learning procedures, global cost functions, or prior knowledge of the dynamic environment. When a large initial tracking error occurs, bioinspired back-stepping controllers can eliminate the speed jump. The current challenges and future directions are discussed in the chapter (Ranjbar-Sahraei et al., 2015).

Talk about the coordination of teams in multi-robot systems that are modeled after the social behaviors of insects, such as ants and honeybees (Gleadow et al. 2019). In particular, we examine the cases of pheromone signaling processes, robot foraging inspired by bees, and robot coverage inspired by ants. We also present some of our bioinspired algorithms to address these issues. Robot foraging refers to the problem of scavenging the environment for food or supplies, while robot coverage refers to the difficulty of deploying a robotic swarm in the environment to optimize the sensor coverage of the environment. Mukherjee et al. (2023) discusses about a robotic platform, which can gain from distributed energy in the following ways: Higher energy density, reduced design complexity, better body dynamics, and operational dependability. This is the first complete analysis that highlights the advantages of bioinspired distributed energy in robotics by focusing on distributed energy and presenting accessible or tuned energy storage and energy-harvesting methods.

Zheng et al. (2013) elaborate about algorithms that draw inspiration from the collective behavior of social colonies and/or natural biological evolutions have shown to be effective and increasingly popular in recent times. We provide an overview of the most recent developments in bioinspired optimization techniques in the realm of sustainable energy development in this study. These techniques include swarm intelligence, artificial neural networks (ANNs), evolutionary algorithms, and their hybridizations. The literature evaluated in this work addresses prospective future research trends and presents the state-of-the-art technology as of now.

The robotic application is presented in Russo et al. (2016). This study is intended to monitor the environment in data centers. Owing to the high energy density that data centers handle, environmental monitoring plays a critical role in managing humidity and air temperature throughout the entire facility. This helps to maximize the life cycle of internet of things (IT) devices, prevent hardware failures, and enhance power efficiency. Modern data center monitoring systems rely on environmental sensor networks to continuously gather data on humidity and temperature. These methods do not scale well in huge contexts and are still somewhat pricey. In this chapter, an alternative to environmental sensor networks based on mobile robots (Ni et al. 2016) that are autonomous and have environmental sensors installed is presented. A centralized cloud robotics platform allows for autonomous navigation and control over the robots.

4.3 BIOINSPIRED DATA PROCESSING IN ROBOTICS

Data-processing approaches that resemble biological systems have emerged from robotics and biology. Modern robotics relies on ANNs, which are modeled after brain neural networks (Kopacek 2016). These networks interpret data like the human brain, allowing robots to see patterns, make choices, and gain experience (Mukherjee et al., 2023). The merging of robotics and biology has led to advanced data-processing methods that imitate biological systems.

Neural networks and evolutionary algorithms inspired by natural selection have changed robotics (Kondoyanni et al. 2022). These algorithms mimic mutation, selection, and reproduction to optimize robot behavior and design. Evolutionary algorithms (Figure 4.1) enable robots to adapt to diverse and dynamic situations by repeatedly improving performance across generations. Swarm robots also mimic natural group behaviors like ants' and bees' synchronized motions. Swarm robotics employs self-organization and decentralized control to let several robots achieve tasks that would be hard for one. Biology and robotics have created many bioinspired-processing systems with unique benefits and capabilities. Science is expanding robotics and preparing for intelligent, adaptable, and autonomous robots by studying natural design principles. These bioinspired processing approaches may lead to greater robotics advances, which will affect society, business, and technology (Shi et al. 2015).

The consequences of these bioinspired robots in many fields and their future revolutionary potential are examined in this chapter: The robots' applications and pros and cons to show how they are changing industries, improving lives, and shaping technology.

A greater understanding of bioinspired processed data-driven mechanisms (Figure 4.2) in robotics reveals a world of boundless possibility, where biology and machines will usher in a new era of innovation and advancement.

4.4 IMPACTS ON HEALTHCARE

Healthcare is a prospective market for bioinspired processed data-driven robots. Robots are helping with surgeries, monitoring vital signs, and companioning seniors, but their impact goes beyond these activities.

4.4.1 Remote patient monitoring

Bioinspired robots with advanced sensors and data processing can remotely monitor patients' health. This is especially useful for chronically unwell

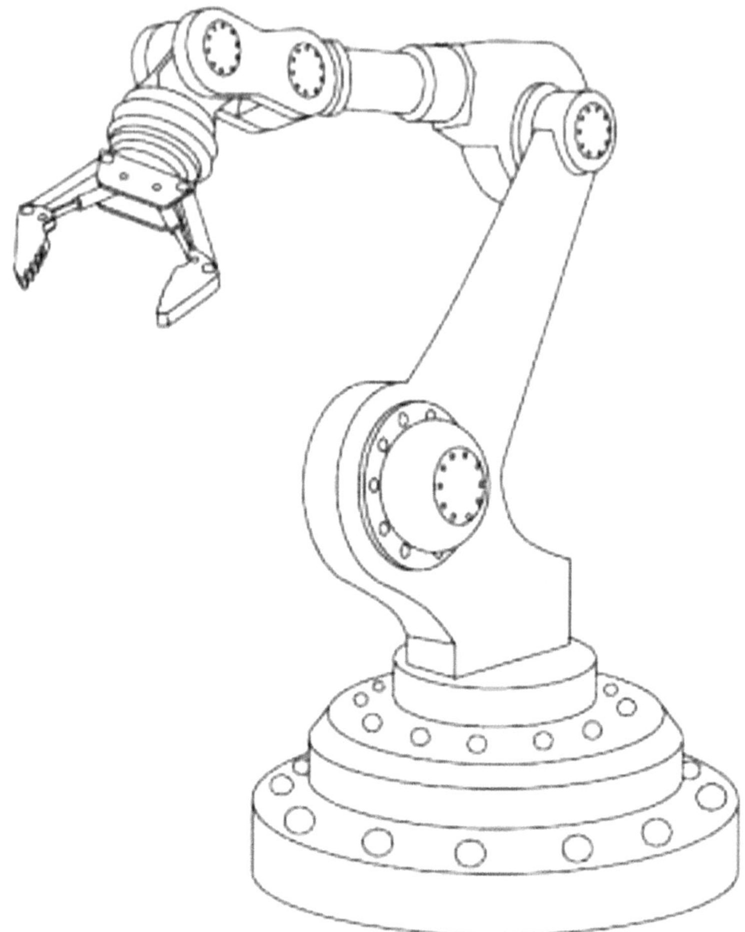

Figure 4.1 Bio-inspired algorithm.

people who need medical monitoring. These robots can monitor glucose, blood pressure, and heart rate to alert doctors about any abnormalities as shown in Figure 4.3. This enables early intervention and personalized treatment. Geographical and clinical constraints have been eliminated by telehealth and remote patient monitoring, which has increased the applicability of traditional clinical practice. Will healthcare become more valuable as a result of this? Will the provision of improved patient experiences be accompanied by an improvement in clinical outcomes? Will technology improve the sustainability of healthcare delivery? Mobile phones, in their

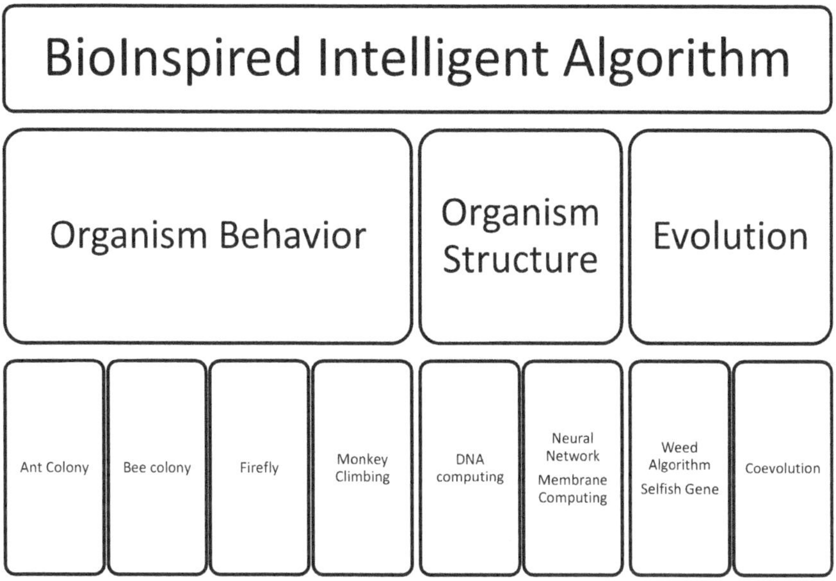

BioInspired Intelligent Algorithm
Organism Behavior
Organism Structure
Evolution
Ant Colony
Bee colony
Firefly
Monkey Climbing
DNA computing
Neural Network
Membrane Computing
Weed Algorithm
Selfish Gene
Coevolution

Figure 4.2 Bio-inspired processes data-driven mechanism.

Neural Network Based
Evolutionary Algorithm Based
Optimization
Search
Design
Bio-inspired PDDM
Swarm Intelligence Based
Reinforcement Learning
Control/Planning

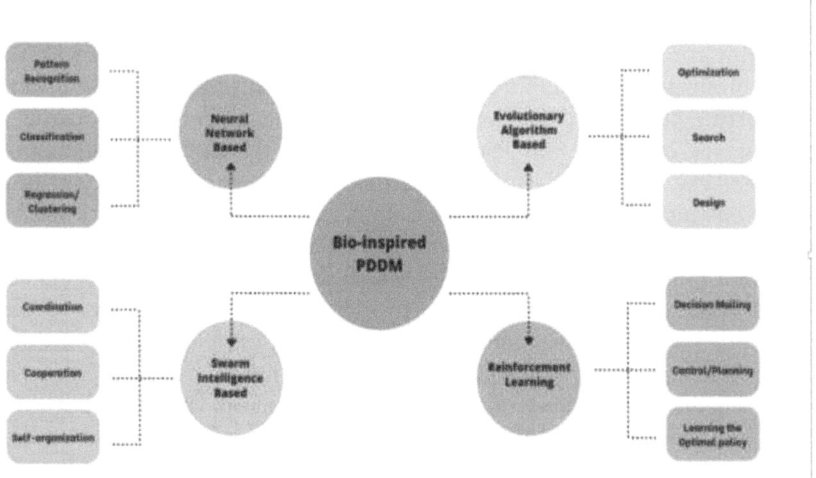

Figure 4.3 Remote patient monitoring.

current form, have just been around for the last 20 years, despite the fact that the majority of us feel as though we could not work, live, or play without them. But as technology and capabilities continue to progress, healthcare has been impacted and medicine is now more accessible—first in our homes and then in our wallets (Jeddi & Bohr, 2020).

4.4.2 Telemedicine and remote consultations

Telemedicine's popularity is making bioinspired robots crucial for remote doctor–patient consultations. These robots can behave like healthcare workers, allowing them to examine patients and prescribe medication remotely (Fukuda et al. 2018).This helps in underserved or rural areas where medical treatment is scarce. The themes of remote consultation versus traditional or classic physical consultation are covered in this systematic review. Research has demonstrated its significance both before and after the coronavirus disease 2019 (COVID-19) pandemic, particularly in the context of caring for patients who were infected. It has been discovered that while remote consultation is required, it needs to be advancement above the current setup. Both patients and general practitioners (GPs) are satisfied with teleconsultations (Picozzi et al. 2023) as a way to cut down on trips, particularly in tight situations.

4.4.3 Mental health support

These robots use motion sensors and feedback mechanism data to personalize rehabilitation programs for each patient, monitor progress, and provide real-time counsel and feedback to maximize results. In addition to physical care, bioinspired robots are providing companionship and mental health support to elderly or cognitively impaired patients. These robots can make patients feel less alone and healthier by including them in social interactions, memory exercises, and reminiscence therapy. The lack of diversity in the situations analyzed and the methodological flaws in the research done so far restrict how broadly the findings can be applied. However, social robot interventions typically show favorable benefits on patients with mental health issues, despite the existence of some impediments to their deployment (e.g., technical challenges, unsuitable location, and staff opposition). More research with higher methodological standards is required to fully comprehend the uses and advantages of social robots in mental health services (Gunderman et al., 2022).

With the growing integration of artificial intelligence (AI) into our daily lives, it is critical to comprehend how engaging with a humanoid robot may increase the probability that humans will attribute purposeful agency to the robot.

4.4.4 Drug discovery and development

The process of finding and developing new drugs is very costly and time-consuming; it can take anywhere from 12 to 18 years, and it typically costs between $2 and $3 billion. With only 10% of medication proposals moving on to clinical development due to slim odds of success, pharmaceutical companies need to stop further erosion of their profit margins. The use of robots to automate steps in the drug discovery and development pathway facilitates quicker analysis and hit production, which helps to shorten the time and expense required to get a medicine from the laboratory to the clinic. Improved medical care and drug discovery are being advanced by bioinspired robots.

These robots can find drug candidates, forecast their safety and efficacy simulate biological processes, and analyze massive datasets to speed up drug development. This could revolutionize the pharmaceutical industry and lead to more effective and precise therapies for several diseases. Robotics has many benefits when it comes to complicated, repetitive jobs. It eliminates human mistake, increases productivity by doing such tasks considerably faster, and can even conduct tests in parallel. AI and robotics are well suited to enhancing all of these factors, as drug discovery is frequently hampered by high prices, protracted delays, and high failure rates.

4.5 ADVANCEMENTS IN AGRICULTURE

A new generation of bioinspired, data-driven robots is changing farming. These self-sufficient machines plant, irrigate, harvest, and assess soil and crop health. These robots analyze sensor and camera data to maximize resource allocation, waste minimization, and agricultural yields. One such example is shown in Figure 4.4 (Navas et al. 2021). The use of agricultural robots in digital agriculture has many benefits for farming productivity. Robotics has fascinated industry and research since the 1950s, when industrial robots were invented. The field of agricultural robotics has experienced rapid evolution because of advancements in computer science, sensing, and control approaches. For a wide range of application scenarios, they now rely on a number of cutting-edge technologies. In reality, significant advancements have been achieved by fusing the tactics of execution, control, decision-making, and observation.

Agriculture robots are meant for farming production. Due to their importance in the robot family, they frequently have advanced vision, control, decision-making, and execution. Even in difficult, dangerous, and complex situations, they accomplish production targets accurately and efficiently (Cheng et al., 2023).

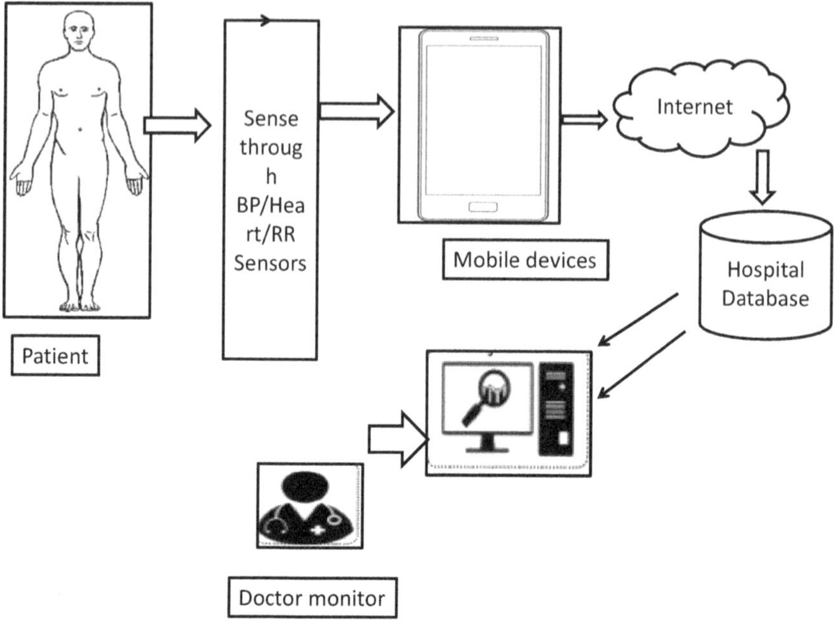

Figure 4.4 Soft gripper robotic arm used in farming.

4.5.1 Precision agriculture and resource management

Precision agriculture uses data analytics and cutting-edge technologies to maximize agricultural productivity. Data on weather, crop health, and soil conditions is needed to make informed planting, fertilizing, and harvesting decisions. This data-driven strategy helps farmers maximize productivity while minimizing costs and environmental impact. Precision farming is being transformed by bioinspired robots with superior sensors and data processing. Farmers optimize irrigation, fertilization, and insect management with these robots' real-time soil, moisture, and crop health analyses. Sustainable agriculture relies on machines that maximize yields while conserving resources and the environment. Robots automate several tasks in precision agriculture, increasing production. Let us look at many key areas where robots are changing farming.

Robots with advanced imagery can identify and selectively eliminate weeds without chemicals. This strategy improves crop health and reduces farming's environmental impact.

1. For better germination, agricultural robots can accurately plant seed at the proper depth and spacing. This precision boosts crop yields and plant growth.

2. 24/7 robotic crop harvesting technologies reduce physical labor. Automation boosts productivity, reduces losses, and allows farmers to harvest crops at their ripest stage.

3. Sensor-equipped robots can measure crop health, nutrient levels, and soil moisture in real time. To maximize crop productivity, this data is examined to guide irrigation, fertilization, and pest management decisions.

4.6 ENVIRONMENTAL MONITORING AND CONSERVATION

Robots must protect and monitor the environment utilizing bioinspired techniques and data. They can detect pollution and habitat degradation, collect environmental data, and negotiate tough terrain. Real-time data processing allows these robots save biodiversity and ecosystems swiftly. Robotic systems provide new insights into the earth and its environmental processes, therefore scientists are increasingly adopting them to collect data. Modern robots are studying deep waters, climatic variables, pollution, algal blooms, and distant volcanoes; over the past 20 years, aerial, marine, and terrestrial robotic systems for environmental monitoring have developed. The study also discusses cooperative robotic teams, adaptive sampling, robot-wireless sensor network (WSN) interaction, and model-aided path planning as potential methods for large-scale environmental monitoring. These innovations increase robotics' accuracy and efficiency in environmental process assessment at unprecedented sizes. Following multiple major natural disasters and ecological damage, such as nuclear meltdowns, oil spills, vast forest fires, hurricanes, floods, tsunamis, and volcanic eruptions, intensive and continued environmental monitoring is needed.

We need a lot of exact spatial and temporal distributed data to understand and quantify environmental health, processes, stressor reactions, and trajectories. Meteorologists monitor air pressure, temperature, and airflow to anticipate weather. Environmental scientists track contaminant transport in air and water. To model and track operations, airflow and dangerous gases like CO_2 and NOx must be monitored. Ecologists study systems to identify, classify, and track living species in their habitats and monitor historical physical and chemical amounts (Dunbabin, 2012).

4.7 WILDLIFE CONSERVATION AND ANTI-POACHING EFFORTS

Conservation projects use bioinspired robots to stop poaching, protect endangered animals, and monitor the environment. These robots can count animals, patrol protected regions, and be vigilant for poaching and destruction. These robots provide real-time data and monitoring to enforce conservation regulations and respond to threats, protecting biodiversity and sensitive ecosystems. Robotic AI is one of the "top three emerging technologies being used in conservation" that helps park rangers prevent endangered species' poaching by automating wildlife data collection to successfully monitor and safeguard enormous African areas that have only been patrolled, for which robotics and AI may be the answer.

4.8 MARINE EXPLORATION AND OCEAN CONSERVATION

Marine research and conservation benefit from bioinspired robots. These robots can map underwater terrain, investigate deep-sea environments, and observe marine life using their cameras and sensors. These robots assess ocean contaminants, temperature, and acidity to help researchers study marine ecosystems and human impact. Unmanned underwater vehicles (autonomous underwater vehicle [AUVs] or remotely operated vehicle [ROVs] are called underwater robots (Cui et al. 2023). To explore and sample aquatic ecosystems, AUVs and ROVs can be equipped with sensors, cameras, and other instruments. Due to their ability to explore new depths and conditions, underwater robots have transformed our vision of the ocean.

Since the 1950s, underwater robots have collected data for mapping the seafloor, investigating marine habitats, and monitoring the environment. First AUV "Self-Propelled Underwater Research Vehicle" studied underwater acoustics and submarine wakes. Since then, underwater robots' "intelligent" skills have allowed them to change shapes and accomplish many functions.

Stanford University scientists constructed the OceanOne underwater robot model, a humanoid AUV that surveys coral reefs and benthic systems like humans. OceanOne can perform prolonged human tasks with its wrists and fingers at depths inaccessible to humans. It is usually used for ecological sampling, but its data might assist scientists to determine how vulnerable a system is to human influences (Ducker 2023).

4.9 DISASTER RESPONSE AND ENVIRONMENTAL REMEDIATION

Natural disasters always strike without warning, causing widespread damage and risking lives. Every second matters, thus cutting-edge technology is critical to speeding rescue efforts. Disaster response and environmental cleanup require bioinspired robots. These robots can travel dangerous terrain, analyze infrastructure damage, and find emergencies. Rescuers employ these robots to concentrate resources and manage cleanup and recovery, limiting environmental harm and restoring nature. Their measurements include air quality, pollutants, and structure integrity. Robotics are essential for disaster relief as natural disasters become more frequent and unpredictable. AI and drones are vital for real-time assessment, search and rescue, and data processing. Human–robot collaboration has the greatest potential to reduce mortality rates, allocate resources, and speed up recovery. By using these technologies, governments and aid organizations can improve disaster response. Robotics will transform disaster response beyond imagination. Robots can speed up and focus natural disaster response, improving emergency operations.

Robotics aids human responders with speedy evaluation, search-and-rescue, dangerous jobs, resource allocation, and data collection. Adopting this technology would decrease risks, save lives, and strengthen the future.

1. Drones: Using the air to support relief efforts: Unmanned aerial vehicles (UAVs), or drones, have fast become disaster relief tools from technology. Relief efforts benefit from drones with advanced software, cameras, and sensors for situational awareness, damage assessment, and search and rescue. Due to their versatility and ability to reach remote areas, they are invaluable disaster response tools. Lessons learned concerning drones' disaster response benefits are as follows:

   Quick and affordable: Drones survey broad areas in hours or days, reducing reaction time. This helps find urgent issues.

   Enhanced safety: Drones let first responders assess risky areas without risking individuals.

   Real-time data capture: High-resolution cameras and sensors allow drones to collect and broadcast videos, photographs, and other data in real time, aiding resource allocation and decision-making.

   Mapping and damage assessment: Drones can create exact maps and 3D models of damaged areas to help emergency responders assess damage and make preparations.

   Drones with thermal imaging and AI algorithms can locate survivors in disaster zones and direct rescuers, possibly saving lives.

2. Robotics with AI: The intelligence providing efficient disaster relief: Disaster assistance relies on AI to assess vast amounts of data and provide useful information for decision-making. AI systems can foresee disasters and optimize resource allocation, giving rescue teams real-time knowledge. Here are a few of the main benefits of AI for disaster relief:

   Early warning systems: Based on historical data and weather trends, AI algorithms can predict disasters and alert authorities.

   Optimal resource distribution: By monitoring social media, sensors, and satellite imagery, AI may identify emergency zones. This allows for the most efficient distribution of aid and resource allocation.

Quick damage assessment: AI-powered data analysis and picture identification help quickly assess disaster damage for a more focused response.

Intelligent infrastructure restoration: Through aerial photography and structural damage data analysis, AI algorithms help specialists restore vital infrastructure swiftly and prioritize activity.

Smart decision-making: AI algorithms speedup infrastructure restoration and prioritize activity using aerial photos and structural damage data analysis.

4.10 ADVANCEMENTS IN AUTOMATION AND MANUFACTURING

Bioinspired processed data-driven robots are changing production and automation. These robots use powerful algorithms to do complex tasks, process data in real time, and adapt to changing conditions. Industries profit from increased output, lower prices, and improved quality control. These robots are improving assembly line and warehouse management productivity and global competitiveness.

In some industries, assembly line robots are becoming more common. These robots improve manufacturing productivity and uniformity by repeating processes quickly and precisely. A typical assembly line robot does pick-and-place, welding, painting, and palletizing.

4.11 CHALLENGES AND FUTURE DIRECTIONS

Bioinspired robots are still in their infancy, and they face several challenges and opportunities for future research and development. The field of *emerging robots using bioinspired processes* is shaped by several obstacles as well as intriguing potential paths forward. Simplifying design complexity, attaining strong control, guaranteeing energy efficiency, addressing safety and bio-compatibility, and resolving moral conundrums are some of the challenges (Massé et al. 2017). Downsizing, transdisciplinary cooperation, bioinspired learning, long-term stability, and ethics will drive biohybrid robot development. A new era of innovation is anticipated as biology, engineering, and ethics come together to blur the lines between artificial and natural systems.

REFERENCES

Cheng, Chao, Jun Fu, Hang Su, and Luquan Ren. (2023). "Recent Advancements in Agriculture Robots: Benefits and Challenges." *Machines* 11, no. 1, 48. https://doi.org/10.3390/machines11010048

Cui, Zhongao, Liao Li, Yuhang Wang, Zhiwei Zhong, and Junyang Li. (2023). "Review of Research and Control Technology of Underwater Bionic Robots." *Intelligent Marine Technology and Systems* 1, no. 7, 1–28.

Ducker, James. (2023). Underwater Robots in Marine Research. Available at www.azorobotics.com/authors/megan-craig.

Dunbabin, M. and Marques L. (2012). "Robots for Environmental Monitoring: Significant Advancements and Applications." *IEEE Robotics & Automation Magazine* 19, no. 1, 24–39. doi: 10.1109/MRA.2011.2181683

Fukuda, Toshio, Fei Chen, and Qing Shi. (2018). "Special Feature on Bio-Inspired Robotics." *Applied Sciences* 8, no. 5, 817–824.

Gleadow, Roslyn, Jim Hanan, and Alan Dorin. (2019). "Averting Robo-Bees: Why Free-Flying Robotic Bees Are a Bad Idea." Edited by Chris Willmott. *Emerging Topics in Life Sciences* 3, no. 6, 723–729.

Gunderman, Anthony, Jeremy Collins, Andrea Myers, Renee Threlfall, and Yue Chen. (2022). "Tendon-Driven Soft Robotic Gripper for Blackberry Harvesting." *IEEE Robotics and Automation Letters* 7, no. 2, 2652–2659.

Jeddi, Zineb, and Bohr Adam. (2020). –"Remote patient monitoring using artificial intelligence." Edited by Adam Bohr and Kaveh Memarzadeh. *Artificial Intelligence in Healthcare*. Academic Press, Denmark, pp. 203–234, ISBN 9780128184387. https://doi.org/10.1016/B978-0-12-818438-7.00009-5

Kondoyanni, Maria, Dimitrios Loukatos, Chrysanthos Maraveas, Christos Drosos, and Konstantinos G. Arvanitis. (2022). "Bio-Inspired Robots and Structures toward Fostering the Modernization of Agriculture." *Biomimetics* 7, no. 2, 69–101.

Kopacek, Peter. (2016). "Development Trends in Robotics." *IFAC-PapersOnLine* 49, no. 29, 36–41.

Li, Junfei, Zhe Xu, Danjie Zhu, Kevin Dong, Tao Yan, Zhu Zeng, and Simon X. Yang. (2021). "Bio-Inspired Intelligence with Applications to Robotics: A Survey." *Intelligence & Robotics* 1, no. 1, 58–83.

Massé, Francis, Alan Gardiner, Rodgers Lubilo, and Martha Themba. (2017). "Inclusive Anti-Poaching? Exploring the Potential and Challenges of Community-Based Anti-Poaching." *South African Crime Quarterly* 60, 19–27.

Mukherjee, Rudra, Priyanka Ganguly, and Ravinder Dahiya. (2023). "Bioinspired Distributed Energy in Robotics and Enabling Technologies." *Advanced Intelligent Systems* 5, no. 4, 1–20.

Navas, Eduardo, Roemi Fernández, Delia Sepúlveda, Manuel Armada, and Pablo Gonzalez-de-Santos. (2021). "Soft Grippers for Automatic Crop Harvesting: A Review." *Sensors* 21, no. 8, 2689.

Ni, Jianjun, Liuying Wu, Xinnan Fan, and Simon X. Yang. (2016). "Bioinspired Intelligent Algorithm and Its Applications for Mobile Robot Control: A Survey." *Computational Intelligence and Neuroscience*, 2016, 1–16.

Picozzi, Paola, Umberto Nocco, Greta Puleo, Chiara Labate, and Veronica Cimolin. (2023). "Telemedicine and Robotic Surgery: A Narrative Review to Analyze Advantages, Limitations and Future Developments." *Electronics* 13, no. 1, 124–142.

Ranjbar-Sahraei, Bijan, Karl Tuyls, Ipek Caliskanelli, Bastian Broeker, Daniel Claes, Sjriek Alers, and Gerhard Weiss. (2015). "Bio-Inspired Multi-Robot Systems." *Biomimetic Technologies*, 273–99.

Russo, Ludovico, Stefano Rosa, Marcello Maggiora, and Basilio Bona. (2016). "A Novel Cloud-Based Service Robotics Application to Data Center Environmental Monitoring." *Sensors* 16, no. 8, 1255–1273.

Shi, Liwei, Maki K. Habib, Nan Xiao, and Huosheng Hu. (2015). "Biologically Inspired Robotics." *Journal of Robotics* 2015, 1–2.

Zheng, Yu-Jun, Sheng-Yong Chen, Yao Lin, and Wan-Liang Wang. (2013). "Bio-Inspired Optimization of Sustainable Energy Systems: A Review." *Mathematical Problems in Engineering* 2013, 1–12.

Chapter 5

IoT's golden age

Pioneering societal change and economic advancements

Aakansha Khanna and Inzimam Ul Hassan

5.1 INTRODUCTION

Internet of Things (IoT)-driven data intelligence is the process of gathering, examining, and drawing conclusions that can be put to use from the vast volumes of data that IoT devices generate. IoT devices have networking features and sensors built in, which enable them to collect and send data over the internet. IoT devices continuously produce enormous amounts of data. Information regarding user interactions, environmental conditions, device status, and more may be included in this data (Ghazal et al., 2021). Due to its size, speed, and variety, it is frequently referred to as "big data". Using several communication protocols like message queuing telemetry transport (MQTT), hypertext transfer protocol (HTTP), or constrained application protocol (CoAP), IoT data is gathered from devices and sent to centralized servers or cloud platforms. A large network of devices, known as an IoT ecosystem, or a single device can both be used to collect data. IoT data is often kept in cloud-based databases or on-premises data lakes. This data can expand quickly and become complex, so it needs to be stored safely and effectively. IoT data frequently has to be preprocessed to extract valuable insights. Data cleansing, standardization, aggregation, and transformation may be necessary for this. For managing data as it comes in, real-time data streaming platforms like Apache Kafka are frequently employed. To find patterns, trends, and anomalies in the data, data intelligence uses a variety of analytics approaches. Descriptive, prescriptive, and predictive analytics can all be a part of this. In this stage, machine learning and artificial intelligence (AI) algorithms are crucial (ABI Research, 2017). To make data more comprehensible for stakeholders, information is frequently represented utilizing dashboards and data visualization tools. Graphs, charts, and real-time displays are a few examples of visualizations (Espinoza et al., 2020). Obtaining actionable insights from data is the main objective of IoT-driven data intelligence. These insights can improve user experiences, operations, efficiency, and decision-making processes. Predictive maintenance is one frequent use of IoT data

DOI: 10.1201/9781003530077-5

intelligence. Organizations can forecast when equipment will need maintenance by evaluating data from industrial IoT (IIoT) devices. Smart city efforts leverage IoT-driven data intelligence to monitor traffic, pollution, energy use, and other factors. Real-time data from environmental sensors can be used to monitor the climate and anticipate disasters. It is important to manage the security and privacy of IoT data. Considerations like data encryption, access controls, and adherence to data protection laws (like General Data Protection Regulation (GDPR)) are crucial (Kelly et al., 2020). As the number of IoT devices increases, scaling becomes a significant difficulty. Platforms for IoT data intelligence must be able to manage datasets that are getting bigger and more complex. Some IoT data processing is migrating to the edge (closer to the devices) to minimize latency and improve real-time decision-making. Edge computing refers to this, and it is particularly pertinent for applications that demand minimal latency.

The IoT-driven data intelligence, in short, entails the gathering, archiving, processing, analyzing, and visualizing of data produced by IoT devices to derive insightful conclusions and promote reasoned decision-making across a range of fields, from industrial settings to healthcare, smart cities, and beyond (Thomas, 2023). With development in technology, more businesses have been harnessing the power of IoT data for efficiency and innovation gains; thus this industry is always changing.

5.2 BENEFITS OF THE IOT

- Significance of Social and Economic: The importance of social and economic considerations cannot be overstated when discussing IoT-driven data intelligence (Eltresy et al., 2020). IoT and data intelligence integration have the ability to address a range of social and economic issues while also opening up new possibilities. The relationship between social and economic variables and IoT-driven data intelligence is as follows.
- Economic Productivity and Efficiency: By streamlining procedures in sectors like manufacturing, agriculture, and logistics, IoT-driven data intelligence can improve economic efficiency. Analysis of real-time data can result in higher production and lower costs.
- Resource Administration: Energy, water, and raw materials may all be managed more effectively because of IoT sensors and data analytics. Businesses and governments may benefit from huge cost savings as a result.
- Supply Chain Improvement: IoT-driven data analytics may increase the visibility and effectiveness of the supply chain (Li et al., 2021), cutting down on waste and assuring on-time delivery. This affects both consumers and businesses economically.

- Prevention-Based Maintenance: For sectors including manufacturing, aviation, and energy production, predictive maintenance, made possible by IoT data, can lower maintenance costs and downtime, having a favorable influence on economic outcomes.
- Market Knowledge and Shopper Behavior: Consumer behavior is affected by social and economic variables. IoT data intelligence can help organizations customize their products and services to fit customer preferences and increase sales by offering insightful information about the market.
- Urban Design and Smart Cities: Intelligent data driven by IoT is essential for building smart cities. It can optimize energy use, transportation networks, and urban infrastructure, which will ultimately save costs and enhance inhabitants' quality of life.
- Healthcare and Social Services: In the healthcare industry, IoT-driven data intelligence can optimize resource allocation, lower costs, and improve patient care. Similar to other sectors, social services can gain from data-driven decision-making for more focused and effective delivery (Li et al., 2021).
- Data Security and Privacy: IoT-driven data intelligence must take into account both economic and social concerns about data security and privacy. Maintaining public trust and preventing economic damage from data breaches require the protection of sensitive data (Kelly et al., 2020).

5.3 IMPACT IOT AND DATA INTELLIGENCE: FOUNDATIONS AND CONCEPTS

5.3.1 Understanding IoT and its components

The impact of IoT and data intelligence is profound and extends across various sectors and industries. These technologies have fundamentally transformed the way we collect, analyze, and utilize data, leading to numerous benefits and challenges. The IoT is a network of physically connected and networked items, including cars, buildings, and other physical objects. These objects include sensors, software, and network connectivity (Ghazal et al., 2021). Through the internet, these devices can gather and exchange data with one another, centralized systems, and cloud platforms. Numerous industries, including healthcare, transportation, manufacturing, agriculture, and smart cities, have used IoT for a variety of purposes. Knowing the foundational elements of IoT is essential for comprehension.

IoT gadgets come with sensors that gather information from the real world. These sensors are capable of measuring a wide range of variables, including pressure, light, motion, humidity, and more. IoT devices frequently feature actuators in addition to sensors, so they can respond to data. Actuators, for instance, can modify a smart thermostat's temperature setting or a smart fan's speed. IoT devices transmit data using a variety of communication protocols. Wi-Fi, Bluetooth, Zigbee, LoRaWAN, cellular networks (3G, 4G, and 5G), and MQTT (a lightweight messaging protocol) are examples of common protocols.

In some IoT setups, data is gathered from many devices and forwarded to the cloud or a central server using gateways (Coetzee & Eksteen, 2011). Data preparation operations can also be carried out via gateways. The processing speed and memory of IoT devices are frequently constrained. Edge computing (Singh, 2023) reduces data transfer and allows for quicker reaction times by processing data locally on the device or at the edge of the network. To store, process, and analyze data, many IoT applications use cloud platforms. Scalability and access to cutting-edge analytics tools are provided by cloud services. The IoT generates significant amounts of data, which must be efficiently and securely stored. Data can be kept in on-site storage systems, data lakes, or cloud databases. IoT systems frequently incorporate user interfaces that let users interact with devices or get data insights. These interfaces can be command-line interfaces, mobile apps, or web-based dashboards.

IoT applications can range from straightforward industrial control systems to complicated home automation projects. Applications determine how data is gathered, handled, and utilized. IoT security is a crucial component. Data privacy (UNESCO Inclusive Policy Lab, 2016) must be upheld while protecting devices and networks from online dangers. Access control, authentication, and encryption are crucial security precautions. IoT device management tools assist businesses in controlling and monitoring their IoT equipment. These platforms have the ability to track device status, update firmware remotely, and resolve problems.

5.3.2 Data intelligence and its role in IoT

For IoT solutions to be successful and effective, data intelligence is essential. It entails gathering, analyzing, interpreting, and using data produced by IoT devices to gain insightful knowledge, guide decision-making, and enhance operations. The first step in developing data intelligence is gathering data from various IoT devices and sensors. Real-time data generated by these gadgets includes sensor readings, ambient information, and gadget status details. To combine and organize data from various sources and make it available for analysis, data integration is crucial (Eltresy et al., 2020). To gain insights from IoT data, data intelligence uses a variety of data analytics approaches. This comprises prescriptive analytics to suggest actions, descriptive analytics to explain historical trends, and predictive analytics to anticipate future events. Complex IoT statistics are routinely analyzed using machine learning and AI algorithms, which can spot trends, outliers, and connections that conventional analysis would miss. IoT-driven data intelligence offers the ability for real-time monitoring and alerts. It makes it possible for businesses to react quickly to crucial situations or occurrences, such equipment failures, security breaches, or environmental changes, that are picked up by IoT devices (Zhang, 2021). Predictive maintenance is one of the main uses for IoT and data intelligence. Organizations are able to forecast when repair is necessary, minimizing downtime and maintenance costs, by studying past data from sensors on machinery and equipment.

Data intelligence aids in the allocation and usage of resources for enterprises. IoT data, for instance, can help farmers conserve water resources by guiding irrigation schedules based on soil moisture levels. Data intelligence can improve user experiences in consumer IoT applications by customizing services and suggestions. Smart home appliances, for instance, can change settings according on user preferences and patterns of activity. Data intelligence may increase the visibility and effectiveness of the supply chain (Yu & Li, 2021). Businesses can use it to track inventory levels, manage the flow of goods, and plan more cost-effective logistical routes; hence it enables businesses to take informed decisions. It also offers useful information that can guide strategic planning, the creation of new products, and operational enhancements. Data intelligence is essential for spotting anomalies and security issues in IoT networks. It can recognize strange behavior patterns that can point to a hack or unauthorized access. It can result in considerable cost reductions for enterprises by streamlining operations, decreasing downtime, and avoiding expensive failures. Thus, data intelligence is a crucial aspect of IoT that enables businesses to understand the enormous amounts of data produced by IoT devices. It improves productivity, decision-making, innovation, and efficiency across a range of sectors and applications. The importance of data intelligence in obtaining value from IoT data is on the rise as IoT expands [CEW: Espinoza et al., 2020)].

5.4 SOCIAL IMPLICATIONS OF IOT-DRIVEN DATA INTELLIGENCE

The IoT-driven data intelligence has far-reaching social implications that impact individuals, communities, and society as a whole. While it offers numerous benefits, it also raises significant concerns. IoT devices continuously collect data from various sources, including homes, workplaces, and public spaces. This raises concerns about personal privacy and the potential for unauthorized surveillance. The vast amount of data generated by IoT devices can be a tempting target for cyberattacks. Ensuring the security of this data is crucial to prevent breaches that could have far-reaching consequences. Access to IoT-driven data intelligence is not uniform across society. The digital divide, where some individuals and communities have limited access to technology, can exacerbate existing inequalities.

The ethical use of IoT data is a complex issue. Decisions about how data is collected, used, and shared can have significant ethical implications, including issues related to discrimination and bias. The automation and optimization made possible by IoT-driven data intelligence can lead to job displacement in certain industries. This shift can have social and economic consequences, particularly for workers in low-skilled jobs (Kelly et al., 2020). As society becomes more reliant on IoT devices and data intelligence, there is a risk of dependency. If these technologies were to fail or be compromised, it could disrupt daily life and essential services. Addressing these social implications requires a multidisciplinary approach involving government regulations, industry standards, ethical guidelines, and public awareness. Balancing the benefits of IoT-driven data intelligence with the protection of individual rights and societal well-being is an ongoing challenge that requires careful consideration and collaboration among various stakeholders.

5.4.1 Transformation of communication and connectivity

Technology breakthroughs are a major force behind the significant and ongoing revolution of connectivity and communication. The way we communicate and stay connected in the modern world has been considerably changed by a number of important trends and innovations. Internet access for everyone, or #1: High-speed internet connectivity has made information and communication more accessible to all people. By bridging the digital divide in many areas, it has made it possible for people all around the world to access internet resources including information, services, and job opportunities. The use of smartphones and other mobile devices has completely changed communication. People who have access to mobile internet can use various applications for work and play, access information, and stay connected while they are on the go. Because 5G networks have far higher speeds, reduced latency, and more capacity, they are revolutionizing communications. IoT, driverless driving, and augmented reality are all made possible by this technology. The IoT has connected billions of objects, from wearables and smart appliances to industrial sensors and autonomous machinery. Through this change, ordinary systems and things are now connected to more than just conventional computing equipment. Global connections are made possible by social media platforms, which have evolved into indispensable communication tools. Additionally, they have changed how information disseminates, enabling the quick broadcast of news and trends. Data and applications are now easily accessible from any location with an internet connection, because of cloud computing. It has revolutionized commercial operations and global collaboration, facilitating remote work. Tools for video conferencing and teamwork like Zoom, Microsoft Teams, and Slack were more widely used as a result of the coronavirus disease 2019 (COVID-19) pandemic. These tools are now necessary for virtual meetings and distant work. Numerous advantages have resulted from these changes in connectivity and communication, including better collaboration, easier access to information, and increased convenience. They do, however, also present issues with cybersecurity, digital literacy, and privacy. These changes will continue to influence how we communicate, work, and live in a more connected world as technology develops.

5.4.2 Impact on healthcare and well-being, and enhancing education and accessibility

Healthcare, well-being, education, and accessibility have all seen substantial developments and benefits as a result of communication and connectivity's revolution. Telemedicine involves remote monitoring, for example: The extensive use of telemedicine, which enables patients to consult with medical specialists remotely, has been made possible by improved connection. For patients in isolated or underdeveloped locations as well as during the COVID-19 pandemic, this has been very helpful. IoT gadgets, wearable health trackers, and remote monitoring technologies are becoming essential for gathering patient data and facilitating ongoing health monitoring. This may result in more proactive treatment and early identification of health problems. Healthcare providers now have easier access to patient data because of electronic health records (EHRs) and platforms for health information sharing. This encourages better treatment that is educated and organized while minimizing mistakes. Researchers now have access to and the ability to analyze enormous volumes of medical data from many different sources, allowing for improvements in illness prevention, drug discovery, and treatment planning.

Algorithms powered by machine learning and AI can spot trends and possible health problems, enhancing diagnostic precision and creating individualized treatment plans. With better internet access, people can now find out more about their health, take care of themselves, and make wise decisions about their well-being. Social media and online groups enable patients to interact with people going through comparable medical issues, offering emotional support and exchanging useful information. Emergency response systems have been enhanced by greater connectivity and communication. During emergencies, first responders can obtain real-time data, coordinate their efforts, and more efficiently allocate resources.

Teletherapy and mental health apps have increased accessibility to mental health support, making it more practical and lowering stigma. Well-being features like stress-tracking and mindfulness activities are frequently seen in wearable technology and mobile applications.

While the evolution of communication and connectivity has greatly benefited accessibility, education, well-being, and healthcare, it is crucial to address issues with digital equity, privacy, and security to make sure that everyone can benefit from these advancements in a secure and safe manner. Let us discuss how the IoT and data intelligence have significantly impacted healthcare.

1. Remote monitoring and telemedicine: IoT gadgets like wearable fitness trackers, smartwatches, and medical sensors enable ongoing observation of patients' vital signs and other health indicators.
   - Real-time data transmission from these devices to healthcare professionals allows for remote monitoring and the early identification of health problems.
   - To make healthcare more convenient and accessible, telemedicine platforms leverage IoT technology to link patients with medical specialists for virtual consultations.
2. Predictive Analytics and Early Disease Detection:
   - Data intelligence tools analyze large datasets from EHRs, IoT devices, and other sources to identify patterns and trends.
   - Machine learning algorithms can predict disease outbreaks, identify at-risk populations, and assist in early disease detection, helping to prevent or manage health crises more effectively.
3. Personalized Medicine:
   - IoT and data intelligence enable the collection of detailed patient data, including genetic information and lifestyle factors.
   - This data can be used to tailor treatments and therapies to individual patients, improving the effectiveness of medical interventions and minimizing side effects.
4. Medication Management:
   - IoT-enabled smart pill dispensers and medication adherence devices help patients take their medications on time.
   - Data intelligence can track medication adherence, helping healthcare providers intervene when necessary.
5. Healthcare Supply Chain:
   - IoT sensors can monitor temperature-sensitive medications and supplies during transportation and storage, ensuring quality control and safety.
   - Data analytics can optimize supply chain management, reducing waste and costs.
6. Health and Wellness Apps:
   - IoT devices and mobile apps for health and wellness enable individuals to take charge of their own health.
   - Data generated by these apps can be used for self-monitoring, setting health goals, and sharing information with healthcare providers (Eltresy et al., 2020).

5.5 ECONOMIC IMPLICATIONS OF IOT-DRIVEN DATA INTELLIGENCE

The economic implications of IoT-driven data intelligence are substantial and touch upon various aspects of businesses, industries, and economies as a whole. IoT-driven data intelligence has the potential to drive economic growth, improve efficiency, reduce costs, and create new business opportunities (Li et al., 2021). IoT-driven data intelligence optimizes processes in various industries, leading to increased efficiency and productivity. For example, predictive maintenance in manufacturing reduces downtime, and smart agriculture enhances crop yields. Predictive maintenance, energy management, and supply chain optimization enabled by IoT data can lead to significant cost savings for businesses (Yu & Li, 2021).

This can include reduced maintenance costs, energy consumption, and inventory management expenses. IoT-driven data intelligence opens up new business models. Companies can offer data-as-a-service, subscription-based analytics, or value-added services based on IoT insights. Businesses can make more informed and data-driven decisions with the insights derived from IoT data.

This can lead to better resource allocation, product development, and market strategies. IoT data intelligence helps reduce energy consumption and promotes sustainability. This is not only environmentally responsible but can also lead to lower energy costs for businesses and individuals (Kg, n.d.). IoT-driven data intelligence can create new markets and opportunities. For example, the market for wearable devices and smart home products has expanded significantly due to IoT innovations.

IoT data acts as a catalyst for innovation. Companies can uncover novel insights and develop new products and services based on IoT-generated data. IoT-driven data intelligence allows businesses to personalize products and services, enhancing the customer experience and potentially increasing sales and customer loyalty. Some industries, such as healthcare and transportation, are subject to regulations that require data collection and reporting. IoT-driven data intelligence can help businesses comply with these regulations efficiently.

While IoT-driven data intelligence offers significant economic benefits, it also raises concerns about data privacy, security, and potential job displacement. Addressing these challenges is essential to realizing the full economic potential of IoT technologies and ensuring equitable benefits for all stakeholders.

5.5.1 Industry disruption and digital transformation

Digital transformation and industry disruption are two concepts that are transforming the corporate landscape in a variety of industries (Sarker,

2022). These phrases describe the significant transformations brought on by technological advances and the use of digital solutions. Industry disruption is the process through which new technology or business models compete with and eventually supplant established markets, industries, or goods. Disruptions are frequently sparked by technologies that are at first viewed as subpar but eventually present special advantages. Innovative business models that redefine value propositions, pricing strategies, or distribution channels may be introduced by new competitors (Thomas 2023). Disruptions can cause major changes in the dynamics of the market, such as alterations in consumer preferences, the rise of new market leaders, and the demise of established players.

In conclusion, successful implementation of digital transformation efforts by creative firms frequently leads to industry upheaval. Digital transformation involves radically altering how firms run and interact with their customers, not merely implementing new technology. Organizations may stay ahead of market changes and spur growth in the digital age by integrating digital transformation with an innovative culture.

5.5.2 Opportunities for new business models

A wide range of new business model prospects across numerous industries are provided by the IoT. IoT refers to the internet-based interconnection of gadgets and sensors that enables data gathering and communication. Create and market IoT products like smart appliances, sensors, and actuators (Zhang, 2021). This includes developing specialized gear for niche markets like manufacturing, healthcare, or the agricultural sector. Provide data analytics services to assist businesses in making sense of the enormous amounts of data produced by IoT devices. Offer suggestions and insights that can be put into practice for process improvement, cost cutting, and better decision-making. Create cybersecurity solutions specifically adapted for IoT networks and devices. There is a rising need for effective security measures as connected devices proliferate.

Develop platforms that provide IoT as a "service", enabling businesses to quickly connect, manage, and analyze IoT devices without making substantial upfront infrastructure investments. Provide solutions that forecast when machines and equipment need maintenance using IoT data, helping businesses save downtime and lower operating expenses. Create and market consumer-focused IoT goods and services, such as wearables, health monitors, and smart home gadgets. Create IoT solutions for precision agriculture, such as crop health management, soil monitoring, and automated farming equipment. Focus on IoT solutions for industrial automation and manufacturing. This entails developing systems for supply chain management, quality assurance, and process optimization. These possibilities only scratch the surface of IoT's potential business concepts. The secret is to pinpoint a niche, resolve a particular issue, and use IoT technology to benefit customers and clients.

5.5.3 Job market and workforce dynamics

In many different industries, IoT is significantly changing the labor market and workforce dynamics. IoT technologies are influencing changes in industry trends, employment positions, and skill requirements. Professionals with knowledge of IoT-related fields, such as IoT architecture, data analytics, cybersecurity, and IoT device development, are in high demand (Dash, 2020). To design, build, and maintain IoT systems, these specialists are necessary (Espinoza et al., 2020). The IoT has made it possible for software developers to build platforms and applications that link and manage IoT devices. Python, Java, and C++ programming expertise are in high demand. IoT generates enormous amounts of data, necessitating the expertise of data scientists and analysts to glean valuable insights. It is advantageous to be familiar with data analytics technologies, machine learning, and data visualization. A major worry is the security of IoT networks and devices. To guard against potential flaws and breaches, cybersecurity experts with knowledge of IoT security are crucial. To build sensors, actuators, and other IoT devices, hardware designers and engineers are required. This includes expertise in hardware development, embedded systems, and electronics.

Urban planning, data analytics, and IoT system integration play a part in the creation of smart cities. Urban infrastructure, energy efficiency, and control of transportation are all growing fields (Coetzee & Eksteen, 2011). The IoT is automating and revolutionizing manufacturing processes. Automation engineers, robotics experts, and process optimization professionals are among the occupations in IIoT. Organizations are getting assistance implementing IoT solutions from consulting firms and systems integrators. There is a high demand for consultants, project managers, and integration specialists. As IoT solutions grow, there is a demand for sales and marketing experts who can explain the benefits of IoT goods and services to customers and organizations.

The workforce must adapt as IoT develops and penetrates more sectors of the economy by learning IoT-related skills and keeping up with current business trends. For those with the necessary skills and knowledge, the job market in IoT-related industries is vibrant and presents a wide range of opportunities.

5.6 SMART CITIES AND URBAN DEVELOPMENT

Smart cities and urban development have become closely intertwined with the IoT as cities seek to leverage technology to enhance efficiency, sustainability, and the quality of life for residents. IoT plays a pivotal role in transforming urban environments by providing data-driven insights and enabling smarter decision-making. IoT sensors are deployed throughout cities to collect data on various parameters, including air quality, traffic flow, noise levels, temperature, and more. These sensors are often embedded in streetlights, utility grids, and public infrastructure. IoT devices continuously gather data from the urban environment. This data is sent to centralized systems for analysis, where it can be used to monitor trends, identify issues, and make informed decisions. IoT-powered traffic management systems use real-time data from sensors to optimize traffic flow, reduce congestion, and improve transportation efficiency. This includes intelligent traffic signals, smart parking solutions, and public transit management. IoT solutions can optimize waste collection routes based on real-time data, leading to cost savings and reduced environmental impact (Sarker et al., 2022). Smart waste bins equipped with sensors signal when they need emptying. IoT plays a role in energy management by monitoring and controlling lighting, heating, and cooling systems in buildings. Smart grids and meters enable better energy distribution and consumption monitoring. IoT-enabled surveillance cameras, gunshot detection systems, and emergency response systems enhance public safety. Data analytics can help predict and respond to emergencies more effectively. IoT solutions can engage citizens in decision-making processes by providing data transparency and involving them in urban development initiatives through digital platforms and mobile apps. The integration of IoT into urban development and management represents a holistic approach to creating smarter, more efficient, and sustainable cities. However, it also raises important considerations related to data privacy, security, and equitable access to technology. Balancing these factors is essential for the successful implementation of IoT-driven smart city initiatives.

5.6.1 Smart infrastructure and resource management

Modern urban development and sustainability efforts depend heavily on resource management and smart infrastructure. Infrastructure and resource management can be made more effective, sustainable, and responsive to the needs of expanding urban populations because of IoT technologies. Real-time monitoring of energy, water, and gas use is done using IoT-enabled smart meters and sensors (Levina et al., 2017). This information aids utilities in better resource management, leak detection, and waste reduction.

The IoT sensors in water distribution systems find leaks and offer crucial information on the quality of the water. Water use in parks, gardens, and farmland is optimized via smart irrigation systems. IoT-enabled garbage cans and collection trucks optimize waste collection routes. Sensors can detect when a bin is full, cutting down on needless pickups and their associated costs. IoT-based traffic sensors and cameras keep track of traffic flow, spot bottlenecks, and instantly change traffic signals. This shortens commutes and consumes less gasoline.

The IoT data is used by smart public transportation systems to enhance passenger services, vehicle maintenance, and route planning. This improves the standard and effectiveness of public transportation. IoT sensors in infrastructure components (such as bridges, roads, and pipelines) can detect wear and deterioration to allow for preventive maintenance. Critical infrastructure's lifespan is extended as a result (Li et al., 2021). IoT sensors track the state of the temperature, pollution levels, and air quality. Cities may use this information to control pollution, make plans for a resilient climate, and maintain a better environment. IoT platforms and mobile apps engage citizens in urban resource management. Residents can report issues, access information, and participate in sustainability initiatives.

5.6.2 Enhancing quality of life and sustainability

The reason behind many IoT deployments is to focus on improving sustainability and quality of life. IoT technologies offer the instruments and information required to improve the livability, effectiveness, and environmental friendliness of urban and rural environments. Remote patient monitoring, wearable medical equipment, and telemedicine are made possible by IoT, making it easier for people to access healthcare services and enhancing overall health results. The elderly can remain independently and safely in their homes for a longer period of time, because of IoT solutions, such as smart home gadgets and monitoring systems, which improve their quality of life. IoT-connected home automation devices increase comfort and convenience. These consist of voice-activated assistants, smart lighting, security systems, and thermostats. IoT-powered mobility options, such ride-sharing applications and real-time transit information, improve commuter efficiency and ease residents' stress and traffic. IoT aids remote learning and online education by giving students of all ages and backgrounds access to educational opportunities and resources. The IoT improves public safety by enabling smart surveillance, emergency response mechanisms, and early disaster warning signals, instilling in the inhabitants a sense of security.

IoT platforms and applications include users in neighborhood activities, promoting a feeling of community and social cohesiveness. IoT devices optimize energy use in homes and buildings, lowering utility costs and having a positive influence on the environment. IoT sensors used in smart waste management systems reduce overflowing trash cans and enhance sanitation, making cities cleaner and more aesthetically pleasing. Building management systems and smart grids powered by the IoT optimize energy use, cut emissions, and support a more sustainable energy infrastructure. IoT sensors and intelligent irrigation systems keep an eye on weather patterns and soil moisture to ensure that water is used effectively in agriculture and landscaping. IoT sensors continuously track air quality, enabling quick response to cut pollution and safeguard public health. IoT devices monitor environmental variables including temperature, humidity, and pollution levels to help with environmental conservation and early problem identification. IoT applications in agriculture boost crop yields while lowering environmental impact, reducing pesticide and fertilizer use, and optimizing resource utilization. IoT supports the infrastructure for electric vehicle (EV) charging and offers EV customers real-time data, encouraging eco-friendly mobility options.

5.6.3 Economic growth and urban innovation

In the framework of IoT, economic development and urban innovation are intertwined, with potential to have a significant impact on cities and regions. IoT technologies present a wide range of options for urban development, efficiency, and economic growth by utilizing the internet to connect physical items and devices for data collection and automation (Sarker, 2022). Skilled labor is needed for data analysis, maintenance, and implementation of IoT deployments in cities. Jobs are made available in a variety of industries, such as technology, data analytics, and telecommunications. By incorporating IoT into their smart city plans, cities can enhance urban services including public safety, waste management, energy, and transportation. These upgrades may draw customers and residents, promoting economic expansion. Hubs and incubators for the IoT inspire startups and business owners to create IoT solutions. These centers encourage innovation, produce jobs, and advance the economy.

Adoption of IoT can lead to growth in traditional areas like manufacturing, agriculture, and utilities. Increasing productivity and competitiveness can be achieved through automation, predictive maintenance, and data-driven decision-making (Levina et al., 2017). Investments in digital infrastructure, such as IoT networks and data centers, can lead to economic growth. These investments enable new business models and assist IoT deployments. IoT-enabled urban infrastructure and smart buildings can boost property prices and encourage real estate development. Through construction, property management, and related services, this can stimulate economic growth. The IoT creates enormous volumes of data that may be evaluated to help people make wise choices. This data-driven strategy can increase economic efficiency, optimize resource allocation, and lower costs. IoT-powered transportation systems can enhance public transit, lessen traffic congestion, and minimize transportation costs for people and businesses, all of which can help the economy develop.

The integration of IoT technologies into urban environments can lead to economic growth through job creation, industry expansion, improved services, and enhanced sustainability. Urban innovation in IoT is an essential driver of economic development, fostering a dynamic and competitive urban landscape.

5.7 ETHICAL AND PRIVACY CONSIDERATIONS

IoT has the potential to greatly benefit society by connecting devices, collecting data, and automating various processes. However, it also raises significant ethical and privacy considerations. IoT devices often collect large amounts of personal and sensitive data. Ensuring that this data is collected only with informed consent is crucial. Data transmitted between IoT devices and cloud servers should be encrypted to prevent unauthorized access. Clear policies regarding how long data is stored and for what purposes should be in place. Users should be fully informed about what data is being collected, how it will be used, and who will have access to it (Espinoza et al., 2020). They should have the option to opt in or opt out of data collection. Clarification on data ownership is important. Users should have control over their data, and IoT companies should not assume ownership or unrestricted access to it. Efforts should be made to anonymize data whenever possible to protect individual privacy. Aggregated, nonidentifiable data should be used for analytics when feasible (Martin, 2022). IoT devices are often vulnerable to cyberattacks due to inadequate security measures. Manufacturers should prioritize security, provide regular updates, and patch vulnerabilities promptly. IoT technologies should be designed to be inclusive, ensuring that all users, including those with disabilities, can use and benefit from them. Consideration of the environmental impact of IoT devices, including their energy consumption and end-of-life disposal, is an ethical concern. Many IoT devices use AI and machine learning algorithms. Ethical considerations around these technologies, such as bias and fairness, also apply to IoT (Sarker et al., 2022). Ethical considerations should be integrated into the design and development of IoT devices from the outset to proactively address potential issues. Addressing these ethical and privacy considerations in the design, deployment, and regulation of IoT technologies is essential to ensure that the benefits of IoT can be realized without compromising individual privacy and societal values. It requires collaboration among stakeholders, including governments, industry, and civil society, to establish a framework that balances innovation with ethical and privacy safeguards.

5.7.1 Data privacy and security challenges

There are several complicated issues with data privacy and security associated with IoT. Due to their ubiquitous use and interconnectedness, IoT devices present certain dangers and vulnerabilities that need to be addressed. Because IoT devices frequently have constrained computer power, they may be more vulnerable to security attacks. Many devices do not have security features built in and do not get regular security updates. The security of data exchanged between IoT devices and cloud servers may be lacking, making it possible for hackers to intercept or alter sensitive data. Inadequate encryption procedures leave data transmissions open to eavesdropping. To safeguard data while it is being transmitted, strong encryption must be used. Malicious actors may take advantage of IoT device default or weak credentials. To prevent unwanted access, appropriate authentication and authorization systems must be in place. IoT devices frequently gather a lot of personal data, including data on location, health, and behavior. Serious privacy violations could occur as a result of unauthorized access or improper use of this data. IoT devices are prime targets for cyberattacks, especially those found in vital infrastructure. Wide-ranging effects of a successful breach can include data theft and infrastructure degradation. As many IoT devices do not have ways to get security upgrades, they eventually become exposed to known security flaws.

Because there are numerous companies involved in the IoT ecosystem, it is difficult to guarantee the security of each component. Security breaches may result from weak supply chain links. Users might not be aware of the privacy implications of IoT devices, including the data they gather and how it is utilized (Coetzee & Eksteen, 2011). Transparency and education are crucial (Dash, 2020). New security vulnerabilities and attack channels are emerging as IoT technology develops, necessitating ongoing vigilance and the adaption of security solutions.

It takes a multifaceted strategy to address these issues, encompassing consumers, manufacturers, regulators, and the cybersecurity community. It entails putting in place strong security procedures, updating hardware frequently, and educating people about the value of data privacy and security in the IoT ecosystem. To guarantee the security and privacy of IoT devices, rules and regulations are also crucially created by governments and industry organizations.

5.7.2 Ethical implications and responsible use

IoT raises a variety of ethical issues and obligations for people, businesses, and society at large. IoT technology must be used responsibly if its advantages are to be realized and potential risks are to be reduced. IoT devices frequently capture enormous volumes of sensitive and personal data. The ethical duty is to seek users' informed consent and limit data gathering to what is required for the function of the gadget. To preserve people's privacy, acquired data must be secured using strong encryption, access controls, and secure storage. Protecting IoT devices from illegal access, data breaches, and other safety risks is crucial. It is the duty of manufacturers to create secure gadgets, release them on schedule, and release security upgrades. Transparency is crucial for IoT operations, data consumption, and business practices. The operation of IoT devices and the utilization of user data should be made explicit to users. IoT technologies should be created with accessibility in mind, making sure that everyone can utilize them and gain from them. When using IoT responsibly, it is important to take sustainability, energy consumption, electronic waste, and other factors into account. AI and machine learning algorithms used in IoT devices should be created and evaluated to reduce bias and guarantee justice, especially in areas like predictive analytics and decision-making. Users must clearly understand who owns and controls the data produced by IoT devices, including the ability to transfer or erase their data. It is crucial to establish responsibility for data breaches, security flaws, and unethical usage of IoT devices. Users, service providers, and manufacturers all need to be aware of their duties. Organizations using IoT solutions have a core ethical obligation to comply with data protection and privacy laws, including GDPR, California Consumer Privacy Act (CCPA), and Health Insurance Portability and Accountability Act (HIPAA). IoT should be created from the beginning with ethics in mind. When creating and implementing IoT solutions, businesses should take into account the potential effects on people and society. It is essential for both consumers and businesses to raise awareness and educate themselves on the ethical implications of IoT technology. The development of IoT ethical standards and guidelines requires cooperation amongst stakeholders, including the government, business, and civil society. IoT ethical considerations are dynamic and constantly changing, need continuing discussion and adaption to handle new problems. To ensure that IoT benefits people and society as a whole while limiting potential harm and hazards, responsible use of IoT technology requires striking a balance between innovation and ethical values [Espinoza et al., 2020].

5.7.3 Legal and regulatory frameworks

Legal and regulatory frameworks related to IoT vary from country to country, and they are continuously evolving to address the unique challenges posed by IoT technology. These frameworks encompass a range of issues, including data privacy, security, liability, and standards. GDPR sets strict rules for the processing of personal data, including data generated by IoT devices (Sarker, 2022). It requires informed consent, data breach notification, and the appointment of data protection officers in certain cases. Similar to GDPR, CCPA grants California residents specific rights regarding their personal data, including the right to know what data is collected and the right to request its deletion. The United Kingdom (UK) and the European Union (EU) are developing security certification schemes for IoT devices to ensure they meet certain security standards. IoT devices often rely on wireless communication networks. Regulatory bodies oversee the allocation of spectrum, standards, and licensing for IoT-related radio frequency usage. Various organizations, such as the International Telecommunication Union (ITU) and the Institute of Electrical and Electronics Engineers (IEEE), develop standards for IoT technology to ensure interoperability and security. Legal frameworks are evolving to determine liability in cases where IoT devices malfunction, cause harm, or are involved in accidents. Questions of liability may involve manufacturers, service providers, or users. Governments are increasingly introducing regulations and standards to address cybersecurity risks associated with IoT devices, including the requirement for regular security updates (Segkouli et al., 2022). Consumer protection laws often come into play when IoT devices fail to meet their advertised capabilities or pose risks to consumers. Data generated by IoT devices often crosses borders. Legal frameworks related to data transfer and cross-border data flow impact IoT deployments. Some regions are exploring ethical frameworks for IoT, particularly related to issues such as bias in AI algorithms and the ethical use of data. Manufacturers, service providers, and users who are involved in the IoT must be aware of and adhere to various legal and regulatory frameworks. Compliance promotes trust in IoT technology while also assisting in the protection of individual rights and privacy, ultimately assisting in the responsible and secure adoption of the technology. Regulatory frameworks will probably continue to change as IoT develops to handle new problems and opportunities in the industry (Zhang, 2021).

5.8 CASE STUDIES: REAL-WORLD APPLICATIONS

IoT technology has been used in a variety of fields and applications to boost productivity, reduce costs, and improve user experiences. Popular IoT in smart home applications include the Nest Thermostat, which is currently owned by Google. To maximize energy economy, it learns user preferences and modifies heating and cooling settings. Farmers utilize IoT sensors to track the health of their crops, changes in weather, and soil quality. The use of fertilizer, irrigation, and planting schedules are all improved by this data. IoT-enabled farm equipment is available from companies like John Deere (Li et al., 2021). Healthcare practitioners can remotely monitor their patients' vital signs and chronic diseases because of IoT technologies like wearable fitness trackers and medical sensors. This may result in earlier intervention and fewer readmissions to the hospital. IoT technology is used in Singapore to monitor traffic, save water, and improve public safety. For example, sensors in trash cans notify when they need to be emptied, reducing the number of trips that garbage trucks need to make. To automate order fulfillment, Amazon deploys hundreds of IoT-enabled robots in its warehouses. These machines quickly transfer product shelves to human workers, speeding up the picking and packing process. Manufacturers utilize IoT sensors to continuously monitor the health of their equipment. They can save downtime and enable timely maintenance by predicting when equipment is likely to fail through data analysis. Siemens offers IoT solutions for cities to optimize energy use, lower emissions, and boost sustainability in general. Cities may improve their energy efficiency by using this technology to assist them make data-driven decisions (Sarker et al., 2022). Amazon Go is a convenience store without a cashier that uses IoT sensors and computer vision to let users buy things and make payments devoid of standard checkout procedures. IoT-connected EV charging stations offer real-time information on charging availability and allow for remote monitoring and maintenance. On the Great Barrier Reef, IoT sensors are set up to keep an eye on factors like water temperature and acidity. This aids in the understanding of the health of the reef and how it adapts to environmental changes. The Edge Amsterdam: The IoT sensors in this smart office building regulate the lighting, temperature, and occupancy to give workers a cozy and energy-efficient working (Espinoza et al., 2020). These case studies emphasize how IoT technology can provide real benefits, such as increased efficiency, cost savings, and improved user experiences, and they show the wide spectrum of IoT applications across many industries. IoT is likely to find even more uses as it develops in fields and sectors that we have not yet considered.

5.8.1 IoT-driven data intelligence in healthcare

IoT-driven data intelligence has significant implications for healthcare, as it can lead to improved patient outcomes, enhanced operational efficiency, and better resource allocation. IoT devices such as wearable fitness trackers, smartwatches, and medical sensors enable continuous monitoring of patient vitals, including heart rate, blood pressure, and glucose levels.

Real-time data is transmitted to healthcare providers, allowing for early detection of health issues and timely interventions, reducing hospital readmissions. Patients with chronic conditions like diabetes can benefit from IoT-enabled devices that track glucose levels and medication adherence (Levina et al., 2017).

Physicians can access historical data to make informed decisions about treatment plans, medication adjustments, or lifestyle recommendations. Smart pill bottles and medication dispensers equipped with IoT sensors remind patients to take their medications and record their adherence.

Healthcare providers can monitor patient compliance and intervene when necessary. IoT-enabled wearables and home monitoring systems can automatically detect falls or abnormal vital signs, triggering immediate alerts to caregivers or emergency services.

This reduces response times and improves the chances of a positive outcome in emergencies. IoT sensors can be attached to medical equipment, such as infusion pumps, wheelchairs, and defibrillators, to track their location, status, and maintenance needs. This improves asset utilization, reduces losses, and ensures equipment availability when needed. Hospitals and pharmaceutical companies use IoT to monitor the temperature and condition of sensitive medications and vaccines during transportation.

Real-time data helps maintain product quality and ensures patient safety. IoT sensors and beacons in healthcare facilities track the movement of patients, staff, and equipment. This data can be analyzed to optimize the layout of hospital departments, reduce wait times, and allocate resources efficiently (Segkouli et al., 2022). Machine learning algorithms can analyze vast amounts of IoT-generated healthcare data to predict disease outbreaks, patient readmissions, and resource needs. This helps healthcare organizations allocate resources efficiently and proactively address health challenges. Protecting patient data is crucial. IoT healthcare systems must adhere to strict security and privacy standards, such as HIPAA in the United States, to ensure the confidentiality and integrity of sensitive information. IoT-driven data intelligence is revolutionizing healthcare by providing real-time insights, enabling early interventions, reducing costs, and improving patient care. However, it also raises important ethical and privacy considerations that must be carefully addressed to ensure the responsible and secure use of IoT technology in healthcare.

5.8.2 Smart energy management systems

Smart Energy Management Systems (SEMS) leverage IoT technology to optimize the generation, distribution, and consumption of energy. These systems play a crucial role in improving energy efficiency, reducing costs, and enhancing sustainability in various sectors. IoT sensors and smart meters are deployed to collect real-time data on energy consumption, grid performance, and environmental conditions. These sensors can monitor electricity, gas, water, and other utility usage. IoT devices rely on communication networks, such as cellular, Wi-Fi, or low-power wide-area networks (LPWANs), to transmit data to centralized systems or the cloud. Advanced analytics and AI algorithms process the massive volume of data generated by sensors and meters. These algorithms identify patterns, anomalies, and opportunities for optimization (Coetzee & Eksteen, 2011). SEMS often include control systems that can adjust energy generation, distribution, and consumption based on real-time data and insights (Dash, 2020). This can involve smart grid management, demand response, and load balancing. Users, who may be consumers, facility managers, and utility providers, can access user-friendly interfaces and dashboards to monitor energy consumption, set preferences, and receive recommendations for energy-saving actions. In smart homes, IoT devices like smart thermostats, lighting controls, and energy-efficient appliances help residents monitor and optimize their energy use. Users can remotely control devices and receive recommendations for energy-saving actions. SEMS are widely used in commercial and industrial facilities to reduce energy costs, enhance comfort, and lower carbon footprints. Building automation systems integrates IoT devices to control heating, ventilation, and air conditioning (HVAC), lighting, and other systems based on occupancy, weather, and energy prices. IoT technology helps utilities monitor and manage the distribution grid more effectively. Sensors detect faults and congestion, while smart meters provide real-time data on consumer demand. Demand-response programs use IoT to communicate with appliances and shift electricity usage during peak hours. SEMS optimize the charging and discharging of energy storage systems, such as batteries, to store excess energy during low-demand periods and release it when demand is high or during power outages. IoT-enabled EV charging stations can communicate with EVs to optimize charging schedules and manage grid impact.

Utility providers use SEMS to monitor grid performance, detect leaks or faults, and optimize resource allocation. SEMS are a critical component of the transition to a more sustainable and efficient energy ecosystem. By harnessing the power of IoT technology and data analytics, these systems empower individuals, businesses, and utilities to make informed decisions and contribute to a more sustainable energy future.

5.8.3 Transportation and logistics optimization

The manner that commodities are carried, tracked, and managed has undergone a change due to IoT technologies. IoT solutions in this industry increase productivity, cut costs, boost safety, and give supply chain visibility in real time. Global Positioning System (GPS) and IoT sensors are used to track the whereabouts and state of goods, containers, and vehicles in real-time. The use of resources and route planning are improved with the use of this information. IoT sensors keep an eye on the health of machinery and vehicles. Algorithms for predictive maintenance use data analysis to predict when maintenance is necessary, minimizing downtime and expensive breakdowns (Liu, 2021). IoT technologies monitor driving behavior and fuel use, offering insights for enhancing fuel economy and cutting emissions. IoT sensors in storage and warehouses keep an eye on inventory levels, enabling automatic replenishment requests and lowering the chance of stockouts or overstocking. IoT temperature sensors ensure that perishable commodities are transported and stored at the proper temperature, maintaining product quality and safety. IoT algorithms determine the most efficient routes, saving time and fuel, by taking into account real-time traffic data, weather conditions, and delivery limits (Sarker, 2022). IoT data is utilized to forecast consumer demand, enabling companies to modify their production and inventory plans as necessary. Using IoT data, retailers and transportation companies can create dynamic pricing plans that are dependent on supply, demand, and current circumstances. In-vehicle IoT devices record information on driver behavior, including speed, abrupt braking, and acceleration. Training and safety improvements can be made with the help of this information. IoT solutions support adherence to environmental and safety laws, such as the Hours of Service (HOS) guidelines for drivers. IoT allows for real-time cargo tracking, increasing transparency and providing projected arrival timings. Automated alerts and notifications advise clients of the progress of their delivery and any hiccups. IoT systems can suggest routes that reduce emissions and fuel usage, advancing sustainability objectives. IoT sensors have the ability to spot unauthorized entry or cargo tampering, setting off alerts and notifying authorities in the event of theft or tampering. In addition to increasing efficiency, IoT-driven transportation and logistics optimization also lowers costs, has a smaller negative impact on the environment, and increases customer happiness. The supply chain is becoming more flexible and responsive to the demands of a fast-changing global market as a result of these technologies' ongoing evolution and increasing incorporation of AI and machine learning.

5.9 FUTURE DIRECTIONS AND EMERGING TRENDS

IoT is a dynamic and rapidly evolving field, and several emerging trends and future directions are shaping its development. Edge computing, which involves processing data closer to the source (IoT devices), is gaining prominence. It reduces latency, enhances real-time decision-making, and minimizes the need to send all data to centralized cloud servers. The rollout of 5G networks enables faster, more reliable, and lower-latency communication, making IoT applications like autonomous vehicles, augmented reality, and industrial automation more feasible and efficient. IoT devices are increasingly being equipped with AI and machine learning capabilities (ABI Research, 2020). This enables data analytics at the edge, allowing devices to make intelligent decisions without relying on cloud-based processing. IoT devices are increasingly being equipped with AI and machine learning capabilities. This enables data analytics at the edge, allowing devices to make intelligent decisions without relying on cloud-based processing. Digital twins are virtual replicas of physical objects or systems. IoT-driven digital twins offer real-time insights into the performance and behavior of physical assets, enabling predictive maintenance and optimization. Blockchain technology is being explored for enhancing IoT security by providing secure, tamper-proof data transmission and authentication. This is particularly important in critical applications like supply chain and healthcare. IoT is transforming healthcare through remote patient monitoring, wearable devices, and telemedicine. The future will likely see increased adoption of IoT for personalized medicine and improved patient outcomes. Research into energy harvesting technologies, such as self-powering IoT devices through solar, kinetic, or thermal energy, is ongoing, which could reduce the reliance on traditional power sources. Efforts to establish common IoT standards and interoperability protocols will facilitate seamless integration of diverse devices and ecosystems. The future of IoT promises continued innovation and transformative impact across industries and everyday life. As these trends evolve, addressing challenges related to security, privacy, standardization, and sustainability will be essential to realize the full potential of IoT technology.

5.9.1 Advancements in AI and machine learning

Advancements in AI and machine learning related to IoT are driving significant improvements in IoT applications and capabilities. These technologies enable IoT devices to become smarter, more autonomous, and better at processing and interpreting the vast amount of data they generate. Edge AI refers to the deployment of AI algorithms directly on IoT devices, enabling real-time data processing and decision-making at the edge of the network. This reduces latency and minimizes the need to send data to centralized cloud servers. Federated learning allows IoT devices to collaboratively train machine learning models without sharing sensitive data. This preserves data privacy while still benefiting from collective intelligence. Machine learning algorithms can analyze IoT data streams to detect anomalies or unusual patterns in realtime. This is invaluable for identifying potential security breaches, equipment malfunctions, or quality control issues. IoT cameras and sensors can leverage computer vision algorithms to recognize and interpret visual data (Espinoza et al., 2020). Applications include object detection, facial recognition, and gesture recognition. Reinforcement learning techniques are being used to optimize control systems in IoT devices and autonomous vehicles. These algorithms learn through trial and error, making them suitable for dynamic environments. IoT-connected autonomous vehicles rely on AI for navigation, collision avoidance, and decision-making in complex traffic situations. AI can analyze behavioral data from IoT devices to gain insights into user habits and preferences, which can inform product design and marketing strategies (Sarker et al., 2022).

5.9.2 IoT and data intelligence in emerging industries

IoT and data intelligence are playing pivotal roles in the transformation and growth of emerging industries. These technologies are reshaping operations, enabling data-driven decision-making, and driving innovation in various sectors. IoT sensors, drones, and satellite data are used to monitor soil conditions, crop health, and weather. Farmers can optimize irrigation, fertilization, and pest control, leading to increased crop yields and resource efficiency. IoT devices and wearables allow healthcare providers to remotely monitor patients' vital signs and chronic conditions. Data intelligence provides early warnings and helps tailor treatment plans. Smart hospitals use IoT for asset tracking, patient flow optimization, and ensuring medical equipment is properly maintained. Manufacturers are integrating IoT sensors into machinery and equipment to monitor performance, predict maintenance needs, and improve production efficiency. Data intelligence in manufacturing predicts when equipment is likely to fail, reducing downtime and maintenance costs. IoT enables the monitoring and management of electricity grids, improving grid stability and integrating renewable energy sources seamlessly (Li et al., 2021). IoT sensors optimize the performance of solar panels and wind turbines by adjusting their orientation and tracking weather conditions (ABI Research, 2020). IoT sensors and data intelligence inform urban planners about traffic patterns, environmental conditions, and infrastructure performance, leading to more sustainable and efficient cities. Smart city applications include video analytics for surveillance, emergency response optimization, and predictive policing. IoT connectivity via satellites is opening up new possibilities in agriculture, environmental monitoring, and asset tracking, especially in remote or inaccessible areas. In emerging industries, quantum computing coupled with IoT and data intelligence could revolutionize optimization problems, materials discovery, and cryptography. The combination of blockchain and IoT enhances data security and trust in emerging industries like supply chain management, healthcare, and energy trading. In the education sector, IoT devices and data analytics enhance personalized learning, campus management, and student engagement. These examples highlight the transformative potential of IoT and data intelligence across various emerging industries. As these technologies continue to advance, we can expect further innovation and growth in these sectors, leading to more sustainable, efficient, and data-driven operations.

5.9.3 Anticipated social and economic impact

On society and the economy, IoT is anticipated to have a significant and far-reaching impact. Its influence is felt in a variety of industries, including manufacturing, smart cities, healthcare, and agriculture. IoT devices in healthcare offer remote monitoring and individualized treatment, enhancing the quality of life for elderly patients and patients with chronic diseases (Segkouli et al., 2022). Remote patient monitoring and telemedicine services, particularly in underserved areas, improve accessibility to healthcare, lower healthcare costs, and offer prompt treatments. IoT makes a positive impact on environmental sustainability by reducing waste, monitoring pollution levels, and optimizing resource utilization. Environmental aims are supported by smart grids, effective transportation systems, and precision agriculture. IoT sensors improve public safety by providing real-time monitoring and emergency response in smart cities and public infrastructure. Applications like traffic management, disaster detection, and monitoring fall under this category. The IoT-enabled assistive technology and gadgets increase accessibility for persons with disabilities, enabling greater involvement in society and the workforce. IoT devices in smart homes offer convenience, energy savings, and increased security, helping to create a cozier and safer living space. Personalized learning opportunities are provided by IoT in schools and colleges, which enhances student engagement and performance. IoT optimizes business operations, lowers unproductive time, and boosts output, resulting in cost savings and economic expansion. The development of IoT industries, such as software development, data analytics, and device manufacturing, leads to the creation of jobs and business prospects. IoT technology may promote economic growth in emerging nations by enhancing infrastructure, healthcare, education, and agriculture. IoT data analytics produce insights that help businesses make better decisions, enhance customer experiences, and add value to their operations. IoT presents both social and economic benefits (Thierer & O'Sullivan, 2015), but it also has its drawbacks, such as issues with data privacy, security, and ethics. To guarantee that the promised benefits of IoT are realized while minimizing potential hazards, it is imperative to address these challenges.

5.10 CONCLUSION

In conclusion, the integration of IoT-driven data intelligence into various aspects of society and the economy is poised to bring about transformative changes with far-reaching implications. IoT-driven data intelligence in healthcare, smart homes, and assistive technologies is improving the quality of life for individuals, particularly those with chronic conditions or disabilities. Remote patient monitoring and telemedicine are increasing healthcare accessibility, reducing costs, and providing timely care, particularly in remote or underserved areas. IoT is helping to address environmental challenges by optimizing resource use, reducing waste, and enabling more sustainable practices in agriculture, transportation, and energy management. IoT technologies in smart cities and infrastructure are enhancing public safety through real-time monitoring, emergency response systems, and efficient traffic management. IoT-driven assistive technologies are promoting accessibility and inclusivity, allowing individuals with disabilities to participate more fully in society. IoT-enhanced learning experiences in education are improving student engagement and outcomes, contributing to better-prepared future generations. IoT-driven data intelligence is optimizing industrial processes by reducing downtime and enhancing productivity, resulting in cost savings and economic growth. The growth of IoT industries is creating jobs in manufacturing, software development, data analytics, and related fields. IoT-enabled supply chains are becoming more efficient, reducing costs, minimizing waste, and benefiting both producers and consumers. IoT is fostering economic development in emerging markets by improving agriculture, infrastructure, healthcare, and education. IoT is enabling innovative business models based on subscription services, data monetization, and predictive maintenance, generating new revenue streams. IoT-driven energy management is reducing consumption, lowering costs, and enhancing energy efficiency, benefiting both businesses and consumers. IoT systems for disaster management and recovery are increasing regional resilience to natural disasters and reducing economic impacts. IoT is fostering innovation in multiple industries, enhancing global competitiveness and driving economic growth through technological advancements. Data intelligence from IoT devices allows improved business decisions and customer experiences and increased value creation for organizations. Despite the remarkable potential benefits, it is crucial to address challenges related to data privacy, security, and ethical considerations associated with IoT and data intelligence. Ensuring responsible and secure adoption will be essential to fully realize IoT's positive socioeconomic impacts while mitigating potential risks and vulnerabilities. Overall, the future is promising as IoT-driven data intelligence continues to drive innovation and enhance the well-being of individuals and the prosperity of economies worldwide.

REFERENCES

ABI Research. (2017). *AI and Machine Learning for the IoT*. Retrieved from www.abiresearch.com/blogs/2020/09/18/ai-and-machine-learning-iot/

Coetzee, L., & Eksteen, J. (2011, May). The Internet of Things—Promise for the future? An introduction. In *2011 IST-Africa Conference Proceedings* (pp. 1–9). IEEE.

Dash, S. P. (2020). The impact of IoT in healthcare: Global technological change & the roadmap to a networked architecture in India. *Journal of the Indian Institute of Science, 100*(4), 773–785.

Eltresy, N. A., Dardeer, O. M., Al-Habal, A., Elhariri, E., Abotaleb, A. M., Elsheakh, D. N., … Abdallah, E. A. (2020). Smart home IoT system by using RF energy harvesting. *Journal of Sensors, 2020*, 1–14.

Espinoza, H., Kling, G., McGroarty, F., O'Mahony, M., & Ziouvelou, X. (2020). Estimating the impact of the Internet of Things on productivity in Europe. *Heliyon, 6*(5), e03935.

Ghazal, T. M., Hasan, M. K., Alshurideh, M. T., Alzoubi, H. M., Ahmad, M., Akbar, S. S., …Akour, I. A. (2021). IoT for smart cities: Machine learning approaches in smart healthcare—A review. *Future Internet, 13*(8), 218.

Kelly, J. T., Campbell, K. L., Gong, E., & Scuffham, P. (2020). The Internet of Things: Impact and implications for health care delivery. *Journal of Medical Internet Research, 22*(11), e20135.

Levina, A. I., Dubgorn, A. S., & Iliashenko, O. Y. (2017, November). Internet of things within the service architecture of intelligent transport systems. In *2017 European Conference on Electrical Engineering and Computer Science (EECS)* (pp. 351–355). IEEE.

Liu, Y. (2021). Intelligent analysis platform of agricultural sustainable development based on the Internet of Things and machine learning. *Acta Agriculturae Scandinavica, Section B—Soil & Plant Science*, 71(8), 718–731.

Martin, R. (2022). *Impacts of Internet of Things on Society*. Retrieved from www.mytechmag.com/impacts-of-iot-on-society/

Sarker, I. H. (2022). Smart City Data Science: Towards data-driven smart cities with open research issues. *Internet of Things*, 19, 100528.

Segkouli, S., Fico, G., Vera-Muñoz, C., Lecumberri, M., Voulgaridis, A., Triantafyllidis, A., … Votis, K. (2022, May). Ethical decision making in IoT data driven research: A case study of a large-scale pilot. *Healthcare*, 10(5), 957). MDPI.

Singh, S. (2023). The future of IoT: Emerging trends and technologies in the Internet of Things. Retrieved from https://ifacet.iitk.ac.in/knowledge-hub/internet-of-things/the-future-of-iot-emerging-trends-and-technologies-in-the-internet-of-things/

Thierer, A., & O'Sullivan, A. (2015). *Projecting the Growth and Economic Impact of the Internet of Things*. Retrieved from www.mercatus.org/research/policy-briefs/projecting-growth-and-economic-impact-internet-things

Thomas, M. (2023). IoT in Healthcare: 16 Examples of Internet of Things Healthcare Devices and Technology. Built In. https://builtin.com/articles/iot-in-healthcare]

UNESCO Inclusive Policy Lab. (2016). *Data Privacy and the Internet of Things*. Retrieved from https://en.unesco.org/inclusivepolicylab/analytics/data-privacy-and-internet-things

Zhang, C. (2021, March). Intelligent Internet of Things service based on artificial intelligence technology. In *2021 IEEE 2nd International Conference on Big Data, Artificial Intelligence and Internet of Things Engineering (ICBAIE)* (pp. 731–734). IEEE.

Chapter 6

Impactful future

Harnessing IoT for socio-economic data intelligence

*Padma Nandanan, Hariharan R, and
Shyam Shankar Menon*

6.1 INTRODUCTION

The Internet of Things (IoT) is a technology that enables the connection of various devices and systems to the internet, allowing them to exchange data and interact with each other. The concept of IoT has been around for a long time, but it was not until the early 2000s that the term "Internet of Things" was coined by Kevin Ashton, a British technology pioneer.

The origins of IoT can be traced back to as early as 1960, but significant development started in the early days of the internet when a group of students at the Carnegie Mellon University connected a vending machine to the internet in the early 1980s with the purpose being to monitor the availability of cold drinks and to determine whether the drink was cold enough. The project was the first step toward making everyday objects "smart" by connecting them to the internet.

The concept of IoT has evolved significantly since then as it allows participants in IoT to interact without human intervention. The guiding essence of the IoT is enabling the computers to synthesize information without assistance from humans (Gubbi et al., 2013; Hsu and Yeh, 2016; Li et al., 2019). With the development of sensors, wireless communication technologies, and cloud computing, IoT has become a reality by getting separate pieces of technology to work together and rapidly expanding into various industries, including healthcare, transportation, manufacturing, etc. By leveraging IoT, these industries are able to collect and analyze vast amounts of data in real-time, which has the potential to unlock significant value and improve operational efficiencies. For example, in healthcare, IoT devices are used to monitor patients remotely and provide real-time data to healthcare providers, allowing for better and more personalized care. In transportation, IoT devices are used to monitor and manage traffic, reducing congestion and improving safety. In manufacturing, IoT devices are used to monitor and optimize production processes, reducing downtime and improving efficiency. The performance of SMART factories have increased tremendously through IoT as the latter ensures improved product

DOI: 10.1201/9781003530077-6

customizations, role integration of customers, companies and suppliers, flexibility in production and increased sustainable development (Haller et al., 2009; Shrouf et al., 2014).

The impact of IoT can be felt in various aspects of our lives. For instance, smart homes have become increasingly popular, with consumers using IoT devices to control everything from lighting to thermostats. Smart cities are also emerging, with connected sensors and devices being used to manage everything from traffic flows to energy usage (Madakam et al. 2015).

Overall, the history of IoT is a testament to the power of innovation and the ability of technology to transform the way we live and work across the world. Hoshi et al. (2017) in their paper emphasize on the continuing problem of aging population in Japan. They have highlighted the role of IoT in making farming convenient for the aging population by improving the food production output in Japan.

The objective of this paper is to give a historical background on IoT, its impact on the social, economic, and sustainability, and its effect on individuals, businesses, and society at large.

6.2 THE ALPHA

The history of IoT (Figure 6.1) is marked by several significant milestones with exponential growth since the beginning of the 21st century. While IoT has come a long way, its origins can be traced back to the 1960s when the concept of interconnected devices was first envisioned. In 1964, Carl Steinberg predicted that computers would be interwoven into every industrial product.

The first IoT device, a toaster that could be controlled over the internet, was developed in 1990 by John Romke. It took more than 25 years to build this device. Additionally, Carnegie Mellon University students also connected a Coca-Cola machine to the internet at the same time.

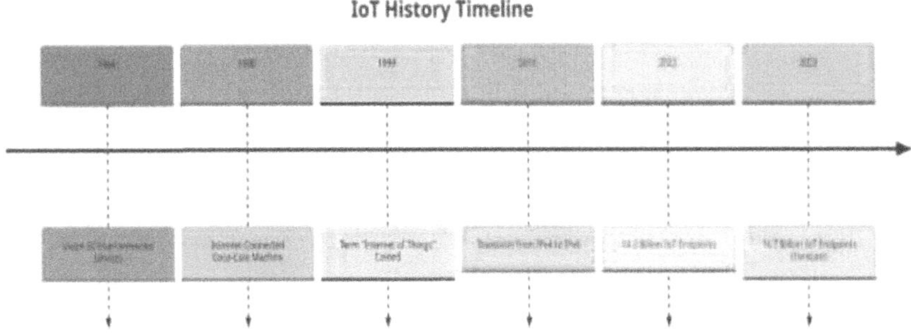

Figure 6.1 IoT history timeline.

The term "Internet of Things" was coined by Kevin Ashton, a British technology pioneer, in 1999. He introduced the term during a presentation for Procter & Gamble, where he discussed linking radio-frequency identification (RFID) tags in the company's supply chain to the internet and this marked a significant milestone in the growth and development of IoT.

The IoT Analytics "State of IoT—Spring 2023" report shows that the number of global IoT connections grew by 18% in 2022 to 14.3 billion active IoT endpoints. In 2023, IoT Analytics expects the global number of connected IoT devices to grow another 16%, to 16.7 billion active endpoints. While 2023 growth is forecasted to be slightly lower than it was in 2022, IoT device connections are expected to continue to grow for many years to come.

In recent years, the IoT industry has experienced rapid growth, with the number of devices expected to reach over 24 billion by 2030. The future of IoT will be shaped by advancements in 5G, artificial intelligence, and advanced analytics. Key developments include the adoption of 5G networks, improvements in smart cars and road safety, the rise of smart cities, increased business adoption, advancements in artificial intelligence for data insights, edge computing for local data processing, and the emergence of massive IoT (Mahmood et al. 2020). These innovations promise to enhance the capabilities of clients across various industries and contribute to the development of technologies that improve our lives.

6.3 IMPACTS UNLEASHED

6.4 SOCIAL IMPACTS

The drastic effect that IoT has on human life and the magnitude of the changes that it introduces results in it being considered a social innovation (Shin, 2017). The impact of IoT on social factors is significant and multifaceted. IoT has had both positive and negative impacts.

6.4.1 Positive impacts

1. *Convenience*: IoT gadgets have increased daily ease. For instance, smart home technologies enable individuals to remotely regulate their equipment, lighting, and temperature controls using their mobile devices, leading to a more comfortable standard of living. For example, Amazon Echo devices, powered by Alexa voice assistant, enable users to control smart home devices, get news updates, and perform various tasks through voice commands. Google Nest offers a range of smart home devices such as thermostats, cameras, and doorbells, as well as Google smart home speakers with Google assistant.

2. *Safety and security*: IoT has a notable part in improving safety and protection measures. Home protection systems, wearable technologies, and broad city surveillance systems are some illustrations of IoT applications that can minimize accidents and discourage criminal behavior. In emergency situations, IoT devices can assist in prompt alerting and response to mitigate risks and provide assistance to those in need. For example, Ring is company known for smart video doorbells and security cameras that allow users to monitor their homes remotely. Arlo offers a range of wireless security cameras designed to provide comprehensive home security solutions.

3. *Healthcare*: IoT is revolutionizing the health sector through wearable health trackers, distantly monitored patients, and intelligent medical tools. These advancements can enhance medical services availability, facilitate early diagnosis of health problems, and elevate patient care. For example, Fitbit is a popular wearable technology company that produces fitness trackers and smartwatches, helping users monitor their health and fitness data. Medtronic is a medical device company that manufactures IoT-enabled devices, such as continuous glucose monitors and insulin pumps for diabetes management.

4. *Efficiency*: Various operations ranging from the managing of supply chains to transport systems can be made more productive with IoT. This could result in conservation of resources, financial savings, and the promotion of sustainability for the future. IoT-enabled transportation systems can optimize traffic flow and reduce congestion, leading to smoother and more efficient travel experiences. For example, Siemens offers a variety of IoT solutions, such as MindSphere, an IoT operating system that enables industrial companies to optimize operations and improve efficiency. Schneider Electric provides IoT-enabled energy management and automation solutions for homes, buildings, and industries, helping to optimize energy usage and reduce waste.

5. *Environmental impact*: IoT can aid in the supervision and control of environmental elements, like waste management, air purity, and water consumption, thereby fostering a greener and more sustainable society. Smart grids can optimize energy distribution and consumption, reducing waste and carbon emissions. IoT sensors can monitor and manage water usage, leading to conservation and improved water resource management. Smart agriculture and environmental monitoring applications can help optimize fertilizer and pesticide usage, leading to reduced environmental impact. Google's Nest smart thermostats learn users' habits and optimize energy usage, helping to reduce energy consumption and lower utility bills. Aclima designs and deploys IoT-based air quality monitoring systems that provide

real-time data on air pollution, helping communities and businesses make informed decisions about their environmental impact.

6.4.2 Negative impacts

1. *Privacy concerns*: The widespread use of IoT devices has raised significant privacy concerns. Many of these devices collect data about users' behaviors and activities, leading to potential data breaches and invasions of privacy.

   Devices like Amazon Echo, Google Home, and Apple Home Pod are always listening for wake words to activate voice assistants. There have been instances where these devices inadvertently recorded private conversations and sent the recordings to random contacts or store them on the company's servers. This raises concerns about who has access to these recordings and how they may be used.

   Smart toys connected to the internet can collect data on children's activities, preferences, and even conversations. In 2017, the smart toy company Cloud Pets leaked personal data, including voice recordings, of millions of users due to poor security practices. This raised concerns about the privacy of children and the potential misuse of their data.

2. *Security risks*: IoT devices can be vulnerable to cyberattacks if not properly secured. Hacked IoT devices can be used in large-scale attacks, and personal information can be compromised. Ensuring strong data protection measures and implementing robust security protocols are crucial to addressing these concerns and maintaining trust in IoT systems.

   In 2016, the Mirai Botnet, which comprised thousands of infected IoT devices like routers and cameras, launched a massive, distributed denial-of-service (DDoS) attack against the Dyn DNS provider. This attack caused widespread internet outages, affecting major websites like Twitter, Netflix, and Reddit.

   In 2017, the Food and Drug Administration (FDA) confirmed vulnerabilities in St. Jude Medical's (now Abbott) IoT-enabled cardiac devices, such as pacemakers and defibrillators. These vulnerabilities could have allowed hackers to gain unauthorized access and control over the devices, posing a potential risk to patients' health and safety.

3. *Job disruption*: Automation and IoT technology can lead to job displacement in certainindustries, as some tasks become automated. This can have social and economic implications.

   The IoT devices like smart shelves, automated checkout systems, and inventory management solutions have disrupted the retail sector, reducing the need for manual labor in stocking shelves and managing inventory. This has led to concerns about job losses in retail positions like cashiers and stock clerks (Ng and Wakenshaw 2017).

The IoT technologies like autonomous vehicles, Global Positioning System (GPS) tracking, and fleet management systems have the potential to disrupt the transportation and logistics industries. As these technologies become more widespread, job roles such as truck drivers, taxi drivers, and delivery personnel may be at risk.

4. *Digital divide*: The digital divide refers to the gap between individuals, households, businesses, and geographic areas with access to modern information and communication technologies (ICT) and those without. Not everyone has equal access to IoT technology due to factors like affordability and digital literacy. This can exacerbate social inequalities.

   The effective use of IoT devices often requires digital literacy and technical skills. Individuals without access to education or training in these areas may face difficulties in using and benefiting from IoT technologies, further exacerbating the digital divide. This was experienced during the coronavirus disease (COVID) pandemic where students without access to smart phones and tablets were at a disadvantage as compared to others.

5. *Ethical concerns*: IoT raises ethical considerations regarding data ownership, consent, and potential inequality. The collection and use of personal data by IoT devices and systems raise questions about who owns the data, how it should be protected, and how individuals' consent should be obtained.

   IoT devices often collect and transmit data without users' explicit consent or awareness. This raises ethical questions about data ownership, informed consent, and users' rights to control their personal information.

   The IoT devices and algorithms can inadvertently perpetuate biases and discrimination, particularly if the data used to train these systems is biased. For example, facial recognition technologies have been found to be less accurate in identifying people of color, which raises concerns about fairness and potential misuse of these technologies.

6.5 ECONOMIC IMPACTS

IoT has significant economic implications, both positive and negative which cannot be overlooked.

6.5.1 Positive impacts

1. *Increased efficiency*: IoT-enabled automation and real-time monitoring leads to improved efficiencies and associated cost savings as well as increased productivity. The sectors where IoT is widely used include manufacturing, logistics, and agriculture.

   The IoT devices like drones, precision farming equipment, and connected sensors have revolutionized agriculture by enabling farmers to monitor crop health, soil conditions, and environmental factors more accurately. This has led to more efficient use of resources like water, fertilizers, and pesticides, resulting in higher crop yields and reduced environmental impact.

   The IoT technologies like GPS tracking, fleet management systems, and autonomous vehicles have improved efficiency in transportation and logistics by optimizing routes, reducing fuel consumption, and automating various tasks. This has led to cost savings and increased productivity for businesses in these sectors.

2. *New business models and opportunities*: IoT opens new business models and revenue streams. It has enabled subscription-based services, data analytics, and targeted advertising. The data is obtained from IoT encourages innovation and the development of new products and services.

   The IoT devices can collect real-time data on product usage, enabling companies to offer usage-based pricing models. For example, insurance companies can use telematics devices in vehicles to track driving behavior and offer personalized, usage-based insurance premiums based on actual driving habits.

   IoT enables companies to shift from selling products to offering them as services. For example, instead of selling industrial equipment, companies can lease the equipment and provide ongoing maintenance and monitoring services through IoT connectivity. This model can generate recurring revenue and foster long-term customer relationships.

3. *Better decision-making*: IoT provides access to vast amounts of data, which can be analyzed to make betterinformed decisions, optimize processes, and enhance resource management. IoT sensors and devices collect real-time data on production processes, equipment performance, and product quality. By analyzing this data, manufacturers can optimize production schedules, identify bottlenecks, and make informed decisions on maintenance and resource allocation, resulting in improved efficiency and reduced costs.

   The IoT technologies like GPS tracking, fleet management systems, and connected vehicles provide data on vehicle performance, traffic conditions, and fuel consumption. By analyzing this data, transportation and logistics companies can benefit in route optimization, fleet maintenance, and driver performance, resulting in reduced costs and increased efficiency.

4. *Enhanced customer experience*: IoT allows businesses to provide personalized experiences, anticipate customer needs, and improve customer satisfaction. This leads to increased brand loyalty and higher customer retention rates.

    The IoT devices can collect and analyze data on individual customer preferences and behavior, enabling businesses to offer personalized product recommendations, targeted promotions, and tailored experiences. For example, retailers can use beacon technology to send personalized offers and discounts to customers' smartphones as they enter a store.

    The IoT technologies have enabled the development of connected cars that offer a range of services to improve the driving experience. Features like real-time traffic updates, remote diagnostics, and predictive maintenance can make driving safer, more convenient, and more enjoyable for consumers.

5. *Economic growth*: The widespread adoption of IoT technologies can boost economic growth by improving productivity, fostering innovation, and creating new business opportunities.

    IoT has given rise to new business models, such as product-as-a-service, data-driven services, and platform-based models, creating new revenue streams and opportunities for growth. These innovative business models can drive economic growth by fostering entrepreneurship and the development of new market segments.

    Researchers of McKinsey Global Institute in their study estimated the likely economic impact of IoT technologies to be $2.7 to $6.2 trillion per year by 2025.According to Thierer and Castillo (2015), the manufacturing and healthcare industries would have the maximum economic growth through the various IoT modeling.

    Another interesting study by Morgan Stanley forecasts that driverless cars will save the US economy $1.3 trillion per year,once they fully penetrate the market, while saving the world another $5.6 trillion a year. The following specific areas of economic gains were identified in Table 6.1 (Thierer & Castillo, 2015).

6.5.2 Negative impacts

1. *Regulatory challenges*: The rapidly evolving IoT landscape presents regulatory challenges, as governments struggle to keep pace with technological advancements. Unclear or restrictive regulations can hinder innovation and economic growth.

Table 6.1 Economic impact of IoT on Automobile business

Economic gains	Area
$507 billion	Productivity gains
$488 billion	Prevented accident costs
$158 billion	Fuel cost savings
$138 billion	Productivity gains from congestion prevention
$11 billion	Fuel cost savings from congestion prevention

Source: Thierer and Castillo (2015).

IoT devices collect vast amounts of data, raising concerns about the privacy and security of personal information. Regulators must balance the need to protect user privacy with the desire to foster innovation in the IoT space. This has led to the development of data protection regulations like the European Union's General Data Protection Regulation (GDPR) and the California Consumer Privacy Act (CCPA).

The IoT devices can be vulnerable to cyberattacks, potentially leading to data breaches, unauthorized access, and the disruption of critical infrastructure. Regulators face the challenge of developing and enforcing security standards for IoT devices and networks to mitigate these risks. Examples include the US's IoT Cybersecurity Improvement Act and the European Union's Cybersecurity Act.

2. *Job displacement*: Automation and increased efficiency through IoT can lead to job displacement in certain sectors like manufacturing and customer service. Workers in these sectors may need to acquire new skills to stay relevant in the job market.

   The IoT-enabled automation and robotics have significantly impacted manufacturing jobs, particularly in assembly and production lines. IoT devices can perform tasks more efficiently and accurately than humans, leading to job displacement for workers performing repetitive or manual tasks.

   The IoT devices like chatbots and virtual assistants can automate customer service tasks, potentially displacing jobs in call centers and support roles. Automated support systems can handle customer inquiries more efficiently, reducing the need for human customer service representatives.

6.6 SUSTAINABLE IMPACT

The role of IoT in improving energy efficiency, increasing the share of renewable energy, and reducing environmental impacts of the energy use cannot be overlooked (Motlagh et al., 2020). The IoT has the potential to significantly impact sustainability both positively and negatively. Some of the positive and negative impacts of IoT on sustainability are described in the following:

6.6.1 Positive impacts

1. *Energy efficiency*: IoT devices and sensors can monitor and optimize energy consumption in buildings, industrial processes, and homes. This can lead to reduced energy waste and lower greenhouse gas emissions. For example, smart thermostats can learn user preferences and optimize heating and cooling schedules to save energy. The IoT and Big Data plays a vital role in the development of energy-efficient and intelligent vehicle (EEIV). This in turn will, in a big way, address energy, environment, and safety issues (Aris et al., 2015).

   The IoT technologies enable the development of smart grids that can intelligently manage energy distribution, optimize power generation, and balance demand and supply. Smart grids can integrate renewable energy sources, enable real-time monitoring of power usage, and respond to fluctuations in demand, resulting in more efficient and sustainable energy management.

   The IoT-enabled smart lighting systems can automatically adjust lighting levels based on occupancy, time of day, and natural light availability. This can result in significant energy savings, as lights are only used when needed and at the optimal intensity.

2. *Resource management*: IoT can help monitor and manage resources such as water, electricity, and raw materials more effectively. Smart grids can optimize power distribution and reduce outages, while smart water systems can detect leaks and optimize water usage.

   The IoT sensors can be deployed in water infrastructure systems to monitor water usage, detect leaks, and optimize irrigation schedules. This can lead to more efficient water management, reduced waste, and lower water consumption. Smart irrigation systems, for example, can adjust watering schedules based on soil moisture levels, weather forecasts, and plant needs, which reduces water wastage and improves agricultural productivity.

   The IoT technologies can improve inventory management by providing real-time data on stock levels, location, and movement. IoT-enabled devices like RFID tags and smart shelves can help businesses optimize inventory levels, reduce waste, and streamline supply chain operations (Motlagh et al. 2018).

3. *Waste reduction*: IoT can help reduce waste production by enabling real-time monitoring of waste levels, optimizing waste collection routes, and improving recycling processes. Smart agriculture can minimize the use of fertilizers and pesticides, reducing environmental pollution.

The IoT technologies, such as smart refrigerators and connected supply chain management systems, can help reduce food waste by monitoring product freshness, tracking expiration dates, and optimizing inventory levels. Retailers and consumers can use this information to make informed decisions on food consumption and purchases, thus reducing spoilage and waste.

The IoT-enabled waste management systems, such as smart bins equipped with sensors to monitor waste levels, can help optimize waste collection routes and schedules, reducing fuel consumption and CO_2 emissions. Cities can also use data from IoT devices to implement recycling programs, reduce litter, and improve the overall waste management process.

4. *Transportation*: IoT can improve transportation sustainability through real-time traffic monitoring, route optimization, and electric vehicle charging infrastructure. This can lead to reduced fuel consumption, traffic congestion, and emissions.

   The IoT technologies can improve the efficiency and sustainability of public transportation systems by providing real-time data on vehicle location, passenger load, and schedules. This information can be used to optimize routes, reduce waiting times, and encourage the use of public transportation, ultimately reducing the number of private vehicles on the roads and their associated emissions.

   The IoT-enabled smart parking systems can help drivers find available parking spaces more efficiently, reducing the time spent searching for parking and the associated fuel consumption and emissions. Smart parking solutions can also optimize the use of existing parking infrastructure, leading to more efficient use of urban space.

5. *Environmental monitoring*: IoT sensors can collect data on air quality, water quality, soil conditions, and wildlife habitats. This can help identify pollution sources, monitor ecosystem health, and inform environmental policies and conservation efforts.

   The IoT-enabled air quality sensors can be deployed in urban and industrial areas to monitor air pollutants, such as particulate matter (PM), nitrogen oxides (NOx), and volatile organic compounds (VOCs). Real-time data from these sensors can help governments and organizations identify pollution hotspots, implement mitigation measures, and track progress toward air quality goals.

   The IoT devices, such as camera traps and tracking collars, can be used to monitor wildlife populations, track animal movements, and detect poaching activities. This information can inform conservation efforts and help protect endangered species and their habitats.

6.6.2 Negative impacts

1. *E-waste*: The rapid growth of IoT devices can contribute to an increase in electronic waste (e-waste). Many IoT devices have short lifespans and may not be easily recyclable, leading to a buildup of e-waste in landfills and potential environmental pollution.

   Many IoT devices are designed to be difficult or impossible to repair, with components that are soldered or glued together. When these devices malfunction or become outdated, they are often discarded rather than repaired or upgraded, contributing to e-waste.

   The recycling infrastructure for electronic devices, particularly IoT devices, is often inadequate or nonexistent. This can result in a large portion of discarded IoT devices ending up in landfills or being improperly disposed of, leading to environmental pollution and increased e-waste.

2. *Energy consumption*: While IoT can optimize energy use in many cases, the devices themselves require energy to operate. The growing number of IoT devices, along with the data centers and communication networks needed to support them, can lead to increased overall energy consumption.

   Many IoT devices, such as smart speakers, home automation systems, and security cameras, consume energy even when they are not actively in use. This standby power consumption can contribute to increased energy usage, particularly when numerous IoT devices are deployed in homes and businesses.

   The increasing use of IoT devices generates vast amounts of data, which needs to be stored, processed, and analyzed. This has led to a growing demand for data centers and cloud computing services, which can consume significant amounts of energy for cooling and powering servers.

3. *Privacy and security*: IoT devices can collect vast amounts of data, raising concerns about user privacy and data security. Poorly secured devices can be exploited by hackers, leading to potential data breaches and negative consequences for individuals and organizations.

   Some fitness trackers and health apps have been found to inadvertently leak users' personal information, such as location data, workout routines, and health metrics. This can expose users to privacy risks and potential misuse of their sensitive data.

   The sheer volume of data collected by IoT devices can enable companies to aggregate and analyze user data in ways that may not be immediately apparent to users. This can lead to invasive profiling, targeted advertising, and other potentially undesirable uses of personal information.

4. *Digital divide*: The benefits of IoT may not be equally accessible to everyone. Developing countries and poor communities may lack the necessary infrastructure and resources to adopt IoT technologies, potentially widening the gap between the "haves" and "have-nots."

5. *Resource extraction*: The production of IoT devices requires raw materials, such as rare earth metals and minerals. The extraction of these resources can have negative environmental and social impacts, including habitat destruction, water pollution, and human rights violations.

Cobalt is a critical component of lithium-ion batteries, which are used in a wide range of IoT devices, including smartphones, laptops, and electric vehicles. Most of the world's cobalt supply comes from the Democratic Republic of Congo, where mining activities have been associated with environmental degradation, water pollution, and human rights violations, including child labor.

Rare earth elements, such as neodymium, dysprosium, and yttrium, are used in various electronic components, including magnets, batteries, and displays. The extraction and processing of rare earth elements can generate toxic waste, contribute to air and water pollution, and cause soil degradation. China, which produces

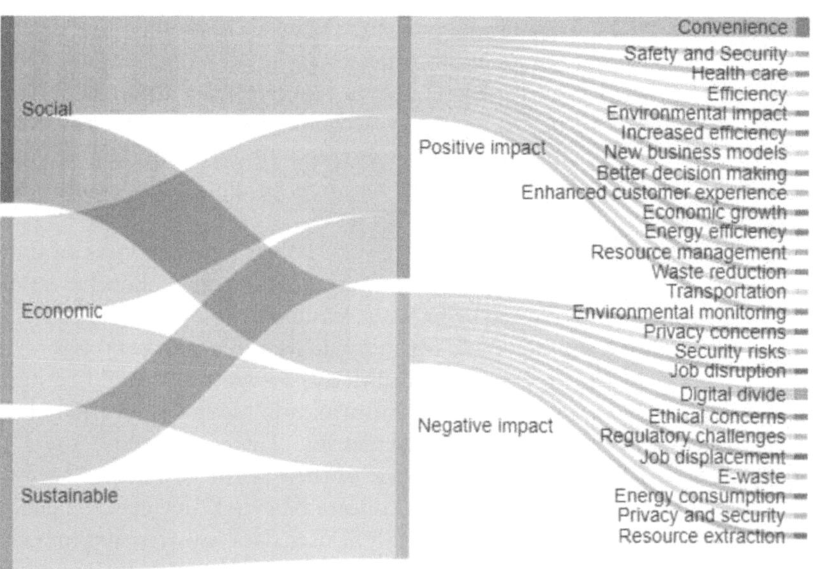

Figure 6.2 Social, economic, sustainable impacts.

the majority of the world's rare earth elements, has faced significant environmental challenges related to rare earth mining and processing.

In conclusion, the IoT has the potential to significantly improve sustainability across various sectors. However, it also comes with potential negative impacts that need to be considered and addressed. Balancing the benefits and drawbacks of IoT technologies requires responsible design, implementation, and regulation to ensure a sustainable and equitable future (Figure 6.2).

6.7 METAMORPHIC SEGMENTS

IoT has had a significant impact on individuals, businesses, and society as a while. By connecting everyday objects to the internet and enabling them to communicate and share data, IoT has transformed the way we live, work, and interact with our environment.

6.8 INDIVIDUALS

The IoT devices like smart home systems, wearable devices, and personal assistants have made daily life more convenient by automating tasks and providing personalized experiences. Wearable devices and health monitoring systems promote improved health and well-being by helping individuals make better-informed decisions about their health and fitness. IoT devices like smart security systems, connected smoke detectors, and real-time location tracking can help improve personal safety and security. IoT has made it easier for individuals to stay connected with their friends and family, leading to an increased sense of community and belonging.

The IoT devices, such as glucose monitors and insulin pumps, have transformed the management of chronic diseases like diabetes. By providing real-time feedback on blood sugar levels and automating insulin delivery, these devices can help individuals maintain better control of their condition and prevent complications.

The devices, such as fall detection systems, medication reminders, and remote health monitoring, can support elderly individuals in maintaining their independence and ensuring their safety. These IoT technologies can also provide peace of mind to caregivers and family members.

Wearable devices, such as fitness trackers and smartwatches, have enabled individuals to monitor their physical activity, heart rate, sleep patterns, and other health metrics. These devices can help users make informed decisions about their exercise routines, nutrition, and overall well-being, potentially leading to healthier lifestyles and improved outcomes.

6.9 BUSINESSES

Businesses experience improved efficiency by adopting IoT, which enables automation of processes, monitoring of equipment and optimization of resources, leading to efficiency and cost savings. IoT devices collect vast amounts of data, which can then be analyzed, giving businesses valuable insights for decision-making and strategic planning. The data associated with IoT offers unbounded potential for organizations to obtain valuable insights (Dwivedi et al., 2017; Hashem et al., 2015). Using the data from the IoT devices, new business models such as remote monitoring and predictive maintenance, where it is possible to reduce personnel and initiate maintenance much before a failure occurs. The improvements in timeliness and the sheer magnitude of data provided by IoT results in improved operational planning and increased ability to react quickly to previously unforeseen events (Brous et al., 2020), which enhances workflows of the organization. IoT allows businesses to offer personalized services and experiences that leads to increased customer engagement and loyalty. It has transformed supply chain management by providing businesses with better visibility and control over their inventory, transportation, and logistics processes. IoT-enabled devices, such as RFID tags, GPS trackers, and sensors, can help businesses optimize their supply chains, reduce waste, and increase overall efficiency. IoT technologies have been used by retailers to enhance customer experiences, streamline operations, and gather valuable insights into consumer behavior. Examples include smart shelves that monitor inventory levels, beacon technology for personalized promotions, and smart shopping carts that facilitate contactless payments.

6.10 SOCIETY

IoT has enabled the development of smart cities, where various infrastructure components like traffic management, waste management, and energy distribution are connected and optimized for efficient urban living. IoT technologies monitor and manage environmental conditions, leading to more sustainable practices and reduced environmental impact. Public safety can be improved by having IoT devices monitor public spaces, detect hazards, and alert authorities to potential emergencies. IoT has created new industries, job opportunities, and spurred economic growth by driving innovation and creating new markets.

The IoT technologies can play a crucial role in disaster management and response by providing real-time data on weather conditions, infrastructure damage, and the movement of people. This information can assist authorities in making informed decisions about resource allocation, evacuation planning, and recovery efforts (Shenkoya and Dae-Woo 2019).

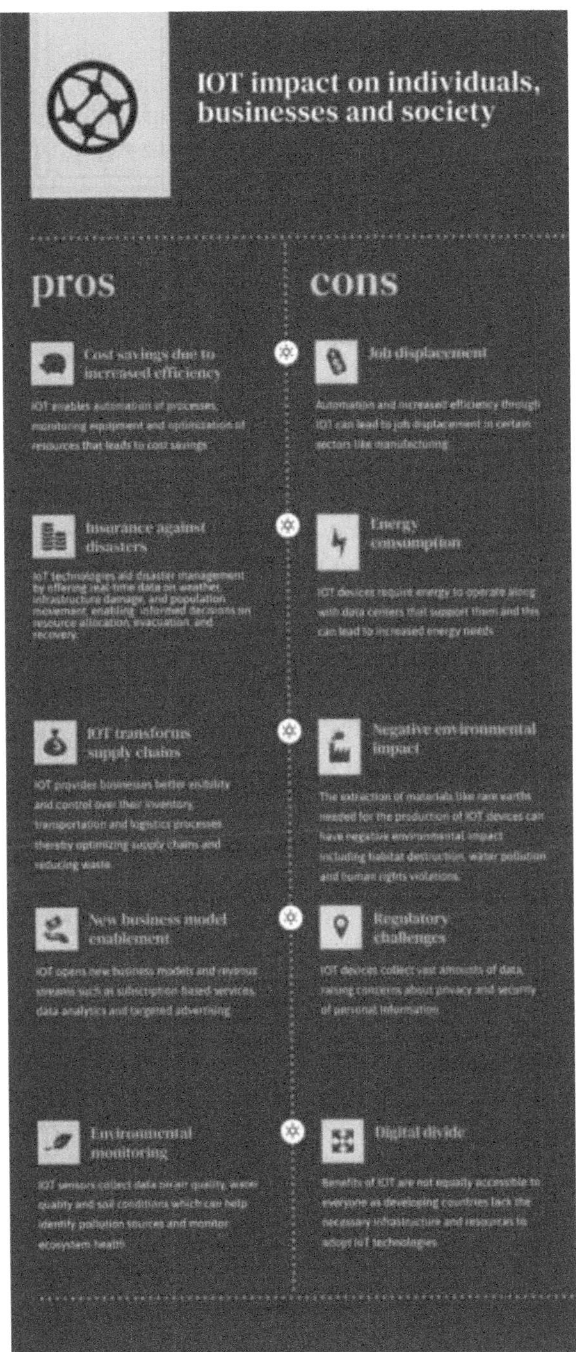

Figure 6.3 Summarized impact.

The IoT devices, such as surveillance cameras, gunshot detection systems, and body-worn cameras, can help improve public safety and support law enforcement efforts. These technologies can provide valuable data for crime prevention, emergency response, and evidence collection.

The IoT technologies are being integrated into educational settings to enhance learning experiences and outcomes. Examples include adaptive learning systems that personalize instruction, classroom management tools that monitor student engagement, and virtual reality applications that provide immersive learning experiences.

Despite the numerous benefits, IoT has also raised concerns about privacy, security, and data ownership. As IoT continues to expand, it is crucial to address these issues and develop frameworks that ensure the responsible and ethical use of IoT technologies.

6.11 CONCLUSION

In conclusion, the IoT has undeniably transformed the way individuals, businesses, and societal functions, and has offered numerous benefits as well as posing certain challenges. On the one hand, it has enhanced efficiency, convenience, and connectivity, promoting innovation and economic growth. On the other hand, it has raised concerns regarding privacy, security, and digital inequality. As we continue to embrace IoT technologies, it is crucial to strike a balance between harnessing their potential and addressing the accompanying risks. By fostering collaboration among stakeholders, implementing effective regulations, and promoting digital literacy, we can ensure that the socio-economic impact of IoT leads to a more connected, inclusive, and sustainable future for all (Figure 6.3).

REFERENCES

Aris, I. B., Sahbusdin, R. K. Z., & Amin, A. F. M. (2015, September 8). Impacts of IoT and big data to automotive industry. *In 2015 10th Asian Control Conference: Emerging Control Techniques for a Sustainable World, ASCC 2015* (pp. 1–5). Kota Kinabalu, Malaysia: IEEE. https://doi.org/10.1109/ASCC.2015.7244878

Brous, P., Janssen, M., & Herder, P. (2020). The dual effects of the Internet of Things (IoT): A systematic review of the benefits and risks of IoT adoption by organizations. *International Journal of Information Management*, 51, 101952. Elsevier Ltd. https://doi.org/10.1016/j.ijinfomgt.2019.05.008

Dwivedi, Y. K., Janssen, M., Slade, E. L., Rana, N. P., Weerakkody, V., Millard, J., & Snijders, D. (2017). Driving innovation through big open linked data (BOLD): Exploring antecedents using interpretive structural modelling. *Information Systems Frontiers*, 19(2), 197–212. https://doi.org/10.1007/s10796-016-9675-5

Gubbi, J., Buyya, R., Marusic, S., & Palaniswami, M. (2013). Internet of Things (IoT): A vision, architectural elements, and future directions. *Future Generation Computer Systems*, 29(7), 1645–1660.

Haller, S., Karnouskos, S., & Schroth, C. (2009). The Internet of Things in an enterprise context. In: Domingue, J., Fensel, D., & Traverso, P. (eds.), *Future Internet – FIS 2008. FIS 2008. Lecture Notes in Computer Science*, vol 5468. Springer, Berlin, Heidelberg. https://doi.org/10.1007/978-3-642-00985-3_2

Hashem, I. A. T., Yaqoob, I., Anuar, N. B., Mokhtar, S., Gani, A., & Khan, S. U. (2015). The rise of "big data" on cloud computing: Review and open research issues. *Information Systems*, 47, 98–115.

Hoshi, T., Yasuba, K., Kurosaki, H., & Okayasu, T. (2017). Ubiquitous environment control system: An Internet-of-things–based decentralized autonomous measurement and control system for a greenhouse environment. In *Automation in Agriculture—Securing Food Supplies for Future Generations*. https://doi.org/10.5772/intechopen.71661

Hsu, C. W., & Yeh, C. C. (2016). Understanding the factors affecting the adoption of the Internet of Things. *Technology Analysis & Strategic Management*, 29(9), 1089–1102. https://doi.org/10.1080/09537325.2016.1269160

Li, S., Zhao, S., Yang, P., Andriotis, P., Xu, L., & Sun, Q. (2019). Distributed consensus algorithm for events detection in cyber-physical systems. *IEEE Internet of Things Journal*, 6(2), 2299.

Madakam, S., Ramaswamy, R., & Tripathi, S. (2015). Internet of Things (IoT): A literature review. *Journal of Computer and Communications*, 3, 164–173. http://dx.doi.org/10.4236/jcc.2015.35021

Mahmood, Y., Kama, N., Azmi, A., & Ya'acob, S. (2020). An IoT based home automation integrated approach: Impact on society in sustainable development perspective. *International Journal of Advanced Computer Science and Applications*, 11(1), 240–250. https://doi.org/10.14569/ijacsa.2020.0110131

Motlagh, N. H., Khajavi, S. H., Jaribion, A., & Holmstrom, J. (2018). An IoT-based automation system for older homes: A use case for lighting system. In *Proceedings of the 2018 IEEE 11th Conference on Service-Oriented Computing and Applications (SOCA)* (pp. 1–6). Paris, France.

Motlagh, N. H., Mohammadrezaei, M., Hunt, J., & Zakeri, B. (2020). Internet of Things (IoT) and the energy sector. *Energies*, 13(2), 494. https://doi.org/10.3390/en13020494

Ng, I. C. L., & Wakenshaw, S. Y. L. (2017). The Internet-of-Things: Review and research directions. *Journal of Research in Marketing*, 34(1), 3–2.

Shenkoya, T., & Dae-Woo, C. (2019). Impact of IoT on social innovation in Japan. *Asia Pacific Journal of Innovation and Entrepreneurship*, 13(3), 341–353. https://doi.org/10.1108/APJIE-06-2019-0040

Shin, D. (2017). An exploratory study of innovation strategies of the Internet of Things SMEs in South Korea. *Asia Pacific Journal of Innovation and Entrepreneurship*, 11(2), 171–189.

Shrouf, F., Ordieres, J., & Miragliotta, G. (2014). Smart factories in industry 4.0: A review of the concept and of energy management approached in production based on the Internet of Things paradigm. In 2014 *IEEE International Conference on Industrial Engineering and Engineering Management* (pp. 697–701). Selangor, Malaysia. https://doi.org/10.1109/IEEM.2014.7058728

Thierer, A., & Castillo, A. (2015). *Projecting the growth and economic impact of the internet of things*. George Mason University, Mercatus Center, June, 15.

Chapter 7

Cultural evolution in the digital age

IoT implementation and technology readiness in organizational context

Channi Sachdeva and Veer P. Gangwar

7.1 INTRODUCTION

Combining bioinspired concepts and contemporary technology has created new opportunities in organizational dynamics in the digital transformation era (Thrift, 2006). In the context of organizations, "Bio-Inspired Cultural Evolution in the Digital Age" takes readers on an enthralling trip to investigate the meeting point between the disruptive power of the Internet of Things (IoT) and the evolutionary wisdom of nature (Mourtziset al., 2022). It provides a comprehensive review of the complex interplay between bio-inspiration, IoT deployment, and technological preparedness, setting the tone for the following engrossing investigation. We compare two subsamples, one from rich nations and the other from developing countries, to investigate the impact of economic circumstances on e-business value (Zhu et al., 2004).

The distinction between the culture and social structure is necessary, and it is claimed that the analysis of organizational life's culture cannot be based on a single universal concept. The occurrence of cultural change frequently triggered strong emotional responses. The reactions were often intense in nature (Smollan & Sayers, 2009; Meek, 1988). The use of biodiversity benefits was anticipated to not only fund conservation goals but also to support the sustainable development of countries (Morgera & Tsioumani, 2010). Addressing and adoption of biodiversity is important due to climate changes (Visseren-Hamakers & Kok, Eds., 2022). The use of AI technology impacts individual's drive and their competence in performing complex tasks efficiently (Uren & Edwards, 2023).

A key feature in the fast-paced, technologically advancing digital era is the merging of cultural evolution with the incorporation of cutting-edge technology into organizational structures. Numerous companies have implemented new policies, products, or procedures to tackle pollution, reduce consumption, and enhance relationships with the community and stockholders (Linnenluecke & Griffiths, 2010). Development of skills to manage changes and address resistance within organization is crucial

(Warrick, 2023). Two interrelated components are leading this change: the broad use of the IoT and the general notion of technological readiness in organizational settings. Biodiversity loss is not adequately comprehended or recorded (Chandra & Idrisova, 2011). The advancement in technology continues to build on technological progress (Reding & Eaton, 2020). New system capabilities are usually built upon the successful outcomes of previous advanced technology research and development tasks (Mankins, 2009). Simultaneously, in light of our digital growth, the complex interplay between biodiversity and technological advancement adds another level of complexity that calls for careful and comprehensive analysis (Aubry et al., 2022).

Recognition of the fact that the biological world has long been a source of inspiration for creativity, adaptability, and evolution is crucial as we enter this field. With its complex web of linked devices, the IoT has reshaped the operational dynamics of organizations in addition to changing how we view and engage with our environment. Managing environment dynamics relies heavily on the crucial roles of digital leadership, along with IT capabilities and organizational learning (Mollah et al., 2023). The influence of IoT on organizational cultures is significant, ranging from process optimization and connectivity enhancement to real-time data-driven decision-making. The exploration, manipulation, and synthesis of biological systems have been undergoing a revolution due to advancements in new technologies that create, analyze, and utilize large amount of data (Halewood et al., 2018). The adoption of enterprise resource planning is significantly influenced by factors such as technology readiness, size, perceived barriers, and production and operations improvement (Pan & Jang, 2008). Technology readiness—which includes an organization's ability to adapt and smoothly integrate these cutting-edge technologies—takes center stage as enterprises traverse this technological frontier. We explore how companies may use bioinspired techniques to adapt, grow, and evolve in the digital age by drawing on the abundance of natural occurrences (Barrett, 1995), in order to identify contributing factors, such as strategic planning (Li et al., 2016) depending on their success or failures. In an increasingly technologically integrated world, the IoT is a major force that is upending whole businesses and upending established norms. Managers must recognize that the growing role of technology and customer coproduction has reduced the impact of the human touch (Antwi et al., 2021). Adopting an appropriate strategy for each specific project is crucial for its success (Shenhar et al., 2005). It lays the groundwork for integrating IoT into the organizational culture and provides an overview of its revolutionary capacity. An increasing amount of scholarly works has recognized HR expertise as a crucial component of technology deployment and preparedness (Aboelmaged, 2014).

At the same time, the digital age makes us consider how technology and biodiversity interact. Our technological endeavors inexorably collide with natural ecosystems as they transform the world around us. It is essential to comprehend how technology affects biodiversity to design a responsible and sustainable future. In this age of cultural change, weighing the advantages of technological advancement against a dedication to environmental stewardship becomes crucial.

This investigation aims to disentangle the complex web of cultural development in the digital era, emphasizing IoT adoption, technological preparedness in corporate settings, and their consequences for biodiversity. In order to improve strategic abilities, people need to adopt a mindset of continuous learning, foster an atmosphere of creativity, and implement effective practices for developing talent and knowledge (Sachdeva & Grover, 2024). The traditional approaches used for industrial automation are no longer sufficient to meet the evolving technological and business requirements (Leitão et al., 2016). We hope to traverse the changing terrain of our digital culture by exploring these interrelated domains and provide insights into how businesses might balance technology advancement with environmental responsibility in this period of unparalleled change. Such organizational changes are being measured and evaluated by different assessment tools given by Holt et al. (2007).

We shall investigate the workings of this revolutionary combination. We'll explore the symbiotic link between bioinspired cultural development and the IoT, providing a multifaceted view of how this interaction will impact business, technology, and society in the future (Näyhä, 2020). By careful analysis, we hope to highlight the shifting paradigms in a technologically advanced world and demonstrate the significant influence of this convergence on organizational evolution and preparedness. Industry 4.0 practitioners and senior management from various organizations comprehend the preparedness factors that must be taken into account before Industry 4.0 technology is implemented (Antonyet al., 2023).

7.2 TECHNOLOGY READINESS: A FOUNDATION FOR IOT ADOPTION

This chapter examines the complex relationship between technological readiness, IoT adoption, and the development of an organizational culture that is biodiversity-conscious in an era where the delicate balance of biodiversity is threatened and digital transformation is reshaping the way organization's function. Our planet's survival depends on biodiversity, yet environmental degradation and climate change are posing growing threats to it (Mourtzis et al., 2022). The readiness of technologies embodies a combination of psychological factors that together influence an individual's inclination to adopt new technologies (Parasuraman & Colby, 2015). Organizations are embracing IoT technology as an effective instrument for biodiversity preservation at the same time that they realize how important it is to play this role. The use of IoT technology is important in various areas, bringing together businesses, platforms, and industries (Nezhad et al., 2023). Nevertheless, organizational culture and technological readiness are equally important for the effective integration of IoT into biodiversity conservation initiatives (Barrett, 1995). The future will benefit from cleaner solutions offered by new technologies that will also aid in preventing and resolving issues created by current technologies (Orozco & Grundmann, 2022). This chapter explores the complex link that exists between IoT adoption and technological readiness in the framework of an organizational culture that places a high priority on biodiversity protection. Anticipating the challenges in organizational culture is crucial because organizational culture efforts are typically intensive and time-consuming. Companies that can foresee these challenges may experience a more seamless shift into the fourth industrial revolution (Tortorella et al., 2023). It looks at how promoting an atmosphere that appreciates and incorporates an understanding of biodiversity affects the effective implementation of IoT technology. We reveal how IoT technologies are being used by organizations that value biodiversity to monitor ecosystems, save endangered species, and support environmental sustainability. The capacity and willingness of a corporation to adopt and use pertinent new technical resources is indicated by its level of technology readiness (Vize et al., 2013). The most important factor was recognized as skill in technology and ability to resolve issues (Balouei et al., 2022).

A key precondition for IoT adoption in an organizational setting is technological preparedness. IoT needs a strong technology basis for effective integration as it represents a paradigm shift in how businesses gather, handle, and use data. These are some essential elements of IoT adoption preparedness in an organizational setting.

1. **Network Infrastructure:** To accommodate IoT devices, an enterprise needs a strong and expandable network infrastructure. This comprises a dependable network design, wireless access points, and high-speed internet connectivity. A dependable network is necessary to send the massive amounts of data generated by IoT devices to on-site or cloud servers.
2. **Data Processing and Storage:** Enterprises must have the ability to effectively store and handle the vast volumes of data produced by IoT devices. This might entail a hybrid strategy, on-site data centers, or cloud-based storage options. Requirements for adequate data processing skills include strong servers and analytics software to get valuable insights from the data provided by the IoT.
3. **Security and Privacy:** IoT creates new security and privacy problems in terms of security and privacy. It is vital to guarantee the security of IoT devices and the data they gather. To guard against such breaches entails putting encryption, access control, authentication, and

ongoing monitoring into place. The occurrence, impact, and management of information privacy have been influenced by the evolution of information and communication technologies, despite its existence long before their emergence (Bélanger & Crossler, 2011). Adherence to data privacy laws is also essential, particularly in industries such as healthcare and banking (Musyaffi et al., 2022).

4. **Device Management:** To oversee and control IoT devices, organizations need to have a framework in place. This covers maintenance, firmware upgrades, device provisioning, and troubleshooting. Reliability and security are maintained in the IoT ecosystem through effective device management.

5. **Knowledge and Education:** The company ought to spend money on hiring and educating employees who are knowledgeable about IoT technology. This covers not just IT specialists but also staff members who are aware of how IoT data is used in business.

6. **Cost Management:** Adopting IoT may be expensive, so businesses need to have a well-defined plan in place for handling expenses. Budgeting for software, hardware, data transfer, and continuing maintenance is part of this.

7. **Regulatory Compliance:** Businesses need to be aware of and follow any local, industry-specific, or state laws about security, privacy, and IoT data. There may be financial and legal repercussions for noncompliance.

In conclusion, a fundamental prerequisite for the effective implementation of IoT in an organizational setting is technological preparedness. It includes all of the components required to support IoT adoption and make use of its potential to enhance decision-making and company operations, including infrastructure, security, data management, integration, scalability, expertise, cost management, and compliance.

7.3 IOT IMPLEMENTATION IN ORGANIZATIONAL CONTEXT

IoT technology used in an organizational setting can improve biodiversity conservation efforts as well as technology preparedness. Businesses may better monitor and manage biodiversity by integrating IoT devices and sensors into ecosystems. These tools help in conservation decision-making by gathering information on habitat changes, wildlife behavior, and environmental conditions. In addition, IoT promotes technological preparedness by enhancing data gathering and analysis, streamlining resource distribution, and enabling quick reaction to environmental hazards (Narkhede et al., 2023). This dual strategy supports global environmental goals and corporate social responsibility by fostering a healthy balance between technological progress and biodiversity preservation. The progression of human societies is fueled by advancements in technology (Tverskoi, Babu, & Gavrilets, 2022).

1. **Management and Protection:** The management and protection of biodiversity can benefit greatly from the application of IoT (IoT) in an organizational setting. Ecosystems and natural surroundings may be monitored, safeguarded, and preserved with the use of IoT technologies. IoT may be used to promote biodiversity in an organizational setting in the following ways:

2. **Wildlife Monitoring:** You may keep an eye on the whereabouts and activities of wildlife by using IoT sensors like GPS tracking tags, acoustic monitoring devices, and cameras. Organizations may use this data to monitor threatened species, spot trends, and evaluate how human activity is affecting their ecosystems.

3. **Monitoring of the Habitat:** IoT sensors may be used in ecosystems to keep an eye on parameters including air quality, water quality, temperature, and humidity. These sensors can assist organizations in monitoring ecosystem health and quickly responding to pollution incidents or changes.

4. **Early Warning Systems:** With IoT technology, natural calamities such as landslides, floods, and forest fires can have early warning systems. Organizations may take prompt action to safeguard biodiversity and ecosystems by recognizing and alerting them to these occurrences in real time.

5. **Smart Land Management and Agriculture:** By optimizing water and land usage, IoT sensors may be employed in agricultural settings to mitigate the adverse impacts of farming practices on biodiversity. Precision farming allows you to monitor soil conditions and use less water, herbicides, and fertilizer.

6. **Control of Invasive Species:** The early identification and control of invasive species that pose a danger to local biodiversity can be facilitated by the use of IoT sensors and data analytics. More efficiently, organizations may put policies in place to eliminate or control invasive species.

7. **Studies on Bird and Insect Migration:** By monitoring the migratory paths, breeding sites, and hazards associated with migration, IoT-based sensors can assist organizations in better understanding the movements of these species. Conservation activities can benefit from this data.

8. **Water Resource Management:** Rivers, lakes, and seas may all be kept an eye on with the use of IoT sensors. This aids in determining how pollution and climate change affect aquatic ecosystems and direct efforts to preserve water quality and save marine life.

9. **Ecosystem Restoration:** IoT may be used by organizations working on these initiatives to track their performance and development. Sensors can monitor the expansion of local plant species, evaluate the condition of the soil, and offer guidance for adaptive management.

10. **Collaboration and Data Sharing:** IoT technology facilitates collaboration and data sharing between organizations and other stakeholders, including government agencies, non-profit conservation organizations, and researchers. An ecosystem with shared data can offer a more thorough picture of ecosystem health and biodiversity.

11. **Public Engagement and Education:** The IoT has the potential to improve biodiversity-related public engagement and education initiatives. By making real-time sensor data available to the public via applications and websites, conservation concerns may be better understood and raised to the public's attention.

When used in an organizational setting, IoT deployment can enable businesses to take prompt, well-informed action to preserve and maintain biodiversity. Additionally, technology can make data-driven strategies possible to tackle the difficult problems linked to protecting endangered species and sustaining thriving ecosystems. However, when deploying IoT solutions in natural ecosystems, businesses must take data privacy, security, and ethical issues into account.

7.4 CULTURAL EVOLUTION IN THE DIGITAL AGE

In the digital era, cultural development is a major influence on how to see biodiversity and technical preparedness (Mourtzis et al., 2022). The term "cultural evolution" in the context of the digital era describes how civilizations, cultures, and customs adjust and transform in response to the quick developments in digital technology and communication. Differences between subcultures in organization are observed at various levels of the hierarchy (Cooke & Rousseau, 1988). The change in leadership forms is necessary for change in culture in all aspects of life (Burnes, 2004). This phenomenon has had a significant influence on many facets of human civilization, such as the production, dissemination, and consumption of knowledge, art, and information. These are some salient features of the digital age's cultural growth. The modern era of technology has brought significant changes, instability, unpredictability, and intricate conditions in organizational settings (Pfaff et al., 2023).

1. **Environmental Education and Awareness:** Social media and digital platforms have increased public knowledge of conservation initiatives and biodiversity. The classification of inventions should be based on

their impact on individuals and society (Ghaleb et al., 2021). A cultural change toward appreciating and preserving biodiversity may be fostered by educating people about ecosystems, species, and conservation tactics.

2. **Citizen Science and Data Sharing:** Digital tools enable citizen science initiatives, where individuals contribute to biodiversity research by collecting and sharing data. This collective action bolsters technological readiness by providing vast datasets for analysis, leading to more informed conservation decisions.

3. **International Cooperation:** The digital era encourages international cooperation between scientists, decision-makers, and environmentalists. Technology-enabled collaboration is essential for tackling biodiversity issues and fostering technological preparedness via knowledge exchange.

4. **Technological Solutions:** Big data analytics, artificial intelligence, and the IoT all help in managing and monitoring biodiversity. Our technical readiness to identify and mitigate environmental risks is improved by these technologies (Vize et al., 2013).

5. **Difficulties to Overcome:** Misinformation and digital disparities are only two of the difficulties presented by the digital era. To guarantee that everyone may benefit from technological preparedness and contribute to the preservation of biodiversity, these gaps must be closed.

6. **Information Distribution:** The dissemination of information has been completely transformed by digital technologies. The quick global dissemination of information, news, and ideas is made possible by the internet and social media platforms. This has both advantages and disadvantages as it makes it easier to distribute and manipulate information.

7. **Digital Media and Art:** Virtual reality experiences, music, and digital art are some of the new artistic mediums that have emerged with the advent of the digital age. It is now much easier for makers and artists to access a worldwide audience and try out new media.

8. **Globalization of Culture:** The spread of culture has been made easier by digital technologies. Due to the ease with which people from different areas of the world may now access and engage in the cultures of other places, customs and practices have blended.

9. **Online Communities:** Based on common identities and interests, online communities and social networks have grown. These groups frequently have a big impact on how cultural norms and values are developed and transmitted.

10. **Disinformation Challenges:** One of the major effects of the digital age has been the proliferation of fake news and disinformation, which may drastically alter people's cultural views and beliefs.

In the digital age, cultural development is a complicated and multidimensional process that is reshaping how we communicate, share information, express ourselves, and advance as a society. To optimize the advantages of the digital age while minimizing any possible disadvantages, it presents both possibilities and difficulties that need cautious thought and navigation. The digital age has transformed our cultural approach to biodiversity and technological readiness by facilitating awareness, data sharing, collaboration, and the application of advanced technologies. Balancing the benefits of digital evolution with the challenges it presents is essential for ensuring a sustainable future for both biodiversity and technological advancement.

7.5 THE INTERPLAY BETWEEN TECHNOLOGY READINESS AND IOT IMPLEMENTATION

Organizational culture, biodiversity, IoT deployment, and technological preparedness all interact in complex ways. Organizations can efficiently monitor and protect biodiversity because of IoT and technology preparedness. For their integration to be successful, though, a culture of environmental stewardship and data-driven decision-making is essential. New advancements in technology are causing significant changes in the field of system integrators, offering an efficient solution and delivering numerous positive results (Gupta, Kumar, & Karam, 2020). For their integration to be successful, though, a culture of environmental stewardship and data-driven decision-making is essential. The design concept stage is corresponded to by the system readiness level of technology as has been demonstrated (Kalashnikova et al., 2019). The culture of an organization influences its adoption of IoT for conservation and its dedication to sustainability. Employee adoption of green practices and technology uptake may both be accelerated by a supportive culture. On the other hand, by encouraging an ecological conscience and a stronger feeling of duty towards biodiversity preservation, technology and data from IoT devices may likewise impact and affect the culture of an organization. To effectively conserve and maintain biodiversity, it is imperative that technological preparedness and IoT adoption work together. IoT and technological preparedness work in tandem to improve ecosystem and natural environment monitoring, preservation, and protection. This is how they communicate.

1. **Data-Driven Conservation:** Organizations may make well-informed decisions for the preservation of biodiversity by utilizing real-time data on ecosystems and wildlife provided by IoT devices. The infrastructure and skills necessary to manage and evaluate this data efficiently are ensured by technological readiness.
2. **Rapid Reaction:** Organizations that are equipped with technology can react quickly to environmental hazards. Real-time monitoring,

early warning systems, and aiding in the prevention or mitigation of environmental catastrophes are made possible by IoT.

3. **Efficiency and Cost-Effectiveness:** Including IoT in biodiversity projects may improve cost-effectiveness, minimize manual data collecting, and allocate resources optimally. The effective implementation of IoT solutions is facilitated by technological preparedness.

4. **Sustainability Reporting:** The reporting of environmental sustainability targets is aided by IoT data. A key factor in guaranteeing the dependability and accuracy of this reporting is technological preparedness.

5. **Technology Readiness as a Foundation:** As was previously said, technology readiness provides the fundamental framework and skills required to successfully deploy IoT solutions. It consists of data storage, security, integration, network infrastructure, and knowledge. Before implementing IoT devices and sensors for applications related to biodiversity, organizations must establish the requisite technological framework.

6. **IoT Sensors and Devices:** Organizations can use IoT sensors and devices in the field once they have determined that they are technologically ready. These gadgets gather data in real time on the health of ecosystems, wildlife behavior, and environmental factors. Data on temperature, humidity, the quality of the air and water, the migrations of animals, and other topics can be included in this data.

7. **Data Processing and Analysis:** To identify patterns and insights, the data that IoT devices gather is processed and examined. This is when processing power, analytics tools, and data storage become more important in terms of technological preparedness. Organizations may use advanced analytics to make well-informed decisions concerning the protection of biodiversity.

8. **Rapid Reaction and Early Warning:** IoT technology can offer early warning systems for invasive species, natural catastrophes, and other risks to biodiversity. Rapid reaction mechanisms utilizing the technologically ready infrastructure to warn and mobilize resources for mitigation and protection can be triggered by this data.

9. **Cooperation and Data Sharing:** Researchers, governmental bodies, and non-profit organizations are just a few of the stakeholders that exhibit how technological preparedness and the IoT interact through data sharing and cooperation. A more thorough understanding of biodiversity is possible because of the infrastructure's technological preparedness.

10. **Public Engagement and Education:** Websites and apps that are designed to be technology-ready can be used to make IoT-generated data available to the general public. This increases public awareness

of the value of protecting natural ecosystems and involves them in conservation efforts to conserve biodiversity.

11. **Adaptive Management:** To support conservation efforts, organizations can leverage IoT data and technological ready skills for adaptive management. To optimize their efficacy, plans and treatments must be modified based on data.

12. **Privacy and Ethical Issues:** There are privacy and ethical issues to take into account when IoT devices gather data in natural settings. Protections for data security, privacy, and ethical usage should all be part of the infrastructure for technological preparedness.

13. **Optimizing Resource Allocation:** Organizations may deploy their resources more effectively when they are equipped with data analytics-ready technology. Where there is the most need, as indicated by the statistics, they may focus conservation efforts first.

To put it simply, the relationship between IoT implementation and technology readiness enhances an organization's ability to comprehend, preserve, and manage biodiversity. It blends technological infrastructure with cultural flexibility to form a potent force for ecological preservation and sustainability.

7.6 IMPACT ON ORGANIZATIONAL CULTURE: A MULTIFACETED ANALYSIS

The essence of culture comprises fundamental beliefs and shared understandings of the current state of affairs (Cameron, 2008). The examination of the impact on organizational culture is complicated and reflects the intricate interaction that exists between the values, beliefs, and behaviors of an organization and the different elements that shape it. When considering organizational culture from the perspectives of technological readiness and biodiversity, there are many different aspects to consider. A culture of innovation and environmental responsibility may be promoted by integrating technology, such as the IoT, for biodiversity protection (Thrift, 2006). It promotes a data-driven, sustainability-focused way of thinking. It could, however, also provide difficulties, such as opposition to change or worries about data privacy. To effectively exploit technology for biodiversity preservation while retaining a robust and flexible organizational culture, organizations must strike a balance between technological preparedness, ecological awareness, and cultural flexibility (Clausing & Holmes, 2010). In the culture heritage the challenge is clear and full of information that technology is changing the state of art in working in the organization (Ronchi, 2009). Without a doubt, the following is a comprehensive bullet-point summary of the effects on organizational culture concerning biodiversity and technological readiness:

1. **Technology Integration:** An organization's culture may be significantly impacted by the adoption of new technology inside it. It may result in a work environment that values creativity and adaptation and encourages staff members to welcome change and keep up with emerging technologies. Understanding responses and embracing advanced technologies are essential for the productivity of organization and employees (Mahmood, Imran, & Adil, 2023). On the other hand, it may also breed resistance and apprehension about the displacement of technology, encouraging a cautious and reluctant culture. The vision of the leadership, the organization's readiness to change, and training initiatives all influence how much technology is accepted and incorporated into the organizational culture.

2. **Leadership and Vision:** It is impossible to overstate the influence that leadership has on the culture of an organization. An innovative and flexible leadership group may foster a culture of experimentation and adaptability. On the other hand, a culture of complacency may result from executives who oppose change or who are unable to articulate a clear vision for the integration of technology. As executives frequently set the tone and shape employees' attitudes, there must be a strong congruence between their principles and the organization's culture.

3. **Learning and Development:** Technology is essential to an organization's ability to conduct ongoing learning and development. Businesses may promote a culture of learning and development by investing in e-learning platforms, skill development, and information exchange through technology. The capabilities of big data analytics have been discovered to have a significant and positive effect on the performance of the company (Behl, 2022). However, a culture of stagnation and skill obsolescence may occur if learning opportunities are few or if technology is not used effectively for staff growth.

4. **Employee Empowerment:** The degree of employee empowerment that technology provides is another way in which it impacts organizational culture. Employees can be better empowered by modern technology by having greater access to knowledge and autonomy. A culture of cooperation, trust, and shared accountability may result from this initiative. Conversely, if technology is employed for over-control or monitoring, it may foster a distrustful and micromanaged society.

5. **Diversity & Inclusion:** The use of technology can affect how an organization handles these issues. A more inclusive and varied workforce may be made possible via flexible work schedules and virtual collaboration technologies. But if technology is not inclusive or accessible in and of itself, it might unintentionally obstruct attempts to promote diversity and foster an inclusive culture.

6. **Work-Life Balance:** The work-life balance culture of an organization may be impacted by technological improvements, especially in the areas of flexible scheduling and remote work. Businesses that use technology to promote work-life balance may foster a culture of employee well-being and lower stress levels. On the other hand, technology may have a detrimental effect on work-life balance and employee morale if it fosters an "always-on" culture where workers are expected to be available at all times.

7. **Collaboration and Communication:** Within an organization, technology has a big impact on how workers work together and communicate. Real-time communication and cross-functional cooperation, for instance, can be facilitated by the adoption of digital communication technologies. This has the potential to promote an environment of candor, openness, and teamwork. On the other hand, technology may foster a culture of alienation and detachment if it impedes in-person relationships and creates digital silos.

8. **Innovation Culture:** A culture that emphasizes cutting-edge instruments for biodiversity conservation is fostered by embracing technological preparedness.

9. **Environmental Responsibilities:** The adoption of technology promotes an environmentally aware culture and harmonizes organizational principles with sustainability.

10. **Interdisciplinary Collaboration:** Technology implementation demands interdisciplinary cooperation, which promotes a cooperative culture.

11. **Data-Driven Decision-Making:** Organizational norms are shifted towards data-driven decisions for biodiversity conservation by technology preparedness.

In summary, the influence of technology on organizational culture is complex and varies depending on how an organization uses technology, the values it upholds, and the leadership's vision. How effectively technology fits with the organization's values and dedication to accepting change, empowering staff, and cultivating an innovative and adaptable culture will decide how much it helps or hurts culture. Maintaining an ethical and flexible organizational culture while effectively utilizing technology for biodiversity requires striking a balance between these aspects.

7.7 CHALLENGES AND BARRIERS

The integration of the IoT within the framework of cultural transformation in the digital era has the potential to both facilitate and impede organizational change. The adoption and preparedness of IoT technologies bring special dynamics to organizational and cultural environments. In this perspective, the following are some important things to keep in mind when addressing the obstacles and hurdles related to IoT implementation:

1. **Resistance to Change:** It might be difficult to embrace new technology for biodiversity protection as organizational culture frequently rejects disruptive technological developments.
2. **Lack of Technical Skills:** The effective deployment of technology centered on biodiversity may be hampered by a lack of technical expertise inside the company.
3. **Data Security and Privacy Issues:** Organizations need to handle data security and privacy issues as a result of the massive volumes of data that IoT devices gather and communicate. Adhering to data protection standards and guaranteeing the security of sensitive data can be a formidable obstacle. Using technology to collect data for biodiversity projects presents privacy issues that must be resolved with strong data governance procedures.
4. **Change Management:** Effectively managing cultural shifts toward embracing technology for biodiversity can be complex, requiring well-planned change management strategies.
5. **Data Overload:** The abundance of data from technology can overwhelm organizations, necessitating effective data management strategies.
6. **Technological Preparedness:** Before implementing IoT solutions, organizations may have difficulties with their technological preparedness. This covers problems including out-of-date infrastructure, a shortage of qualified staff, and organizational opposition to change.
7. **Interoperability:** As many IoT devices are made by different companies, there may be problems with compatibility and interoperability. The efficacy of various IoT technologies may be hampered by organizations' inability to seamlessly incorporate them into their current systems.
8. **Costs and Return on Investment:** Purchasing IoT devices and infrastructure might require a sizable upfront expenditure. The long-term expenses of maintaining and updating IoT systems, as well as the return on investment, must be carefully considered by organizations.
9. **Regulatory and Compliance Issues:** When it comes to IoT technology, organizations have to traverse a complicated web of rules and laws. There may be legal and reputational repercussions from non-compliance.
10. **Integration with Current Systems:** It can be difficult and time-consuming to integrate IoT technology with legacy systems; this requires careful planning and technical know-how.

11. **Cultural Resistance:** The changes brought about by IoT technology may encounter resistance or a slow rate of adaptation from organizational cultures. It might be difficult to persuade staff members of the advantages and the necessity of change.

7.8 FUTURE TRENDS AND RESEARCH DIRECTIONS

To meet the changing demands of enterprises and society, future trends in organizational culture, technological preparedness, and biodiversity are essential. Companies will place a higher priority on a flexible culture that encourages creativity and diversity. As technology becomes ready, this cultural transformation will be in line with it, with automation, artificial intelligence, and data analytics playing a key role. Decision-making and efficiency will both be enhanced by these technologies. At the same time, as organizations become more aware of their environmental duties, biodiversity preservation will become more important. Ecological methods and preservation initiatives will become essential, emphasizing ethical sources and circular economies. Strategies for improving technological integration, encouraging cultural flexibility, and supporting biodiversity conservation inside organizations should be the focus of future research. By incorporating these trends, companies may become technologically sophisticated, socially conscious, and sustainable, setting them up for success in the future. Numerous future trends and research avenues are emerging as the fields of cultural development in the digital age and IoT applications in organizational contexts continue to develop. The following areas of emphasis will probably influence the terrain in the years to come:

1. **Sustainability and Green IoT:** As environmental sustainability becomes more and more important, academics will investigate how IoT may be used to lessen environmental effects and improve sustainability initiatives. To promote sustainability, this involves researching IoT devices that are energy-efficient, IoT models for the circular economy, and how to incorporate IoT into smart city efforts.
2. **Blockchain for IoT:** There will be more research done on how to apply blockchain technology to improve the security and transparency of IoT systems. The use of blockchain in managing human resources is gaining attention, particularly in the area of recruiting, managing talent, and handling employee data (Chanda & Singh, 2023). This entails investigating how blockchain technology may be used in supply chain management, data integrity, and smart contracts in an organizational setting.
3. **Edge and FOG Computing:** Research on edge and fog computing, which bring data processing and decision-making closer to the IoT devices, will be conducted in the future. This can improve IoT

systems' real-time capabilities while lowering latency and bandwidth needs. It will be essential to comprehend how these technologies affect organizational preparedness.

4. **Human-Machine Interaction:** As IoT devices proliferate, comprehending the dynamics of human interaction with these technologies will emerge as a critical field of study. Investigating the effects on decision-making processes, trust, and human behavior is included in this.

5. **Integration of AI and Machine Learning:** An increasing number of IoT applications are integrating AI and machine learning. Studies will examine how these technologies might improve the prescriptive and predictive powers of IoT systems and influence the formation of organizational culture.

6. **IoT in Education and Learning:** An increasing trend is the incorporation of IoT into educational environments. Scholars will investigate how IoT might improve educational experiences and the need for cultural adjustment in educational settings.

7. **IoT's Cultural Implications:** Scholars will look at how IoT affects society, businesses, and people on a cultural level. This covers how IoT affects cultural practices, norms, and values as well as how businesses might adjust to these changes in culture.

8. **IoT in Wellness and Healthcare:** As IoT becomes more widely used in the field, studies will concentrate on how technology affects patient care, privacy issues, and how healthcare organizations and practices must adapt to different cultural contexts.

9. **IoT in Developing and Rural regions:** Studies will look at how IoT may help with cultural advancement and close the digital gap in underserved and rural regions. Incorporating a strategic approach is necessary for fostering innovation within companies (Borodako et al., 2023). Examining the difficulties and possibilities of implementing IoT technology in various situations is part of this.

10. **Legal and Ethical Considerations:** As IoT technology advances, so too will the legal and ethical environment around it. The legal frameworks for IoT technologies, liability problems, and ethical issues will be the main areas of research.

11. **Longitudinal Studies:** Long-term research tracing the development of IoT technology within organizations and its cultural influence will offer important insights into its longevity and adaptability given the dynamic nature of IoT and cultural change.

7.9 CONCLUSION

The digital era has brought about a significant change in the way we connect with our environment and with each other due to the broad use of the IoT and improvements in technology readiness within organizational contexts. This shift is not just changing how companies run, but it is also directly related to how we should be caring for the environment, especially when it comes to biodiversity.

A new age of connectedness, efficiency, and cooperation has been ushered in by the integration of IoT in organizational settings. Businesses using IoT technology are reporting increased efficiency, instantaneous decision-making, and a more flexible workplace environment. This progress is not without difficulties, though, as issues with data privacy, cybersecurity, and the need for constant adaption highlight how crucial it is to have a robust organizational culture.

The connection between technology and biodiversity simultaneously draws attention to the advantages and disadvantages of our digital endeavors. While technology may greatly aid in monitoring biodiversity and conservation efforts—especially when it comes to cutting-edge technologies and data analytics—it also presents several risks due to pollution and habitat damage. It is critical to strike a balance between environmental sustainability and technological growth.

From the meeting point of these two domains, it is evident that we need to proceed cautiously as we go further into the digital era. It is imperative to adopt technology responsibly, guided by ethical considerations and a dedication to sustainability. Organizations need to cultivate cultures that value flexibility and moral behavior in addition to embracing technical innovation.

To put it simply, the digital age's cultural progress necessitates a responsible and comprehensive strategy. It demands that organizational flexibility, technological advancement, and a diligent endeavor to protect and enhance biodiversity be harmoniously integrated. We can make sure that our cultural development in the digital era favorably impacts both social advancement and environmental well-being by responsibly navigating this complicated landscape.

In conclusion, there are both advantages and disadvantages to integrating the IoT into the cultural shift of the digital era inside an organizational setting. In the digital age, where cultural evolution is closely tied to technological advancements, the successful implementation of IoT within an organizational context is a strategic imperative. By addressing these challenges and embracing the opportunities presented by IoT technology, organizations can position themselves to thrive in an ever-changing cultural and technological landscape. This journey requires a commitment to innovation, adaptability, and a deep understanding of the evolving digital culture that shapes our world today. In the end, this review article advances our knowledge of the complex interplay of technology, culture, and successful organizations in the digital age, offering insightful information to academics, professionals, and decision-makers alike.

REFERENCES

Aboelmaged, M. G. (2014). Predicting e-readiness at firm-level: An analysis of technological, organizational and environmental (TOE) effects on e-maintenance readiness in manufacturing firms. *International Journal of Information Management, 34*(5), 639–651.

Antony, J., Sony, M., & McDermott, O. (2023). Conceptualizing Industry 4.0 readiness model dimensions: An exploratory sequential mixed-method study. *The TQM Journal, 35*(2), 577–596.

Antwi, C. O., Ren, J., Owusu-Ansah, W., Mensah, H. K., & Aboagye, M. O. (2021). Airport self-service technologies, passenger self-concept, and behavior: An attributional view. *Sustainability (Switzerland), 13*(6), 1–18. https://doi.org/10.3390/su13063134

Aubry, S., Frison, C., Medaglia, J. C., Frison, E., Jaspars, M., Rabone, M., ...& Van Zimmeren, E. (2022). Bringing access and benefit sharing into the digital age. *Plants, People, Planet, 4*(1), 5–12.

Balouei Jamkhaneh, H., Shahin, A., Parkouhi, S. V., & Shahin, R. (2022). The new concept of quality in the digital era: A human resource empowerment perspective. *The TQM Journal, 34*(1), 125–144.

Barrett, F. J. (1995). Creating appreciative learning cultures. *Organizational Dynamics, 24*(2), 36–49.

Behl, A. (2022). Antecedents to firm performance and competitiveness using the lens of big data analytics: A cross-cultural study. *Management Decision, 60*(2), 368–398.

Bélanger, F., & Crossler, R. E. (2011). Privacy in the digital age: A review of information privacy research in information systems. *MIS Quarterly, 35*(4), 1017–1041.

Borodako, K., Berbeka, J., Rudnicki, M., & Łapczyński, M. (2023). The impact of innovation orientation and knowledge management on business services performance moderated by technological readiness. *European Journal of Innovation Management, 26*(7), 674–695. https://doi.org/10.1108/EJIM-09-2022-0523

Burnes, B. (2004). Kurt Lewin and the planned approach to change: A re-appraisal. *Journal of Management Studies, 41*(6), 977–1002.

Cameron, K. (2008). A process for changing organization culture. *Handbook of Organization Development, 14*(5), 2–18.

Chanda, P., & Singh, P. (2023, August). Mapping the landscape of blockchain research in human resource management: A bibliometric analysis. In *Proceedings of the 2023 Fifteenth International Conference on Contemporary Computing* (pp. 115–126). Association for Computing Machinery, New York, NY. https://doi.org/10.1145/3607947.3607968

Chandra, A., & Idrisova, A. (2011). Convention on Biological Diversity: A review of national challenges and opportunities for implementation. *Biodiversity and Conservation, 20*, 3295–3316.

Clausing, D., & Holmes, M. (2010). Technology readiness. *Research Technology Management, 53*(4), 52–59. https://doi.org/10.1080/08956308.2010.11657640

Cooke, R. A., & Rousseau, D. M. (1988). Behavioral norms and expectations: A quantitative approach to the assessment of organizational culture. *Group & Organization Studies, 13*(3), 245–273.

Da Silva, L. B. P., Soltovski, R., Pontes, J., Treinta, F. T., Leitão, P., Mosconi, E., ...& Yoshino, R. T. (2022). Human resources management 4.0: Literature review and trends. *Computers & Industrial Engineering*, *168*, 108111.

Ghaleb, E. A., Dominic, P. D. D., Fati, S. M., Muneer, A., & Ali, R. F. (2021). The assessment of big data adoption readiness with a technology–organization–environment framework: A perspective towards healthcare employees. *Sustainability*, *13*(15), 8379.

Gupta, S., Kumar, V., & Karam, E. (2020). New-age technologies-driven social innovation: What, how, where, and why?. *Industrial Marketing Management*, *89*, 499–516.

Halewood, M., Chiurugwi, T., Sackville Hamilton, R., Kurtz, B., Marden, E., Welch, E., ... & Powell, W. (2018). Plant genetic resources for food and agriculture: Opportunities and challenges emerging from the science and information technology revolution. *New Phytologist*, *217*(4), 1407–1419.

Holt, D. T., Armenakis, A. A., Feild, H. S., & Harris, S. G. (2007). Readiness for organizational change: The systematic development of a scale. *The Journal of Applied Behavioral Science*, *43*(2), 232–255.

Kalashnikova, O. V., Petrunina, A. E., Tsygankov, N. S., & Moskalev, A. K. (2019). The level of generalized technology readiness of the Smart House automation systems. *IOP Conference Series: Materials Science and Engineering*, *666*(1). https://doi.org/10.1088/1757-899X/666/1/012063

Leitão, P., Colombo, A. W., & Karnouskos, S. (2016). Industrial automation based on cyber-physical systems technologies: Prototype implementations and challenges. *Computers in Industry*, *81*, 11–25.

Li, W., Liu, K., Belitski, M., Ghobadian, A., & O'Regan, N. (2016). e-Leadership through strategic alignment: An empirical study of small-and medium-sized enterprises in the digital age. *Journal of Information Technology*, *31*, 185–206.

Linnenluecke, M. K., & Griffiths, A. (2010). Corporate sustainability and organizational culture. *Journal of World Business*, *45*(4), 357–366.

Mahmood, A., Imran, M., & Adil, K. (2023). Modeling individual beliefs to transfigure technology readiness into technology acceptance in financial institutions. *SAGE Open*, *13*(1), 21582440221149718.

Mankins, J. C. (2009). Technology readiness assessments: A retrospective. *Acta Astronautica*, *65*(9–10), 1216–1223. https://doi.org/10.1016/j.actaastro.2009.03.058

Meek, V. L. (1988). Organizational culture: Origins and weaknesses. *Organization Studies*, *9*(4), 453–473.

Mollah, M. A., Choi, J. H., Hwang, S. J., & Shin, J. K. (2023). Exploring a pathway to sustainable organizational performance of South Korea in the digital age: The effect of digital leadership on IT capabilities and organizational learning. *Sustainability*, *15*(10), 7875.

Morgera, E., & Tsioumani, E. (2010). The evolution of benefit sharing: Linking biodiversity and community livelihoods. *Review of European Community & International Environmental Law*, *19*(2), 150–173.

Mourtzis, D., Angelopoulos, J., & Panopoulos, N. (2022). A literature review of the challenges and opportunities of the transition from Industry 4.0 to Society 5.0. *Energies*, *15*(17), 6276.

Musyaffi, A. M., Johari, R. J., Rosnidah, I., Respati, D. K., Wolor, C. W., & Yusuf, M. (2022). Understanding digital banking adoption during post- coronavirus pandemic: An integration of technology readiness and technology acceptance model. *TEM Journal, 11*(2), 683–694. https://doi.org/10.18421/TEM112-23

Narkhede, G., Pasi, B., Rajhans, N., & Kulkarni, A. (2023). Industry 5.0 and the future of sustainable manufacturing: A systematic literature review. *Business Strategy & Development, 6*(4), 704–723. https://doi.org/10.1002/bsd2.272

Näyhä, A. (2020). Finnish forest-based companies in transition to the circular economy-drivers, organizational resources and innovations. *Forest Policy and Economics, 110,* 101936.

Nezhad, M. Z., Nazarian-Jashnabadi, J., Rezazadeh, J., Mehraeen, M., & Bagheri, R. (2023). Assessing dimensions influencing IoT implementation readiness in industries: A fuzzy DEMATEL and fuzzy AHP analysis. *Journal of Soft Computing and Decision Analytics, 1*(1), 102–123.

Orozco, R., & Grundmann, P. (2022). Readiness for innovation of emerging grass-based businesses. *Journal of Open Innovation: Technology, Market, and Complexity, 8*(4), 180.

Pan, M. J., & Jang, W. Y. (2008). Determinants of the adoption of enterprise resource planning within the technology-organization-environment framework: Taiwan's communications industry. *Journal of Computer Information Systems, 48*(3), 94–102.

Parasuraman, A., & Colby, C. L. (2015). An updated and streamlined technology readiness index: TRI 2.0. *Journal of Service Research, 18*(1), 59–74.

Pfaff, Y. M., Wohlleber, A. J., Münch, C., Küffner, C., & Hartmann, E. (2023). How digital transformation impacts organizational culture–A multi-hierarchical perspective on the manufacturing sector. *Computers & Industrial Engineering, 183,* 109432.

Reding, D. F., & Eaton, J. (2020). *Science & Technology Trends 2020–2040. Exploring the S&T edge* (pp. 71–73). NATO Science & Technology Organization.

Ronchi, A. M. (2009). *eCulture: Cultural Content in the Digital Age.* Springer Science & Business Media.

Shenhar, A., Dvir, D., Milosevic, D., Mulenburg, J., Patanakul, P., Reilly, R., ... & Thamhain, H. (2005). Toward a NASA-specific project management framework. *Engineering Management Journal, 17*(4), 8–16.

Smollan, R. K., & Sayers, J. G. (2009). Organizational culture, change and emotions: A qualitative study. *Journal of Change Management, 9*(4), 435–457.

Thrift, N. (2006). Re-inventing invention: New tendencies in capitalist commodification. *Economy and Society, 35*(02), 279–306.

Tortorella, G. L., Prashar, A., Carim Junior, G., Mostafa, S., Barros, A., Lima, R. M., & Hines, P. (2023). Organizational culture and Industry 4.0 design principles: An empirical study on their relationship. *Production Planning & Control,* 1–15.

Tverskoi, D., Babu, S., & Gavrilets, S. (2022). The spread of technological innovations: Effects of psychology, culture and policy interventions. *Royal Society Open Science, 9*(6). https://doi.org/10.1098/rsos.211833

Uren, V., & Edwards, J. S. (2023). Technology readiness and the organizational journey towards AI adoption: An empirical study. *International Journal of Information Management*, *68*, 102588. https://doi.org/10.1016/j.ijinfo mgt.2022.102588

Visseren-Hamakers, I. J., & Kok, M. T. (Eds.). (2022). *Transforming biodiversity governance*. Cambridge University Press.

Vize, R., Coughlan, J., Kennedy, A., & Ellis-Chadwick, F. (2013). Technology readiness in a B2B online retail context: An examination of antecedents and outcomes. *Industrial Marketing Management*, *42*(6), 909–918. https://doi.org/ 10.1016/j.indmarman.2013.05.020

Warrick, D. D. (2023). Revisiting resistance to change and how to manage it: What has been learned and what organizations need to do. *Business Horizons*, *66*(4), 433–441.

Xu, J., & Melick, D. R. (2007). Rethinking the effectiveness of public protected areas in southwestern China. *Conservation Biology*, *21*(2), 318–328.

Zhu, K., Kraemer, K. L., & Dedrick, J. (2004). Information technology payoff in e-business environments: An international perspective on value creation of e-business in the financial services industry. *Journal of Management Information Systems*, *21*(1), 17–54.

Ziadlou, D. (2021). Strategies during digital transformation to make progress in achievement of sustainable development by 2030. *Leadership in Health Services*, *34*(4), 375–391.

Ziaei Nafchi, M., & Mohelská, H. (2020). Organizational culture as an indication of readiness to implement industry 4.0. *Information*, *11*(3), 174.

Chapter 8

Bioinspired swarm model
A new tool for devouring the supply chain

Gurwinder Singh and Monika Singh

8.1 INTRODUCTION

In the dynamic landscape of modern technology and robotics, the integration of bioinspired concepts has emerged as a groundbreaking paradigm, redefining the way we perceive and design systems. The symbiosis of biological principles and technological advancements has given rise to innovative approaches, and the focus herein is on the transformative impact this bioinspired swarm model holds for revolutionizing supply chain dynamics. The application of bioinspired swarm intelligence is a promising solution for industries seeking to enhance efficiency, adaptability, and resilience. This model goes beyond conventional supply chain techniques, taking inspiration from the collective behaviors seen in natural swarms, such as those found in social insects and schooling fish. The bioinspired swarm model not only negotiates the intricacies of the supply chain but also offers a fresh viewpoint on improving autonomous decision-making and resource allocation by utilizing distributed energy and data-driven insights.

The objective of this chapter is to present a thorough examination of the theoretical underpinnings, real-world applications, and possible difficulties related to supply chain management's implementation of the bioinspired swarm model. As we set out on this intellectual trip, it becomes clear that the coming together of state-of-the-art technologies and bioinspired methodologies is going to revolutionize robotics and allow for revolutionary breakthroughs in the field. Let's explore the intricacies of this cutting-edge instrument to see how it can eat up the traditional supply chain problems and open the door to a new era of efficiency and flexibility. Briey highlights the need for disruptive innovation by exposing the bloated and inefficient characteristics of the present robotics and energy supply chains. The ubiquitous, bloated, and ineffective character of present systems is a major concern in the current robotics and energy supply chain scene. Conventional methods in these fields frequently suffer from unwieldy frameworks, needless steps, and an inability to adjust to changing needs. The robotics industry struggles with centralized, sophisticated control systems, which limits the ability

    DOI: 10.1201/9781003530077-8

to make decisions on its own and react quickly to unforeseen obstacles. Energy supply chains simultaneously display inefficiencies characterized by resource-intensive methods, constrained distribution systems, and a reliance on non-renewable resources.

The increasing complexity and interconnectedness of global markets increase this inefficiency, demonstrating how out of step traditional methods are with the needs of a rapidly changing technology environment. As the need to improve scalability, reduce environmental impact, and streamline operations grows, so does the urgency for disruptive innovation. In light of this, the application of bioinspired swarm models emerges as a potentially revolutionary and promising approach. These models provide hope for a new era of efficiency, sustainability, and unmatched adaptability in robotics and energy supply chains by questioning the status quo and utilizing the collective intelligence found in natural swarms.

The "bioinspired swarm model" is a revolutionary and ground-breaking data-driven technique that takes its cues from complex and effective natural processes like ant colonies and bird foraging. This model's fundamental goal is to mimic the collective intelligence shown by social animals, where amazing system optimization is achieved through coordinated activities and decentralized decision-making. The implementation of these bioinspired swarm concepts presents a paradigm change in resource management and problem-solving in the context of robotics and energy supply chains.

Like birds' synchronized flight patterns or ants' well-organized foraging paths, these complex systems' complexity is navigated by the bioinspired swarm model, which uses data-driven insights. The paradigm facilitates adaptive behavior by allowing agents or entities inside the swarm to communicate and exchange information in real time, thereby promoting a dynamic response to changing environmental conditions. In addition to improving decision-making efficiency, this collective intelligence makes it possible to optimize tasks seamlessly, from resource allocation and energy distribution to logistical routing and scheduling. This strategy is beautiful because it mimics the robustness and adaptability found in natural swarms by utilizing data-driven algorithms and distributed energy. The bioinspired swarm model, with its decentralized decision-making and collaboration, presents itself as a revolutionary tool that has the potential to completely change our understanding of and approach to optimizing complex systems. The promise for a future in robotics and energy supply chains that is more agile, responsive, and sustainable becomes more and more clear as we delve into the details of this novel method.

8.2 LITERATURE SURVEY

In the field of bioinspired computing and networking, Xiao's book[1] serves as a comprehensive resource exploring various computational approaches inspired by biological systems. Dehghani et al.[2] introduce the Tasmanian devil optimization algorithm, a novel bioinspired method for solving optimization problems. The paper by Alfeo et al.[3] presents "Urban Swarms," a unique approach for autonomous waste management discussed at the International Conference on Robotics and Automation in 2019. Wang and Wang's work[4] delves into the investigation of determinants and peak prediction of CO_2 emissions in China's transport sector using bioinspired extreme learning machines. Miramontes et al.[5] conduct a comparative study of bioinspired algorithms applied to optimize fuzzy systems. Singh and Singh[6] propose a bioinspired approach for virtual machine migration using re-initialization and decomposition based on whale optimization. Vasant et al.[7] explore bioinspired approaches to address a combined economic emission dispatch problem. Jung et al.[8] survey bioinspired resource allocation algorithms and Media Access Control protocol design for mobile ad hoc networks. Alsayyed et al.[9] present the "Giant Armadillo Optimization," a new bioinspired metaheuristic algorithm for solving optimization problems. Reddy and Kiran's book[10] discusses the bioinspired quality of service-aware routing protocols in mobile ad hoc networks. Shu et al.[11] delve into biologically inspired design principles. Khilar et al.[12] focus on AI-based security protocols for resisting attacks on the Internet of Things. Shivadekar et al.[13] propose an efficient multimodal engine for skin disease[14] and explore swarming drones and energy-saving opportunities inspired by nature. Efendi and Wahyudi[15] introduce a bioinspired multi-objective data-privacy-preservation-technique (MO-DPPT) system model for cloud data. Dziedzic[16] identifies fractional order transfer function models using biologically inspired algorithms. Zheng and Sicker's survey[17] provides an overview of biologically inspired algorithms for computer networking. Rathore's study[18] examines bioinspired approaches across various engineering domains. Yachba et al.[19] present a bioinspired solution for regular carpooling problems. Pop et al.[20] propose a particle swarm optimization (PSO)-based method for generating healthy lifestyle recommendations. Charles et al.[21] discuss biologically inspired artificial intelligence for computer games. Stanzin and Murmu[22] employ a bioinspired approach to construct a minimum-spanning tree in cognitive radio networks. Rath et al.[23] highlight swarm intelligence as a solution for IoT technological problems. Klepac[24] explores the use of PSO for profiling from predictive data mining models. Yang et al.'s book[25] provides insights into swarm intelligence and bioinspired computation. Chen et al.[26] discuss bioinspired computation applications in operation management. Sinha et al.[27] propose a multi-agent-based petroleum supply chain coordination approach using co-evolutionary PSO. Leitão et al.[28] investigate bioinspired multi-agent systems for reconfigurable manufacturing systems. Acharjya and Kauser's work[29] reviews and discusses swarm intelligence in solving bioinspired computing problems. Finally, Sarkar et al.[30] explore the application of bioinspired optimization algorithms in food processing.

8.3 PROBLEM

In the contemporary landscape of supply chain management, optimizing various aspects such as resource allocation, logistics, and overall operational efficiency has become increasingly complex. Traditional optimization methods may fall short in addressing the dynamic and interconnected challenges posed by modern supply chain networks. This study aims to leverage the capabilities of bioinspired swarm models, drawing inspiration from nature's decentralized and self-organizing systems, to revolutionize supply chain optimization. The robotics and energy supply chains are plagued by inefficiencies, waste, and a lack of resilience, hindering their sustainability and economic viability. Traditional optimization methods have proven inadequate in addressing these challenges within increasingly complex and dynamic environments.

8.4 OBJECTIVES

- Dynamic Resource Allocation: Construct a swarm model inspired by biology that can allocate resources in the supply chain network in a dynamic manner while adjusting to shifting requirements and limitations.
- Efficient Logistics Optimization: Swarm intelligence-inspired design algorithms can be used to improve the efficiency of logistics operations by taking into account variables including inventory management, truck scheduling, and route optimization.
- Resilience to Disruptions: To handle and recover from disruptions, whether brought on by unanticipated occurrences, supply chain interruptions, or natural disasters, incorporate resilience mechanisms into the swarm model.
- Adaptive Decision-Making: Integrate the swarm model with adaptive decision-making mechanisms to enable the supply chain to autonomously modify plans in response to changes in the environment and real-time data.
- Collaborative Supply Chain Optimization: Facilitate collaboration among different entities in the supply chain through the swarm model, fostering a decentralized yet coordinated approach to optimization.

8.5 THE MATHEMATICAL MODELS FOR BIOINSPIRED SWARM SYSTEMS

Let us consider a particle swarm optimization (PSO) approach for optimizing the supply chain. In this context, we can define the following mathematical model.

The overall objective of the swarm is to minimize the total cost within the supply chain. Let f(x) represent the objective function, where x is the vector of decision variables.

$$f(x) = \text{Total Cost } (x)$$

Each particle in the swarm represents a potential solution (a set of decisions) to the optimization problem. Let P_i denote the position of the i-th particle, and V_i denote its velocity.

$$P_i = [x_{i1}, x_{i2}, \ldots\ldots x_{in}]$$

$$V_i = [v_{i1}, v_{i2}, \ldots\ldots v_{in}]$$

Here, n is the dimensionality of the decision space.

The position and velocity of each particle are updated iteratively based on the following equations:

$$V_{i,j}{}^{t+1} = w * V_{i,j}{}^{t} + c1*r1*(P_{i,j}{}^{t} - X_{i,j}{}^{t}) + c2*r2*(G_j{}^{t} - X_{i,j}{}^{t})$$

$$P_{i,j}{}^{t+1} = X_{i,j}{}^{t} + V_{i,j}{}^{t+1}$$

Here, w is the inertia weight, c1 and c2 are the acceleration coefficients, r1 and r2 are random numbers between 0 and 1, $G_j{}^{t}$ is the best-known position in dimension j across all particles, and $X_{i,j}{}^{t}$ is the jth component of the position of particle i at iteration t.

The optimization process continues until a specified number of iterations is reached or a convergence criterion is met.

Termination Criteria: Number of Iterations

This generic template provides a starting point for a mathematical model using a PSO-inspired approach in supply chain optimization. These applications showcase how the bioinspired swarm model can enhance decision-making and coordination across various stages of the supply chain in the robotics and renewable energy sectors. The model's ability to leverage collective intelligence and adaptability makes it a powerful tool for addressing the complexities inherent in these industries.

The objective functions play a crucial role in guiding the bioinspired swarm model to find optimal solutions that balance various aspects of supply chain management, including cost, revenue, lead times, and environmental impact. The swarm intelligence algorithm iteratively explores the solution space to converge toward solutions that meet these diverse objectives.

Minimizing costs is a fundamental goal in supply chain management. It involves optimizing various cost components such as shipping costs, manufacturing costs, and other associated costs. This objective function considers decision variables related to shipping costs, manufacturing costs, and other costs. The total cost is calculated by summing the product of quantities, relevant costs, and corresponding decision variables. The goal is to find a combination of decision variables that minimizes the overall cost of the supply chain as given in Algorithm 1.

Algorithm 1: Objective Function to Minimize Cost

---

1. Procedure Objective Function Minimize Cost (position)

2. Extract decision variables from the particle's position

3. shipping_costs gets position [:num_dimensions]

4. manufacturing_costs gets position [2 x num_dimensions:3 x num_dimensions]

5. other_costs gets position [3 x num_dimensions:4 xnum_dimensions]

$$
\begin{aligned}
total_{cost} = \sum\nolimits_{i=1}^{n} &df[\text{'Number of products sold'}][i] \times df[\text{'Shipping costs'}][i] * shipping_{costs[i]} \\
&+ df[\text{'Production volumes'}][i] \times df[\text{'Manufacturing costs'}][i] \times manufacturing \setminus_{costs[i]} \\
&+ df[\text{'OtherCostColumn'}][i] * other\_costs[i]
\end{aligned}
$$

6. Return the total cost as the objective value

7. Return total_cost

8. End Procedure

Maximizing revenue is crucial for business profitability. It involves optimizing pricing strategies to increase sales revenue. This objective function focuses on decision variables related to pricing strategies. The total revenue is calculated by summing the product of quantities, product prices, and corresponding pricing strategy decision variables. The objective is to find the pricing strategy that maximizes the overall revenue generated as depicted in Algorithm 2 by the supply chain.

Algorithm 2: Objective Function to Maximize Revenue

---

1. Procedure Objective Function Maximize Revenue (position)

2. Extract decision variables from the particle's position

3. shipping_costs gets position [:num_dimensions]

4. pricing_strategy gets position [2 x num_dimensions:3 x num_dimensions]

5. other_costs gets position [3 x num_dimensions:4 xnum_dimensions]

$$
total_{revenue} = \sum\nolimits_{i=1}^{n} df[\text{'Number of products sold'}][i] \times df[\text{'Price'}][i] * pricing\_strategy[i]
$$

6. Return the total revenue as the objective value

7. Return total_revenue

8. End Procedure

Minimizing lead times is essential for improving customer satisfaction and meeting delivery deadlines. It involves optimizing shipping times and production lead times.

As given in Algorithm 3 the objective function considers decision variables related to shipping times and production lead times. The total lead times are calculated by summing the product of quantities, relevant lead times, and corresponding decision variables. The objective is to find a combination of decision variables that minimizes the overall lead times in the supply chain.

Algorithm 3: Objective Function to Minimize Lead Times

---

1. Procedure Objective Function Minimize Lead Times (position)

2. Extract decision variables from the particle's position

3. shipping_costs gets position[3 x num_dimensions:4 x num_dimensions

4. production_lead_times gets position[4 x num_dimensions:5 x num_dimensions

$$total_{lead\_times} = \sum_{i=1}^{n} df['Number\ of\ products\ sold'][i] \times df['Shipping\ times'][i] * shipping\_times[i] + df['Production\ volumes'][i] * df['Manufacturing\ lead\ time'][i] *$$

5. $production\_lead\_times[i]$

6. Return the total lead times as the objective value

7. Return total_times

8. End Procedure

Minimizing environmental impact aligns with sustainability goals. It involves optimizing transportation modes to reduce carbon footprint. This objective function focuses on decision variables related to transportation modes. The total environmental impact is calculated by summing the product of quantities, transportation modes, and corresponding decision variables. The objective is to find transportation strategies that minimize the overall environmental impact of the supply chain which could be possible according to Algorithm 4.

Algorithm 4: Objective Function to Minimize Environmental Impact

---

1. Procedure Minimize Environmental Impact(position)

2. Extract decision variables from the particle's position

3. transportation_modes gets position[5 x num_dimensions:6 x num_dimensions

$$total_{impact} = \sum_{i=1}^{n} df['Number\ of\ products\ sold'][i] \times df['Tansportation\ modes'][i] * transportation\_modes[i]$$

4. Return the environmental impact as the objective value

5. Return total_impact

6. End Procedure

The PSO algorithm is a heuristic optimization technique inspired by the social behavior of bird flocks or fish schools. In the context of supply chain optimization, the PSO algorithm is utilized to find optimal solutions by simulating the movement of particles in a multidimensional search space. Each particle represents a potential solution to the optimization problem, and its position in the space corresponds to a set of decision variables. The algorithm begins by initializing a population of particles with random positions and velocities. These particles explore the solution space iteratively over a predefined number of iterations. In each iteration, the particles adjust their positions and velocities based on their own historical best-known solution (personal best) and the best-known solution among all particles in the population (global best). This exploration-exploitation balance allows the particles to converge toward promising regions in the solution space. The update of particle positions is guided by the inertia of their previous movements, cognitive factors reflecting their historical experiences, and social factors influenced by the global best solution. Through this dynamic process, particles converge toward optimal solutions while maintaining diversity in the search space. In the context of supply chain optimization, the algorithm generated in Algorithm 5 where the objective function to be minimized or maximized is evaluated for each particle's position, representing the overall performance metric such as cost, revenue, lead times, or environmental impact.

Algorithm 5: Particle Swarm Optimization (PSO) Algorithm

PSO Parameters:
num_iterations = 100
w = 0.5 //Inertia weight
c1 = 1.5 //Cognitive weight
c2 = 1.5 //Social weight

Initialization:
Particles gets random_matrix [num_particles, 4 x num_dimensions]
Velocities gets random_matrix [num_particles, 4 x num_dimensions]
Global Best Initialization:
global_best_position gets particles[0, :].copy()
global_best_value gets multi_objective_function (df.iloc[0])
PSO Main Loop:
for iteration : 1 to num_iterations do
    For i : 1 to num_particles do
    Update particle velocity
    velocities[i, :] gets (w x velocities[i, :] + c1 x rand() x (particles[i, :] - particles[i, :])
+ c2 x rand() x global_best_position - particles[i, :])
    Update Particle Position:
    Particles [i, :] gets particles[i, :] + velocities[i, :]
    Evaluate Objective Function:

Algorithm 5: Particle Swarm Optimization (PSO) Algorithm

```
    current_value gets multi_objective_function(df.iloc[i])
    Update Personal Best:
    If current_value < multi_objective_function(df.iloc}[i]) then
    particles[i, :] gets particles[i, :].copy()
    end if
Update Global Best:
    If current_value < global_best_value then
    global_best_value} gets current_value
    global_best_position gets particles[i, :].copy()
    end if
    end for
end for

Extract Optimal Solution:
optimal_solution gets global_best_position
optimal_value gets global_best_value

Print Results:
print("Optimal Solution:", optimal_solution)
print("Global Best Objective Value:", optimal_value)
```

The algorithm converges over iterations, and the final positions of particles represent the optimal solution to the supply chain problem. The PSO algorithm is known for its simplicity, adaptability to various optimization problems, and efficiency in finding near-optimal solutions. Its bioinspired nature makes it particularly well-suited for scenarios where decentralized decision-making and adaptive strategies are beneficial, aligning with the complex dynamics of supply chain networks.

The optimal solution obtained from the PSO algorithm for the given supply chain optimization problem is represented as a vector of numerical values. Each element in the vector corresponds to a decision variable in the optimization problem, and collectively, they define the optimal configuration for the supply chain. The optimal solution vector is as follows: The corresponding global best objective value achieved by this optimal solution is approximately 26495.79. This value represents the overall performance metric (e.g., cost, revenue, lead times, or environmental impact) optimized by the PSO algorithm for the given supply chain problem.

8.6 METHODOLOGY

8.6.1 Dataset overview and Exploratory Data Analysis

The success of any optimization algorithm relies heavily on a comprehensive understanding of the dataset. In this section, we present an overview of the supply chain dataset and perform Exploratory Data Analysis (EDA) to gain insights into its structure and characteristics.

Dataset Overview: The supply chain dataset consists of several key columns, including "Product type," "SKU," "Price," "Availability," "Number of products sold," "Revenue generated," "Customer demographics," "Stock levels," "Lead times," "Order quantities," "Shipping times," "Shipping carriers," "Shipping costs," "Supplier name," "Location," "Lead time," "Production volumes," "Manufacturing lead time," "Manufacturing costs," "Inspection results," "Defect rates," "Transportation modes," "Routes," and "Costs."

Exploratory Data Analysis (EDA)

1. Missing Values: We begin by visualizing missing values in the dataset to identify any gaps or inconsistencies.
2. Summary Statistics: A summary of basic statistics provides insights into the central tendencies and distributions of numeric variables.
3. Distributions of Numeric Variables: Histograms illustrate the distributions of key numeric variables, offering a glimpse into their spread and shape.
4. Pairplot of Numeric Variables: A pairplot helps visualize relationships between numeric variables, aiding in the identification of potential patterns or correlations.
5. Correlation Matrix: The correlation matrix provides a quantitative measure of the relationships between numeric variables.
6. Countplots of Categorical Variables: For categorical variables, countplots reveal the distribution of categories, offering insights into the diversity of the dataset.

These EDA steps serve as a foundation for understanding the dataset's characteristics, guiding the subsequent steps in the optimization process. The optimization objectives, as described in the following sections, are informed by the patterns and trends identified during the EDA phase.

8.6.2 Objective functions

Objective functions play a crucial role in guiding the bioinspired swarm model to find optimal solutions that balance various aspects of supply chain management. The following objective functions are defined based on different optimization goals:

1. Minimize Cost:
    - Decision Variables: Shipping costs, manufacturing costs, and other costs.
    - Objective: Minimize the overall cost of the supply chain.
    - Calculation: Sum of the product of quantities, relevant costs, and decision variables.
2. Maximize Revenue:
    - Decision Variables: Pricing strategies.
    - Objective: Maximize the overall revenue generated by the supply chain.
    - Calculation: Sum of the product of quantities, product prices, and pricing strategy decision variables.
3. Minimize Lead Times:
    - Decision Variables: Shipping times, production lead times.
    - Objective: Minimize the overall lead times in the supply chain.
    - Calculation: Sum of the product of quantities, relevant lead times, and decision variables.
4. Minimize Environmental Impact:
    - Decision Variables: Transportation modes.
    - Objective: Minimize the overall environmental impact of the supply chain.
    - Calculation: Sum of the product of quantities, transportation modes, and decision variables.

8.6.3 Particle swarm optimization algorithm

The PSO algorithm operates with specific settings: it runs for 100 iterations, with an inertia weight of 0.5, cognitive weight of 1.5, and social weight of 1.5. In the beginning, particles and their velocities are initialized randomly, and the initial global best position is determined by the first particle's performance. The algorithm then iteratively updates particle positions and velocities, evaluates the objective function for each particle, and adjusts personal and global best positions accordingly. Finally, the optimal solution is derived from the global best position. This process allows PSO to systematically explore and find optimal solutions based on the specified parameters and the objective function.

8.7 RESULTS AND DISCUSSION

This section presents a comprehensive analysis of the supply chain's performance through diverse visualizations. These findings guide strategic decisions

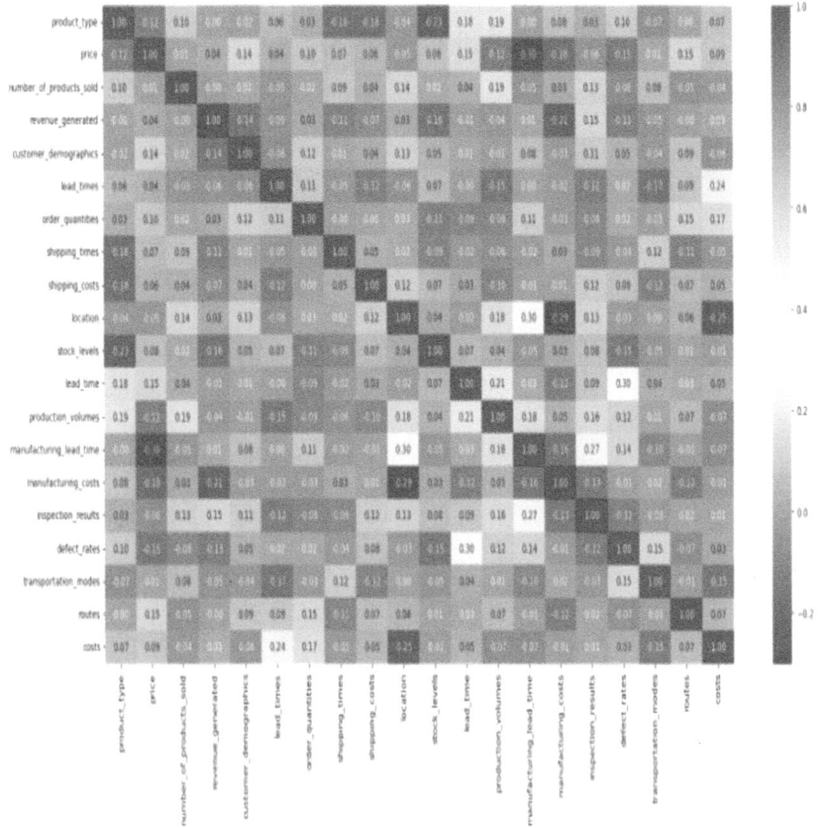

Figure 8.1 Heatmap visualizes the complex interplay of factors influencing our supply chain performance.

for pricing, inventory management, and product offerings, emphasizing the need for a nuanced interpretation based on specific business details.

The heatmap Figure 8.1 has shown correlations between different variables in the dataset, such as how product price is related to the number of products sold or how lead times are related to shipping costs. This information is useful to identify inefficiencies in the supply chain and develop targeted solutions. It visualizes clusters of products or customers with similar characteristics. The heatmap has tracked the changes in key metrics over time, such as the revenue generated or stock levels which is helpful in businesses monitoring their performance and identifying areas for improvement.

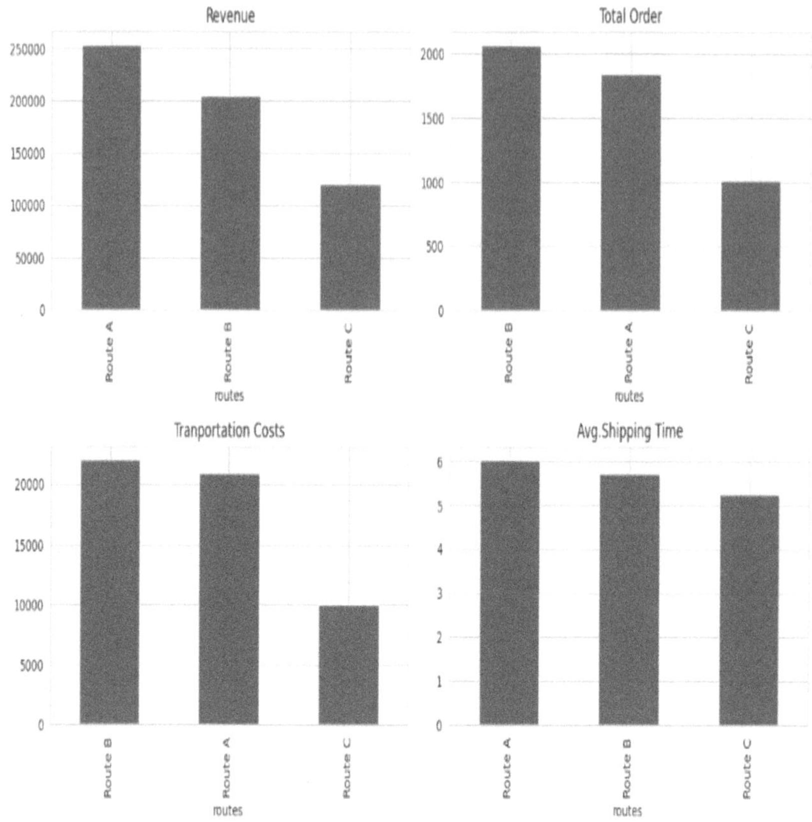

Figure 8.2 Visualizing supply chain metrics and product distribution.

In Figure 8.2, the X-axis represents routes or transportation modes, while the Y-axis represents key performance indicators (KPIs) like revenue, costs, or shipping times. The first four bar charts show the total stock levels, order quantities, manufacturing costs, and revenue generated for each product. Products are sorted based on these metrics, providing insights into their distribution. These visualizations provide insights into the distribution of stock levels, order quantities, manufacturing costs, and revenue across different products in the supply chain.

Figure 8.3 illustrates the relationship between different transportation routes and the corresponding stock keeping unit (SKU) counts. Each bar represents a specific route, and its height corresponds to the number of SKUs associated with that route. The bars are color-coded based on transportation

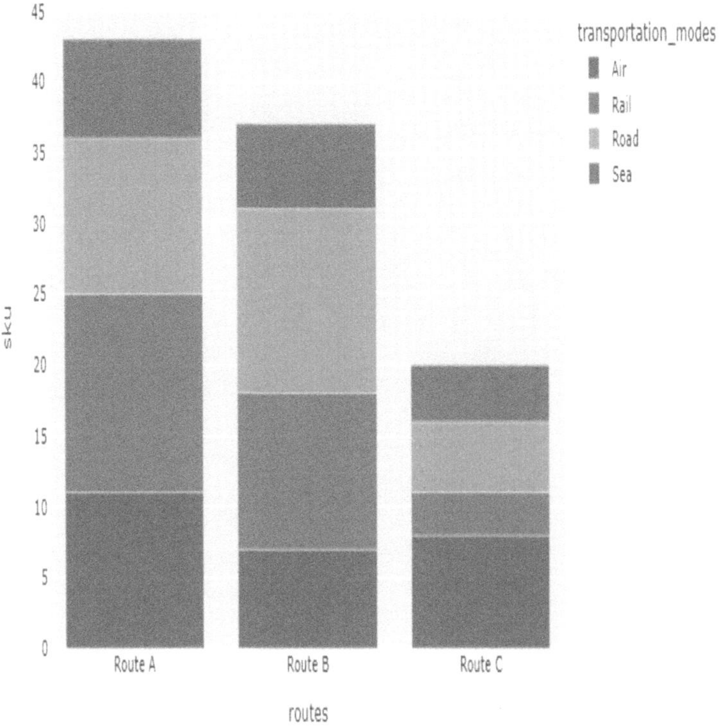

Figure 8.3 Routes by transportation modes.

modes, providing insights into how different transportation methods contribute to the overall distribution across various routes in the supply chain.

Figure 8.4 shows four sets of graphs, each with three plots. Each plot shows the objective function value versus the number of iterations for a particle filter simulation. The simulations were run with 20, 30, and 40 particles, and each set of graphs shows the results for 50, 100, and 150 iterations. The objective function value is a measure of how well the particle filter is able to track the target. In general, a lower objective function value means that the particle filter is doing a better job. The graphs show that the objective function value generally decreases as the number of iterations increases, which means that the particle filter is getting better at tracking the target as it has more time to do so. The graphs also show that the objective function value is generally lower for simulations with more particles, which means that using more particles can improve the performance of the particle filter. However, there are some exceptions to these trends. For example, in

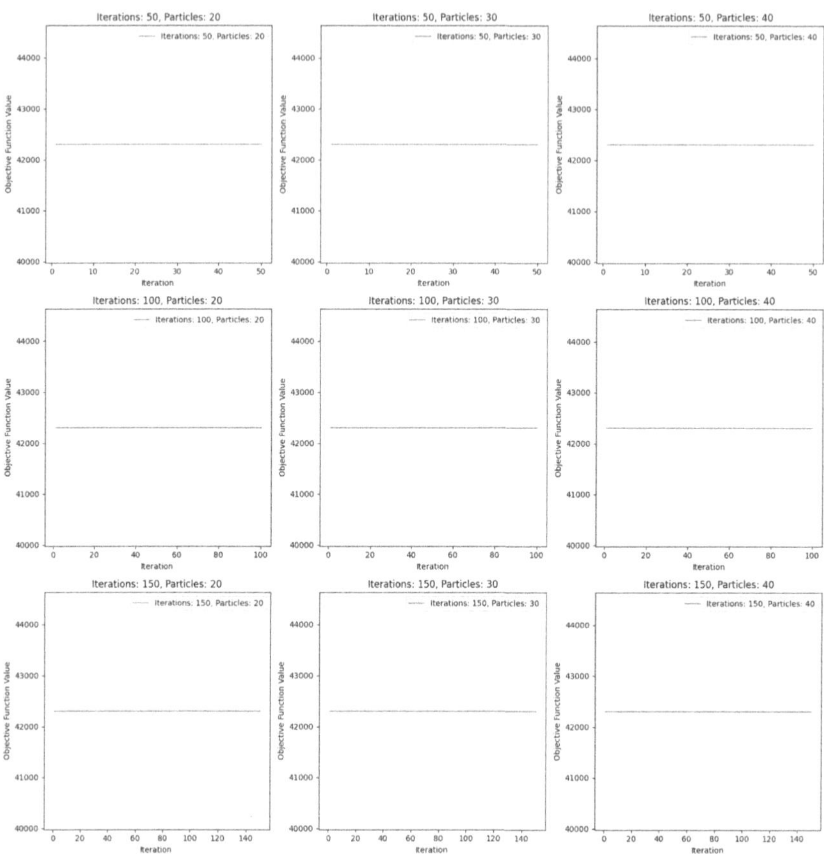

Figure 8.4 Convergence behavior of bio-swarm model with respect to different factors.

the set of graphs for 50 iterations, the objective function value for the simulation with 30 particles is lower than the objective function value for the simulation with 40 particles. This suggests that in some cases, using more particles can actually make the performance of the particle filter worse.

Figure 8.5 consists of scatter plots displaying a retailer's sales data, comparing revenue generated (vertical axis) with different sales metrics (horizontal axis). Each plot focuses on a specific metric: shipping costs, order quantities, and manufacturing costs. Data points on the plots represent the retailer's performance for various products or categories. Blue circles represent all potential solutions, while red circles highlight the "Pareto Front" indicating optimal trade-offs between revenue and the chosen cost metric. Observations include varying product combinations in the top and

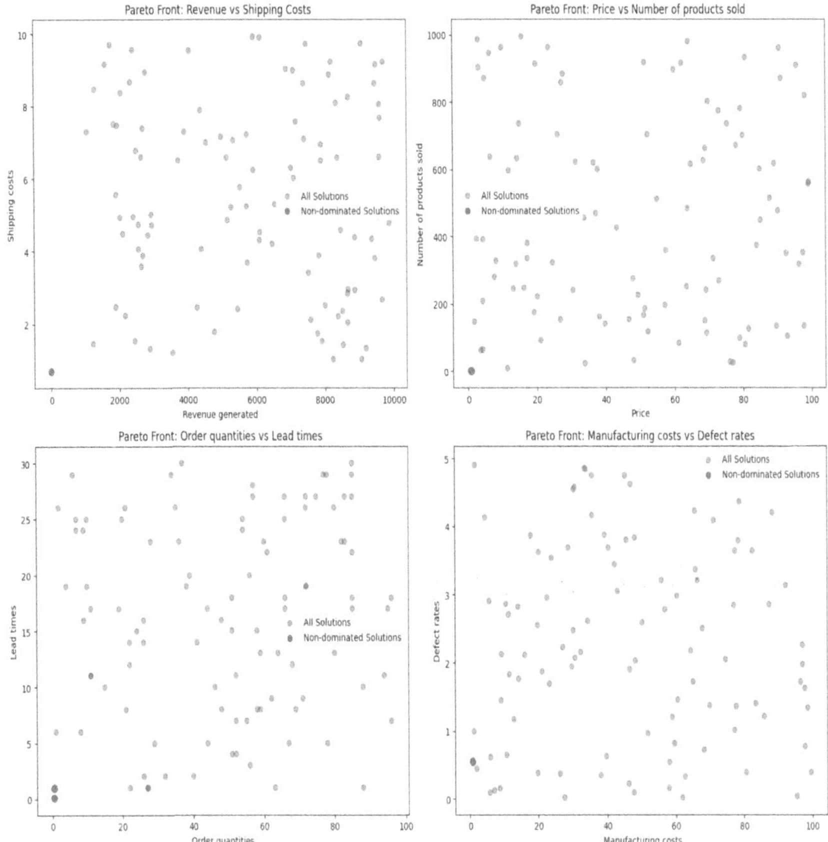

Figure 8.5 Exploring revenue dynamics and cost considerations.

bottom plots, a distinct cluster of order quantities in the middle plot, and trends in the Pareto Fronts, such as the trade-off between higher revenue and increased costs. While more context is needed for precise interpretation, these insights offer a glimpse into the retailer's revenue dynamics and cost considerations.

Figure 8.6 illustrates a Pareto front, presenting trade-offs between revenue and diverse cost/delivery metrics, likely representing an online retailer's product assortment. Each blue circle signifies a specific product or category, positioned based on its performance regarding revenue and a chosen metric (number of products sold, shipping costs, order quantities, or manufacturing costs). The Pareto Front, depicted by a red curve, outlines optimal solutions where revenue enhancement comes with an associated

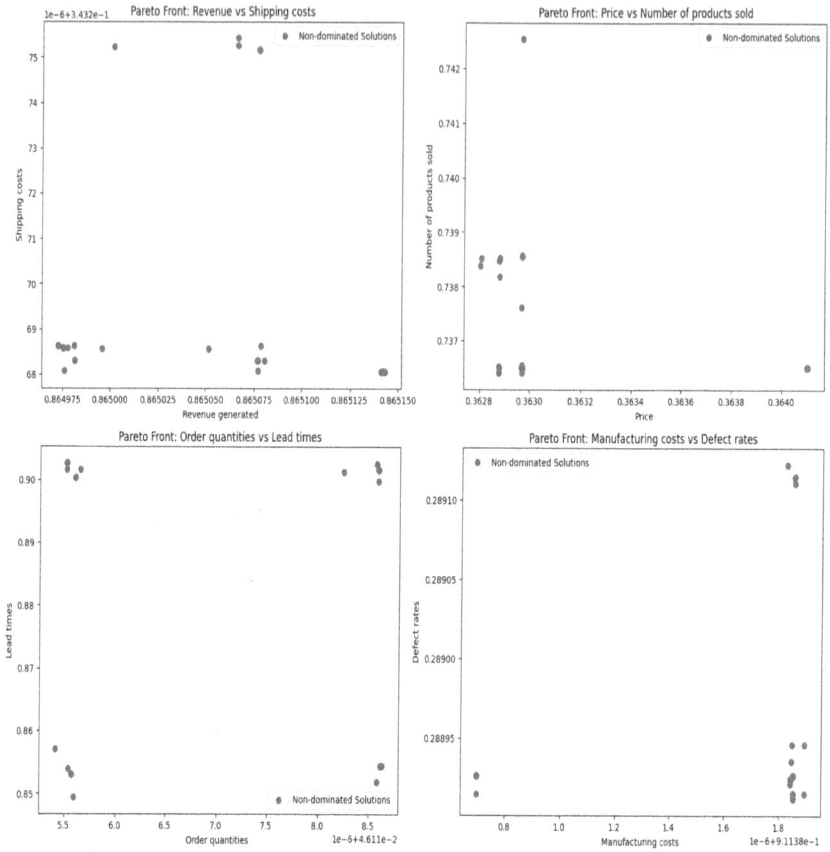

Figure 8.6 Pareto front showcasing trade-offs between revenue and distinct cost/delivery metric.

cost/delivery trade-off. Key observations reveal varied scatterings in the top and bottom plots, indicating different relationships between revenue and costs. The nearly horizontal middle plot suggests the limited impact of order quantities on revenue, potentially due to fixed production constraints. The spread of data points in the bottom plot and the upward slope of the Pareto Front signify a trade-off between revenue and manufacturing costs. This analysis provides valuable insights into the retailer's product mix and guides strategic decisions related to pricing, inventory management, and product offerings. It is crucial to note that a comprehensive interpretation would necessitate additional details about the specific retailer and its products, but these general trends offer meaningful insights into the retailer's business dynamics.

8.8 BENEFITS AND CHALLENGES OF THE SWARM MODEL

The swarm model, inspired by the collective intelligence of animal swarms, is increasingly viewed as a promising approach to optimizing supply chains. Here's how it can drive potential benefits across various aspects:

1. Increased Efficiency:
   - Decentralized Decision-making: Unlike traditional hierarchical models, swarms empower individual actors (entities like suppliers, transportation providers, and warehouses) to make real-time decisions based on local information. This eliminates delays and bottlenecks caused by centralized decision-making processes.
   - Adaptive Route Optimization: Swarms can dynamically adjust routes and delivery schedules in response to real-time changes like traffic congestion, unexpected delays, or product demands. This leads to faster delivery times and reduced transportation costs.
   - Improved Inventory Management: Swarm algorithms can predict demand fluctuations and optimize inventory levels across the supply chain network. This minimizes costs associated with holding excess inventory or experiencing stockouts.
2. Reduced Waste:
   - Collaborative Planning and Forecasting: By sharing information and coordinating plans across the supply chain, swarms can minimize overproduction and optimize resource allocation. This reduces waste generation and disposal costs.
   - Real-time Demand Responsiveness: The ability to adapt to real-time demand changes minimizes unnecessary production and transportation, thus reducing waste associated with unused products or inefficient logistics.
   - Flexible Sourcing and Production: Swarm models can identify alternative suppliers and production facilities based on real-time conditions, thereby reducing reliance on single sources and minimizing waste due to disruptions.
3. Improved Resilience:
   - Fault Tolerance and Self-healing: Decentralized swarms operate without a central point of failure. If one actor in the system

experiences disruption, others can adjust and compensate, maintaining overall supply chain functionality.

- Proactive Risk Management: By analyzing vast amounts of data and identifying potential bottlenecks, swarm models can anticipate disruptions and trigger preventive measures, improving overall supply chain resilience.
- Dynamic Scalability and Adaptability: Swarm systems can easily adapt to changing market conditions, new technologies, or business growth by adding or removing actors from the network, enhancing agility and resilience.

4. Lower Overall Costs:
- Optimized Resource Allocation: The efficient utilization of resources like transportation, inventory, and production capacity leads to significant cost savings across the supply chain.
- Reduced Operational Costs: The swarm's ability to self-optimize routes, adjust schedules, and minimize waste translates to lower operational costs like transportation, storage, and waste disposal.
- Improved Collaboration and Information Sharing: Enhanced transparency and collaboration across the supply chain can lead to better negotiation terms with suppliers, optimize pricing strategies, and reduce inefficiencies, ultimately lowering overall costs.

Implementing the swarm model requires careful consideration of data integration, computational resources, and adaptation to specific supply chain contexts. However, the potential benefits in terms of efficiency, reduced waste, improved resilience, and lower costs make it a compelling approach for enhancing supply chain performance in today's complex and dynamic environments.

8.8 INTEGRATION WITH DISTRIBUTED ENERGY SYSTEMS

Integrating bioinspired swarm models with Distributed Energy Systems (DES) can significantly enhance sustainability and resource utilization in the robotics sector. Here are several ways in which this integration can be beneficial:

1. Energy-efficient Movement and Navigation:
- Inspired by Animal Swarms: Bioinspired swarm models mimic the collective intelligence of animal swarms, promoting decentralized decision-making. This can be applied to robotics for more energy-efficient movement and navigation, where individual robotic entities collaborate to optimize paths and avoid energy-intensive routes.

- DES Integration: Combining swarm intelligence with DES allows robots to make real-time decisions based on local energy availability and demand. This ensures that robots choose energy-efficient paths and modes of operation, contributing to overall energy conservation.

2. Adaptive Energy Harvesting:
   - Swarm Adaptability: Bioinspired swarm models exhibit adaptive behaviors. When applied to robotics, this adaptability can be extended to energy harvesting mechanisms. Robots within a swarm can dynamically adjust their energy harvesting strategies based on environmental conditions, optimizing the use of renewable energy sources.
   - DES Integration: Integrating with DES provides a platform for sharing energy production and consumption data across the swarm. This enables robots to make informed decisions about when and where to harvest energy, ensuring sustained operation and reducing reliance on traditional power sources.

3. Collaborative Charging and Energy Sharing:
   - Swarms for Collaboration: Bioinspired swarms promote collaboration among individual entities. In the context of robotics, this collaboration can extend to sharing energy resources within the swarm. Robots with surplus energy can share it with those in need, promoting overall system efficiency.
   - DES Integration: In a DES, robots can be equipped with bidirectional energy flow capabilities. This allows them to not only consume energy but also share excess energy with other robots in the swarm. Such collaborative charging and energy sharing contribute to a more balanced and sustainable use of energy resources.

4. Decentralized Energy Storage:
   - Swarm-based Storage Strategies: Inspired by swarm behaviors such as fault tolerance and self-healing, robots in a swarm can employ decentralized energy storage strategies. This ensures that energy is distributed across the swarm, minimizing the impact of individual failures.
   - DES Integration: Connecting swarm robots to a DES facilitates the exchange of information about energy storage levels. This information enables robots to make decisions about when to charge or discharge energy, promoting resilience and reliability in energy storage and utilization.

5. Dynamic Resource Allocation:
   - Swarm Adaptability: Bioinspired swarms are known for their dynamic scalability and adaptability. In the context of robotics,

this adaptability can be applied to dynamically allocate resources, including energy.

- DES Integration: Integration with DES provides real-time information about energy availability and demand. Swarm robots can dynamically adjust their tasks and energy consumption based on this information, optimizing resource utilization and contributing to sustainability.

6. Optimized Task Distribution:
- Swarm Efficiency: Bioinspired swarm models excel in optimized task distribution. In robotics, this translates to an efficient allocation of tasks among swarm entities to achieve collective goals.
- DES Integration: Integrating with a DES allows robots to consider energy constraints when distributing tasks. This ensures that tasks are allocated in a way that minimizes energy consumption and maximizes the overall efficiency of the swarm.

8.9 CONCLUSION

In conclusion, this chapter has provided a comprehensive exploration of various aspects related to supply chain optimization and decision-making processes. The discussion covered topics ranging from the formulation of objective functions—such as minimizing costs, maximizing revenue, minimizing lead times, and minimizing environmental impact—to the implementation of the PSO algorithm. Each objective function was carefully examined in the context of decision variables and the calculation process involved. The PSO algorithm, with its defined parameters and initialization steps, was outlined to give readers a clear understanding of its functioning in optimizing supply chain operations. The iterative nature of the PSO main loop, involving the update of particle positions and velocities, evaluation of objective functions, and adjustment of personal and global best positions, was explained as a crucial part of the optimization process.

Furthermore, the extraction of the optimal solution from the global best position was highlighted as the final step in the PSO algorithm. This step underscores the practical applications of the algorithm's outcomes in real-world supply chain scenarios. Overall, this chapter serves as a foundational guide for readers interested in understanding and applying PSO to address complex challenges in supply chain management. The described objective functions, algorithm parameters, and PSO main loop provide a structured framework for optimizing various aspects of the supply chain, contributing to enhanced decision-making and operational efficiency.

REFERENCES

[1] Xiao, Y. (2016). Bio-inspired Computing and Networking. CRC Press.

[2] Dehghani, M., Hubálovský, Š., & Trojovský, P. (2022). Tasmanian devil optimization: A new bio-inspired optimization algorithm for solving optimization problems. IEEE Access, 10, 19599–19620.

[3] Alfeo, A. L., et al. (2019). Urban swarms: A new approach for autonomous waste management. In Proceedings of the 2019 International Conference on Robotics and Automation (ICRA) (pp. 4233–4240). IEEE.

[4] Wang, W., & Wang, J. (2021). Determinants investigation and peak prediction of CO_2 emissions in China's transport sector utilizing bio-inspired extreme learning machine. Environmental Science and Pollution Research, 28(39), 55535–55553.

[5] Miramontes, I., Melin, P., & Prado-Arechiga, G. (2020). Comparative study of bio-inspired algorithms applied in the optimization of fuzzy systems. In Hybrid Intelligent Systems in Control, Pattern Recognition and Medicine (pp. 219–231). Springer.

[6] Singh, S., & Singh, D. (2023). A bio-inspired VM migration using re-initialization and decomposition based-whale optimization. ICT Express, 9 (1), 92–99.

[7] Vasant, P., Banik, A., Thomas, J. J., Marmolejo-Saucedo, J. A., Fiore, U., & Weber, G. W. (2023). Bio-inspired approaches for a combined economic emission dispatch problem. In Human-Assisted Intelligent Computing: Modeling, Simulations and Applications (pp. 3-1–3-38). IOP Publishing.

[8] Jung, J. Y., Choi, H. H., & Lee, J. R. (2018). Survey of bio-inspired resource allocation algorithms and MAC protocol design based on a bio-inspired algorithm for mobile ad hoc networks. IEEE Communications Magazine, 56 (1), 119–127.

[9] Alsayyed, O., et al. (2023). Giant Armadillo Optimization: A new bio-inspired metaheuristic algorithm for solving optimization problems. Biomimetics, 8(8), 619.

[10] Reddy, G. R. M., & Kiran, M. (2016). Mobile Ad Hoc Networks: Bio-inspired Quality of Service Aware Routing Protocols. CRC Press.

[11] Shu, L. H., Ueda, K., Chiu, I., & Cheong, H. (2011). Biologically inspired design. CIRP Annals, 60 (2), 673–693.

[12] Khilar, R., et al. (2022). Artificial intelligence-based security protocols to resist attacks in internet of things. Wireless Communications and Mobile Computing, 2022.

[13] Shivadekar, S., et al. (2023). Design of an efficient multimodal engine for preemption and post-treatment recommendations for skin diseases via a deep learning-based hybrid bioinspired process. Soft Computing, 1–19.

[14] Bawana, N. M. (2021). Swarming Drones: Bioinspiration and Energy Saving Opportunities. New Mexico Institute of Mining and Technology.

[15] Efendi, R., & Wahyudi, R. (2022). Bio inspired based multi-objective data-privacy-preservation-technique (MO-DPPT) system model for cloud data. Int J Circuit Comput Networking, 3(2), 27–43. DOI: 10.33545/27075923.2022.v3.i2a.47

[16] Dziedzic, K. (2020). Identification of fractional order transfer function model using biologically inspired algorithms. In Automation 2019: Progress in Automation, Robotics and Measurement Techniques (pp. 47–57). Springer.

[17] Zheng, C., & Sicker, D. C. (2013). A survey on biologically inspired algorithms for computer networking. IEEE Communications Surveys & Tutorials, 15(3), 1160–1191.

[18] Rathore, H. (2016). Bio-inspired approaches in various engineering domain. In Mapping Biological Systems to Network Systems (pp. 177–194). Springer.

[19] Yachba, K., et al. (2023). A bio-inspired approach to solve the problem of regular carpooling. In Encyclopedia of Data Science and Machine Learning (pp. 2936–2953). IGI Global.

[20] Pop, C. B., et al. (2013). Particle swarm optimization-based method for generating healthy lifestyle recommendations. In 2013 IEEE 9th International Conference on Intelligent Computer Communication and Processing (ICCP) (pp. 15–21). IEEE.

[21] Charles, D., et al. (2007). Biologically Inspired Artificial Intelligence for Computer Games. IGI Global.

[22] Stanzin, T., & Murmu, M. K. (2018). A bio-inspired approach to construct minimum spanning tree in cognitive radio networks. In 2018 International Conference on Communication and Signal Processing (ICCSP) (pp. 0138–0143). IEEE.

[23] Rath, M., et al. (2020). Swarm intelligence as a solution for technological problems associated with Internet of Things. In Swarm Intelligence for Resource Management in Internet of Things (pp. 21–45). Elsevier.

[24] Klepac, G. (2017). Particle swarm optimization algorithm as a tool for profiling from predictive data mining models. In Nature-Inspired Computing: Concepts, Methodologies, Tools, and Applications (pp. 864–892). IGI Global.

[25] Yang, X. S., et al. (2013). Swarm Intelligence and Bio-inspired Computation: Theory and Applications. Newnes.

[26] Chen, T., et al. (2014). Bioinspired computation and its applications in operation management. The Scientific World Journal, 2014.

[27] Sinha, A. K., et al. (2009). Multi-agent based petroleum supply chain coordination: A co-evolutionary particle swarm optimization approach. In 2009 World Congress on Nature & Biologically Inspired Computing (NaBIC) (pp. 1349–1354). IEEE.

[28] Leitão, P., Barbosa, J., & Trentesaux, D. (2012). Bio-inspired multi-agent systems for reconfigurable manufacturing systems. Engineering Applications of Artificial Intelligence, 25(5), 934–944.

[29] Acharjya, D. P., & Kauser, A. P. (2015). Swarm intelligence in solving bio-inspired computing problems: Reviews, perspectives, and challenges. In Handbook of Research on Swarm Intelligence in Engineering (pp. 74–98). IGI Global.

[30] Sarkar, T., et al. (2022). Application of bio-inspired optimization algorithms in food processing. Current Research in Food Science, 5, 432–450.

Chapter 9

A comprehensive study on meta-heuristic optimization approach for maximum power-point tracking in solar power system

Zaiba Ishrat, Kunwar Babar Ali, Taslima Ahmed, and Sriniwas Mishra

9.1 INTRODUCTION

Solar photovoltaic (PV) systems are regarded as more efficient than other renewable energy sources due to their benefits to the environment and economy (Sahin et al. 2020). The cleanest, most abundant, least polluting, and least maintenance-required renewable energy source is solar electricity generation. The absence of moving parts and the absence of noise are the primary benefits of employing a solar PV system. Sun-derived solar irradiation is transformed to electrical power by means of solar PV array. Taking the sources of installed electrical energy in India as a percentage by 30 June 2022, renewable energy sources would account for 28.25% of total energy capacity (MW), and the share percentage is as follows: wind, 35.76%; solar, 50.59%; biomass, 8.95%; waste to energy, 0.42%; and small hydro, 4.29%. The PV cell operates on the electroluminescence effects, which converts light signals directly into electrical signals (CEA 2020; GOI 2022).

Solar panel system (SPS) consists of many PV panels that are linked in parallel and series to achieve required rating. As a result, there is a strong likelihood that partial shading condition (PSC) will develop. This problem arises when the SPS's surrounding environment and irradiance are not uniform. A universal peak and minor peaks are seen on the PV or current power PI plot of SPS under PSC. Numerous traditional maximum power point tracker (MPPT) algorithms have been developed (Sahin et al. 2020) to improve SPS transferrable efficiency under uniform irradiance; however, these traditional procedures are unsuitable for tracking the maximum point under PSC. Therefore, a number of strategies for obtaining the highest power from SPS have been suggested in the literature (Farhat et al. 2015; Hassan et al. 2017; Piegari et al. 2015; Sheik et al. 2016).

The preeminent implementation point for a solar module is known as the MPPT. It is a crucial task to get the most out of the solar power from a solar panel; therefore, an MPPT plays a vital role which ensures the optimum use of a PV system independent of changes in temperature and irradiance. The Global Point (GP) cannot be properly followed under Partial Shading

DOI: 10.1201/9781003530077-9

"

(PS) conditions using conservative MPPT procedures. Advanced MPPT algorithms are requisites to follow the GP to maximize the amount of energy captured by the PV array. This study compares well-known MPPT algorithms that are based on biological inspiration. To illustrate the performance parameter of various optimization strategies under changing weather conditions, a comparative inquiry is set up between three tactics, that is, artificial bee colony (ABC), particle swarm optimization (PSO), and cuckoo search optimization (CSO).

In this review, three mostly used optimization algorithms, namely CSO algorithm (CSOA), PSO algorithm (PSOA), and artificial bee colony algorithm (ABCOA), are brought together and their usefulness and tracking speed convergence to the MPP are evaluated. Section 9.2.1 discusses the algorithm overview, implementation in solar energy, and recent research comparative analysis in terms of major findings and results of CSOA, PSOA, and ABCOA. Section 9.3 discusses the comparative analysis of three swarm-based MPPT optimization techniques in terms of their merits and limitations; and finally conclusions of the paper are drawn, which provides a valuable resource for engineers, scientists, and professionals in the area of solar energy for those interested in the application of soft computing techniques.

9.2 OPTIMIZATION ALGORITHM FOR MPPT

9.2.1 Classification of optimization algorithms

All forms of PV systems are handled using meta-heuristic methods in dynamically changing weather circumstances. This study provides the method demonstration for various strategies that are practicable to execute, along with their benefits, drawbacks, and parameter-based characteristics. Additionally, the literature survey of these techniques contained will be a striking analysis for the specific application. Figure 9.1 categorizes all optimization-based MPPT strategies; the three techniques are explored and contrasted in this literature review. Figure 9.1 shows the classification of algorithms.

9.2.2 Cuckoo search (CS) optimization algorithm (CSOA)

In 2009, Xin-She Yang and Suash Deb introduced the concept of CSOA (Yang and Deb 2009, 2017). The complete family sponging of some cuckoo species, which involves the cuckoos laying their eggs in the nests of host birds of other species, served as its model. There are typically three types of family sponging: intra-specific, obliging, and shell invasion. There are the three idealized guidelines for CSOA as per the sponging activities of cuckoos.

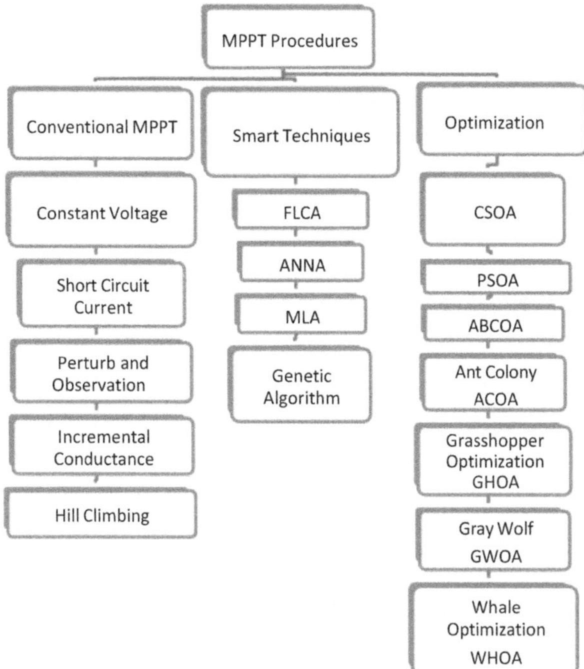

Figure 9.1 Classification of maximum power tracking procedure.

- Every cuckoo deposits one egg at an instant in a shell that is picked at random. This random walk is modeled by levy's flight distribution function.
- The best nest will produce the best eggs for the following generation. This is represented by fitness function (FF).
- The number of nests that are open is set, and the hostess bird set a prospect that the quantity of progeny (set by a cuckoo) it finds is between 0 and 1. The host bird has the option of abandoning its nest or destroying cuckoo eggs if the cuckoo's progeny are found. In either case, given a fixed number of nests, a new nest will be produced with a probability of Px.

The CSOA can be explained in a random code (Yang and Deb 2009, 2017) by using these three rules. For the FF, two variables such as PV array operating voltage and step size Δ (for fine-tuning of GMPP) are used. It is given by Equations (9.1) and (9.2).

$$FF = f(V) \text{ and } Vt+1 \tag{9.1}$$

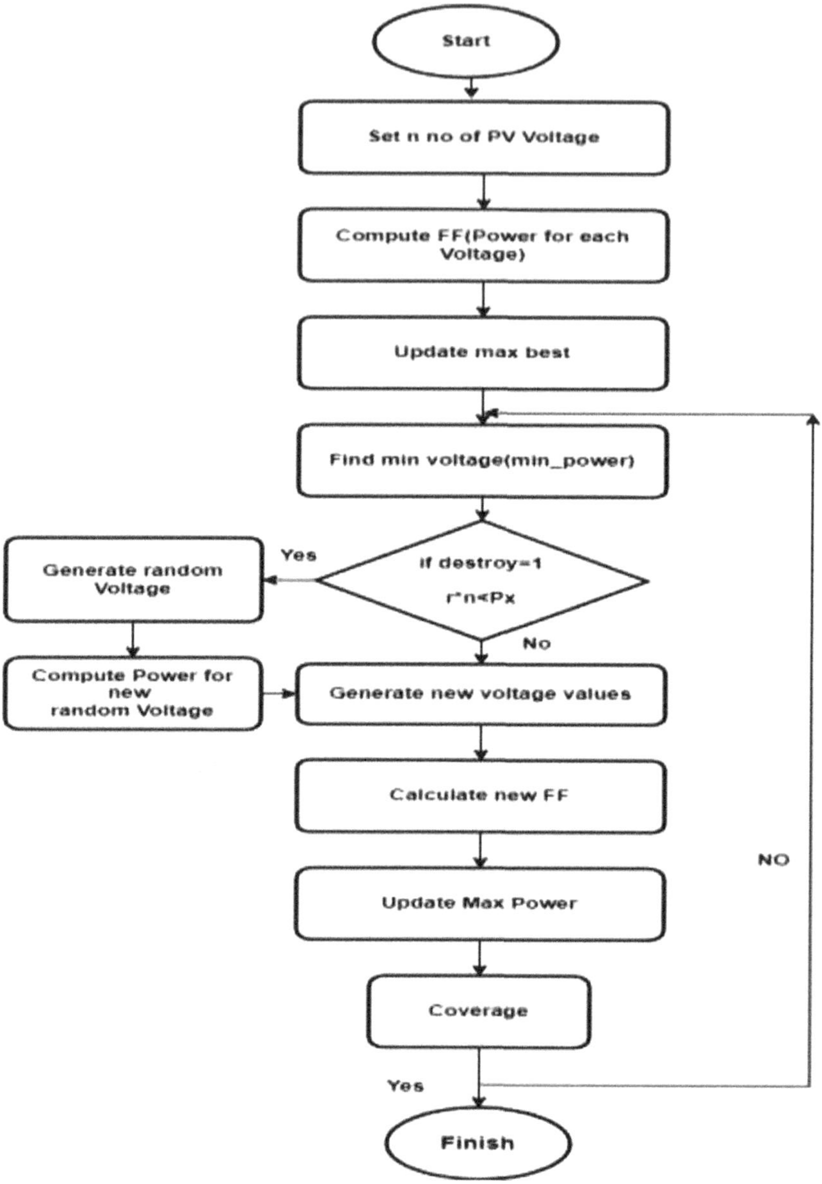

Figure 9.2 Flow chart of CSOA MPPT.

$$i = Vi + \Delta \oplus \text{levy}'\,(\lambda) \tag{9.2}$$

A flow chart of CSOA is shown in Figure 9.2, and Table 9.1 shows the comparison of CSOA MPPT approaches' span.

9.2.2 PSO particle swarm optimization (PSOA)

The fundamental tenet of PSOA is to utilize a population of particles, each of which represents a probable elucidation for the optimization issue. Every fleck's position is updated related to its current velocity, which is then restructured based on optimum situation of the other flecks in the population as well as the particle's own. A combination of the particle's current velocity, its ideal position, and the ideal positions of the other particles in the population are used to update each particle's velocity (Díaz et al. 2021; Pilakkat et al. 2020). Consider a collection of nannometer (Nm) particles (Pi)2<i<Nm, method is based on the following five steps (Hayder et al. 2020). Figure 9.3 displays the flow chart of PSOA, and Table 9.2 shows the comparative analysis of previous work.

Step a: Haphazard situation of every fleck's Pi by Equation (9.3)

$$Pi = \beta,\ 1 \leq i \leq Nm \tag{9.3}$$

where β is a haphazard number [Pinf ...Psup]

Step b: Every fleck discovers its neighborhood optimum place (Popti).

Step c: All flecks should track the global optimum place (PGopt).

Step d: Amendment of every fleck's place using Equations (9.4) and (9.5)
i i

$$\Delta Pi^{(n+1)} = \Omega \times \Delta Pi\ n + r1c1\ (Popti - P^n) + r2c2\ (PGopt - P^n) \tag{9.4}$$

$$Pi\ (n+1) = Pi\ n + \Delta Pi\ n+1 \tag{9.5}$$

where $P^{(n+1)}$ is the fresh fleck's place, P^n is the real place of fleck, $\Delta P^{(n+1)}$ is the perturb quantity to implement at the real place, $\Delta P\ i\ n$ is the perturbation in the prior iteration, Ω is the inertia mass, r1 and r2 are haphazard variable in [0,1], c1 is the cognitive constant, c2 is the public constant, PGopt is the global optimum place of the chief swarm flecks, and Piopt is the neighborhood optimum place of the ith-flecks.

Step e: Reiterate b, c, and d until every fleck's place congregates to the PGopt.

For the SPS, the Pi is the duty duration which should be adjusted to attain the optimum output from SPS. For SPS metric, Equation (9.6) should be

$$\Delta Pi^{(n+1)} = \Omega \times \Delta Pi\ n + \mu(Popti - Pin) + \pounds(PGopt - Pin) \tag{9.6}$$

Table 9.1 Comparison of CSOA MPPT approaches from literature survey

Authors	Sense Parameter	Hardware/Software	Convertor	Findings	Result
Ahmed and Salam (2013)	Power	Matlab/Simulink	Boost	Result compared by P&O method under PSC and shows zero oscillation around MPP under steadystate.	Tracking time (CS) 22 ms for STC and for P&O method 45ms.
Ahmed and Salam (2014)	Power	DSP-TMS320F28035 controller Matlab/Simulink	Buck-Boost	CSOA with GSS result compared with the PSO and CS method shows better tracking accuracy and less tracking time.	Tracking accuracy = 99.77% $\pm$ 0.28 Tracking time = 2.9.5 $\pm$ 0.44
Nugraha et al. (2019)	Voltage and current	Matlab/Simulink	Buck-Boost	Comparing the outcome to the P&O and PSO tactics.	Tracking duration 100 to 250 ms and energy hammering in stable condition is 0.000008%
Mohamed et al. (2019)	Voltage	Matlab/Simulink	Boost	Results are compared with Inc-cond and ANN method and shows better power at MPP under STC and varying weather conditions.	MPP power = 60.4728 W
Basha et al. (2020)	Voltage and current	Matlab/Simulink	SEPIC	The CSOA tracking technique uses the complete search space of IV plot to provide accurate global power point with rapid convergence.	MPP voltage 230 V for first irradiation pattern and for second pattern it is 250 V.

Table 9.1 (Cont.)

Authors	Sense Parameter	Hardware/Software	Convertor	Findings	Result
Farag et al. (2020)	Voltage and current	Matlab/Simulink	Boost	To locate a more efficient global maximum power point, CSOA outperforms PSO.	CS efficiency 99.31% Compared to 99.28% for PSO.
Yang and Deb (2017)	Voltage and current	Matlab/Simulink	Boost	The outcomes are contrasted with P&O tactic and variable P&O tactic and offer a smaller number of iteration and optimal maximum point by easy computational calculation.	Comparing the traditional (P&O) and variable-step P&O algorithm, tracking duration was reduced by 46.42% and 11.76%, respectively.
Rezk et al. (2017)	Voltage and current	Matlab/Simulink	Boost converter	Compare the result of PSO and CS method with conventional technique and shows that CS performs better than other techniques under PSC.	Tracking time 0.24 s Tracking efficiency 99.87–99.94%

where $\pounds + \mu + \Omega = 1$. PGopt is the duty period which related to the universal energy, while Popti considered the duty value of ith flecks, which related to the minor optimum energy produced during n repetition.

9.2.3 Artificial bee colony (ABCOA)

This tactic discusses how bees search for nectar together. There are three categories of bees: employed bees, observer bees, and explorer bees (Bilal 2013). Employed bees find food source and then inform about the location of food to nearby bees by doing waggle dance. A food source must be chosen

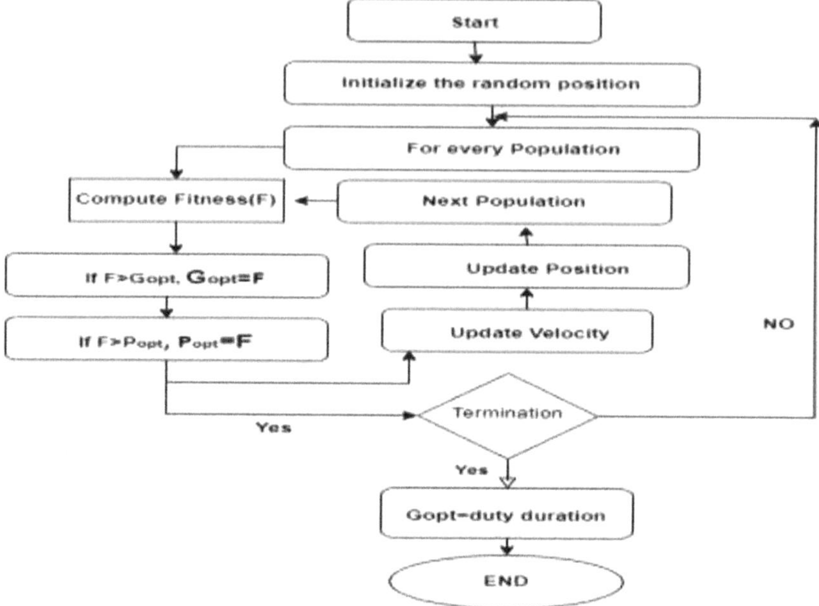

Figure 9.3 Flow chart of PSOA.

by the spectator bee based on the information it has been given (Fanani et al. 2020). The ABC algorithm combines the confined finding method (by hired bees) and the comprehensive finding method to maintain a balance between exploration and exploitation by onlooker and scouts (Karaboga 2010). This is then applied to the PV system's MPPT by adjusting the duty cycle. Figure 9.4 displays the flow chart of ABCOA, and Table 9.3 shows the comparative analysis of literature work.

9.3 CORRELATION TABLE BETWEEN CSOA, PSOA, AND ABCOA

A comparative analysis between all the three discussed algorithms based on the parameters is discussed in Table 9.4.

9.4 CONCLUSION

The three contemporary MPPT algorithms utilized in software and hardware platforms are briefly described in this review article. It deals with MPPT optimization methods that are mostly targeted in partial shading situation. In terms of SPS-MPPT tracking, each of the algorithms, CSOA,

Table 9.2 Comparison of PSOAMPPT approaches from previously works

MPPT Technique	Sense Parameter	Hardware/Software	Convertor	Findings	Result
Hayder et al. (2020)	Voltage and current	Matlab/Simuli nk	Boost	Result shows better performance as compared to ANN-PSO and P- PSO method under varying solar irradiance but temperature is constant.	Accuracy % in steady-state Ass = 99.9980
Ben et al. (2016)	Voltage and current	Matlab/Simuli nk	Buck	Result is compared with P&O and Fuzzy-TS method under different meteorological condition	ηMPPT-99.30 ISE-10-6
Anoop and Nandakumar (2018)	Current	Matlab/Simuli nk	Boost	PSO-OCC tactics is analyzed with classical PSOA.	The anticipated tactic can follow the global MPP precisely and rapidly
Kalaiarasi et al. (2018)	Voltage and current	Dspace 1104 controller Matlab/Simuli nk	Zsource Inverter	Result is compared with P&O method and shows less tracking time to MPP.	Settled time = 0.4 s
Li et al. (2018)	Voltage and current	Matlab/Simuli nk	Buck	Result is compared with firefly MPPT and PO-PSO method and shows better efficiency, less tracking time.	Pave - 134.17 W Efficiency - 99.82% Tracking time - 0.210 (s)
Merchaoui et al. (2018)	Voltage and current	Matlab/simuli nk	Boost	Nonlinear weight distribution with PSO result shows better tracking speed withcomparison to linear weight distribution under PSC.	Avg tracking speed < 0.75 s
Chang et al. (2018)	Voltage and current	PIC18F8720 microcontroller Matlab/Simuli nk	Boost	A weight value is modified as per the gradient and modify in power of P–V plot.	Maximum average power is 35.32, 37.28, 45.55, and 64.73 And tracking speed 0.55, 0.98, 1.12, and 0.67 s.
Díaz et al. (2021)	Voltage and current	Matlab/Simuli nk	Boost	Compared the simulation result of different PSO techniques with P&O method Shows better performance.	Steady-state average efficiency >99.75%

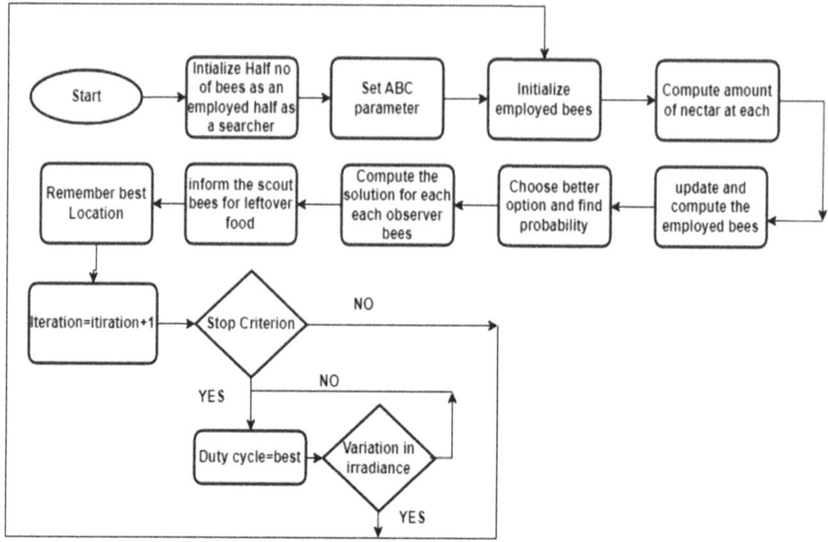

Figure 9.4 Flow chart of ABCOA.

PSOA, and ABCOA, has benefits and shortcomings of its own. Although the PSO method performs well globally, the CS approach is computationally economical and ideal for usage in embedded systems and real-time applications. The ABCOA is less prone to becoming stuck in local optima and is less sensitive to parameter selection. The MPPT tracking problem's specific requirements and the trade-offs between computing efficiency, parameter sensitivity, and optimization performance will determine which solution is best. The application, hardware accessibility, price, convergence speed, accuracy, and system dependability all affect the MPPT option that is selected. Given the significance of MPPT in partial shade conditions, it is clear that there is a large area of study to be done to identify an appropriate MPPT that can increase SPS output efficiency. This review is anticipated to be a very useful tool for all PV system researchers as well as for all sectors that excel at producing energy that is effective, clean, and sustainable for humankind.

CONFLICT OF INTEREST

The authors declare that they have no conflict of interest related to this research work. Additionally, none of the researchers had any connections that could be seen as impacting the research's neutrality or the paper's content on a personal, professional, or financial level.

Table 9.3 Comparison of ABC optimization-based MPPT approaches from previous works

Authors	Sense Parameter	Hardware/ Software	Convertor	Findings	Result
González-Castaño et al. (2021)	Sun insolation and ambient temperature Vref controlling parameter	PLECS RT Box I DSC (TI 28069M)	Boost	The output voltage of SPS under variable weather condition can be controlled by two sequential control loops; one is current (inner loop) controller and other voltage (outer loop) controller along with the ABCOA.	Relative error = 1.72; mean absolute error = 0.4; root mean squared error = 0.42 Mean power = 39.92w; Tracking factor = 98.92; Efficiency = 99.6%
Fanani et al. (2020)	Duty cycle Ppv	PSIM	Zeta	Human psychology optimization (HPO) procedure outcomes correlated with the ABCOA.	Tracking accuracy = 99.95% Average tracking time = 0.0727 s
Benyoucef et al. (2015)	Photon current duty cycle	Matlab/Simulink and Cadence/ PSpice	Boost	Comparison with PSO & PO MPPT	Accuracy = 98.76–99.22% Convergence cycle = 10.4–10.1
Nie et al. (2019)	Duty cycle	Matlab/Simulink	Buck-Boost	Performance of P&O, PSOA, and ABCOA is compared with modified ABCOA.	Modified ABCOA shows better result in terms of tracking efficiency.
Pilakkat et al. (2020)	Voltage, current, duty cycle	Matlab/Simulink	Boost	ABC-PO algorithm with grid-connected solar power system	Efficiency = 99.93; Tracking speed = 0.08s
Pilakkat and Kanthalakshmi (2019)	Voltage, current, duty cycle	Matlab/ Simulink	Boost	ABCPO algorithm for standalone system. Result is compared with ABC, PO, and INC method.	Efficiency >99.5%

Table 9.4 Comparative analysis based on parameters

Parameter	CSOA	PSOA	ABCOA
Inspiration	Procreation behavior of cuckoo	Bird group trying to reach and unknown position	For aging behavior of honeybee
Global optimum solution	Yes	Yes	No
Computational efficiency	Yes	Yes	More than CSOA and PSOA
Computational power need in RTOS	Less	More	Moderate
Parameter sensitivity	Yes	Yes	Moderate
Fitness function dependence	Yes	Yes	Yes
Local optimization stuck susceptibility	More	More	Less
Convergence speed	Moderate	Fast	Moderate
Gradient information need	No	Yes	Yes
Convergence dependency on initial condition	Yes	Yes	No (Sundareswaran et al. 2015)
Complexity	More	Simple	Simple
Control parameter	Less	More	Less

REFERENCES

Abo-Elyousr, F.K., Abdelshafy, A.M., Abdelaziz, A.Y. (2020). MPPT-based particle swarm and Cuckoo Search algorithms for PV systems. In Eltamaly, A., Abdelaziz, A. (eds.), Modern Maximum Power Point Tracking Techniques for Photovoltaic Energy Systems. Green Energy and Technology. Springer, Cham. https://doi.org/10.1007/978-3-030-05578-3_14

Ahmed, J., Salam, Z. (2013). A soft computing MPPT for PV system based on Cuckoo Search algorithm. In 4th International Conference on Power Engineering, Energy and Electrical Drives, pp. 558–562. Istanbul. http://doi.org/10.1109/PowerEng.2013.6635669

Ahmed, J., Salam, Z. (2014). A Maximum Power Point Tracking (MPPT) for PV system using Cuckoo Search with partial shading capability. Applied Energy, 119, 118–130. https://doi.org/10.1016/j.apenergy.2013.12.062

Anoop, K., Nandakumar, M. (2018). A novel maximum power point tracking method based on particle swarm optimization combined with one cycle control. In Proceedings of International Conference on Power, Instrumentation, Control and Computing (PICC), pp. 1–6. Thrissur, India.

Ben Belghith, O., Sbita, L., Bettaher, F. (2016). MPPT design using PSO technique for photovoltaic system control comparing to fuzzy logic and P&O controllers. Energy and Power Engineering, 8, 349–366.

Bilal, B. (2013). Implementation of artificial bee colony algorithm on maximum power point tracking for PV modules. In 2013 8th International Symposium on Advanced Topics in Electrical Engineering (ATEE), pp. 1–4. Bucharest, Romania. https://doi.org/10.1109/ATEE.2013.6563495

Central Electricity Authority (CEA). India. Annual Report 2020–21. https://cea.nic.in/wpcontent/uploads/annual_reports/2021/CEAAnnualReport_fnal.pdf (accessed August 15, 2022).

Chang, L.Y., Chung, Y.N., Chao, K.H., Kao, J.J. (2018). Smart global maximum power point tracking controller of photovoltaic module arrays. Energies, 11, no. 3, 567–582.

Díaz Martínez, D., Trujillo Codorniu, R., Giral, R., Vázquez Seisdedos, L. (2021). Evaluation of particle swarm optimization techniques applied to maximum power point tracking in photovoltaic systems. International Journal of Circuit Theory and Applications, 2021, pp. 1–19. https://doi.org/10.1002/cta.2978

Fanani, M.R., Sudiharto, I., Ferdiansyah, I. (2020). Implementation of maximum power point tracking on PV system using artificial bee colony algorithm. In 2020 3rd International Seminar on Research of Information Technology and Intelligent Systems (ISRITI), pp. 117–122. Yogyakarta, Indonesia. https://doi.org/10.1109/ISRITI51436.2020.9315527

Farhat, M., Barambones, O., Sbita, L. (2015). Efficiency optimization of a DSP-based standalone PV system using a stable single input fuzzy logic controller. Renewable and Sustainable Energy Reviews, 49, 907–920. https://doi.org/10.1016/j.rser.2015.04.123

González-Castaño, C., Restrepo, C., Kouro, S., Rodriguez, J. (2021). MPPT algorithm based on artificial bee colony for PV system. IEEE Access, 9, 43121–43133. https://doi.org/10.1109/ACCESS.2021.3066281

Government of India, Ministry of Power. Power Sector at a Glance All India. https://powermin.gov.in/en/content/power-sector-glance-all-india (accessed August 15, 2022).

Hassan, S.Z., Li, H., Kamal, T., Arifoğlu, U., Mumtaz, S., Khan, L. (2017). Neuro-Fuzzy wavelet based adaptive MPPT algorithm for photovoltaic systems. *Energies*, 10, no. 3, 394. https://doi.org/10.3390/en10030394

Hayder, W., Ogliari, E., Dolara, A., Abid, A., Ben Hamed, M., Sbita, L. (2020). Improved PSO: A comparative study in MPPT algorithm for PV system control under partial shading conditions. Energies, 13, 2035. https://doi.org/10.3390/en13082035

Hussaian Basha, C., Bansal, V., Rani, C., Brisilla, R.M., Odofin, S. (2020). Development of Cuckoo Search MPPT algorithm for partially shaded solar PV SEPIC converter. In Das, K., Bansal, J., Deep, K., Nagar, A., Pathipooranam, P., Naidu, R. (eds.), Soft Computing for Problem Solving. Advances in Intelligent Systems and Computing (vol. 1048). Springer, Singapore. https://link.springer.com/chapter/10.1007/978-981-15-0035-0_59.   https://doi.org/10.1007/978-981-15-0035-0_59

Kalaiarasi, N., Dash, S.S., Padmanaban, S., Paramasivam, S., Morati, P.K. (2018). Maximum power point tracking implementation by dspace controller integrated through Z-Source inverter using particle swarm optimization technique for photovoltaic applications. Applied Science, 8, no. 1, 145–162. https://doi.org/10.3390/app8010145

Karaboga, D. (2010). Artificial bee colony algorithm. Scholarpedia, 5, no. 3, 6915. https://doi.org/10.4249/scholarpedia.6915

Li, H., Yang, D., Su, W., Lü, J., Yu, X. (2019). An overall distribution particle swarm optimization MPPT algorithm for photovoltaic system under partial shading. IEEE Transactions on Industrial Electronics, 66, no. 1, 265–275. https://doi.org/10.1109/TIE.2018.2829668

Merchaoui, M., Saklyl, A., Mimouni, M.F. (2018). Improved fast particle swarm optimization based PV MPPT. In Proceedings of the 9th International Renewable Energy Congress (IREC 2018), pp. 1–6. Hammamet, Tunisia.

Mosaad, M.I., Osama abed el-Raouf, M., Al-Ahmar, M.A., Banakher, F.A. (2019). Maximum Power Point Tracking of PV system Based Cuckoo Search Algorithm; review and comparison. Energy Procedia, 162, 117–126. http://doi.org/10.1016/j.egypro.2019.04.013

Nie, L., Mao, M., Wan, Y., Cui, L., Zhou, L., Zhang, Q. (2019). Maximum power point tracking control based on modified abc algorithm for shaded PV system. In 2019 AEIT International Conference of Electrical and Electronic Technologies for Automotive (Aeit Automotive). Turin, Italy.

Nugraha, D.A., Lian, K.L., Suwarno. (2019, Summer). A novel MPPT method based on Cuckoo Search algorithm and golden section search algorithm for partially shaded PV system. Canadian Journal of Electrical and Computer Engineering, 42, no. 3, 173–182. https://doi.org/10.1109/CJECE.2019.2914723

Piegari, L., Rizzo, R., Spina, I., Tricoli, P. (2015). Optimized adaptive perturb and observe maximum power point tracking control for photovoltaic generation. *Energies*, 8, no.5, 3418–3436. https://doi.org/10.3390/en8053418

Pilakkat, D., Kanthalakshmi, S. (2019). An improved P & O algorithm integrated with artificial bee colony for photovoltaic systems under partial shading conditions. Solar Energy, 178, 37–47. https://doi.org/10.1016/j.solener.2018.12.008

Pilakkat, D., Kanthalakshmi, S. (2020). Single phase PV system operating under partially shaded conditions with ABC-PO as MPPT algorithm for grid connected applications. Energy Reports, 6, 1910–1921. http://doi.org/10.1016/j.egyr.2020.07.019

Pilakkat, D., Kanthalakshmi, S., Navaneethan, S. (2020). A comprehensive review of swarm optimization algorithms for MPPT controller of PV systems under partially shaded conditions. Electronics, 24, 3–14. https://doi.org/10.7251/ELS2024003P

Rezk, H., Fathy, A., Abdelaziz, A.Y. (2017). A comparison of different global MPPT techniques based on meta-heuristic algorithms for photovoltaic system subjected to partial shading conditions. Renewable and Sustainable Energy Reviews, 74, 377–386. https://doi.org/10.1016/j.rser.2017.02.051

Sahin, Z., Ismaila, K.G., Yilbas, B.S., Al-Sharafi, A. (2020). A review on the performance of photovoltaic/thermoelectric hybrid generators. International Journal of Energy Research, 44, no. 5, 3365–3394. https://doi.org/10.1002/er.5139

Sheik Mohammed, S., Devaraj, D., Imthias Ahamed, T.P. (2016). A novel hybrid Maximum Power Point Tracking Technique using Perturb & Observe algorithm and Learning Automata for solar PV system. Energy, 112, 1096–1106. https://doi.org/10.1016/j.energy.2016.07.024

Soufyane Benyoucef, A., Chouder, A., Kara, K., Silvestre, S., Sahed, O.A. (2015). Artificial bee colony based algorithm for maximum power point tracking (MPPT) for PV systems operating under partial shaded conditions. Applied Soft Computing, 32, 38–48. http://doi.org/10.1016/j.asoc.2015.03.047

Sundareswaran, K., Sankar, P., Nayak, P.S.R., Simon, S.P., S. Palani, S. (2015). Enhanced energy output from a PV system under partial shaded conditions through artificial bee colony. IEEE Transactions on Sustainable Energy, 6, no. 1, 198–209. https://doi.org/10.1109/TSTE.2014.2363521

Yang, X.S., Deb, S. (2009). Cuckoo Search via Levy flights. In 2009 World Congress on Nature & Biologically Inspired Computing (NaBIC), pp. 210–214. Coimbatore.

Yang, X.S., Deb, S. (2017). Cuckoo search: State-of-the-art and opportunities. In 2017 IEEE 4th International Conference on Soft Computing & Machine Intelligence, pp. 55–59. Port Louis.

Chapter 10

Fuzzy logic and bio-inspired Ant Colony Algorithm-based technique to find relative desirability in IoT-based healthcare system

Firos A

10.1 INTRODUCTION

In the context of Internet of Things (IoT)-based healthcare systems, a novel approach to assessing the relative desirability of health-related data emerges through the integration of Fuzzy Logic and a Bio-Inspired Ant Colony Algorithm. This hybrid technique leverages the adaptability of Fuzzy Logic to handle uncertainty and imprecision in healthcare data while drawing inspiration from the collective intelligence of ant colonies for optimization. Fuzzy Logic enables a nuanced understanding of diverse health parameters, considering linguistic variables and membership functions. Concurrently, the Ant Colony Algorithm introduces a bio-inspired optimization mechanism, mimicking the collaborative decision-making process observed in ant colonies. This integrated technique enhances the efficiency of evaluating and prioritizing healthcare data within the IoT framework. It empowers healthcare systems to make informed decisions based on a comprehensive analysis of patient information, treatment outcomes, and other critical factors. By combining the strengths of Fuzzy Logic and bio-inspired algorithms, this approach contributes to creating more adaptive, intelligent, and effective IoT-based healthcare systems.

Preference-Leveled Evaluation Functions (PLEF) method can be used in fuzzy Artificial Neural Network (ANNs) systems to evaluate the relative desirability. The goal of this technique is to create a system that can accurately identify abnormality of cells through cells' dataset to alert the patients accordingly. The technique involves Defining Factors and Preference Levels. This approach can help us to make informed decisions when selecting desired instruction and further courses of action based on multiple factors that might have varying degrees of importance. It will identify the factors that are important for evaluating the smart system dataset. These could include factors like rhythm annotations of types (AFIB – atrial fibrillation), (AFL – atrial flutter), (J – AV junctional rhythm), and (N – used to indicate all other rhythms), and more. For each factor, define preference levels that capture the degree of desirability. These levels could be "Highly Desirable,"

DOI: 10.1201/9781003530077-10

"Desirable," "Neutral," "Less Desirable," and "Undesirable." We create membership functions for each factor and each preference level. These functions define the fuzzy boundaries for each preference level. Then we implement a Fuzzy Inference Engine that uses the fuzzy rules to compute the fuzzy output membership degrees for each preference level. A defuzzification process then will convert the aggregated fuzzy preference levels back into crisp values for each preference level using the defuzzification method.

10.1.1 Generative models for IoT-based healthcare systems using Artificial Neural Network

Generative models, particularly those based on ANNs, play a crucial role in shaping the future of IoT-based healthcare systems. These models offer innovative solutions for generating synthetic healthcare data, which can be invaluable for training and validating machine learning (ML) algorithms. In the context of healthcare IoT, where data privacy and security are paramount, generative models can create realistic yet anonymized datasets that facilitate robust algorithm development without compromising patient confidentiality. ANNs, with their ability to learn complex patterns and representations, excel in capturing the diverse and dynamic nature of healthcare information. By training generative models on real-world healthcare data, they can produce synthetic datasets that closely mimic the statistical characteristics of the original data. This synthetic data can be used to enhance the performance of predictive models, anomaly detection systems, and other artificial intelligence (AI) applications within IoT-based healthcare, contributing to improved patient care, diagnosis, and overall system efficiency while adhering to privacy regulations.

Generative models provide innovative solutions for generating synthetic healthcare data, offering a versatile and privacy-preserving approach in the realm of healthcare analytics. These models, including but not limited to Generative Adversarial Networks (GANs) and Variational Autoencoders (VAEs), are adept at capturing the complex patterns and distributions present in real-world healthcare datasets. The generation of synthetic data is particularly valuable in situations where access to large, diverse, and privacy-sensitive datasets is limited. By training generative models on existing healthcare data, these models can produce synthetic counterparts that closely resemble the statistical properties of the original information.

The synthetic data generated by these models maintain the essential features and relationships present in real healthcare data, ensuring that downstream analyses and ML algorithms trained on such synthetic datasets remain robust and effective. Importantly, this synthetic data creation process mitigates privacy concerns by removing personally identifiable information while retaining the underlying patterns crucial for model training. Consequently, generative models pave the way for advancements in healthcare analytics, allowing researchers and data scientists to develop and refine models without compromising individual patient privacy or data security.

Generative models excel in creating realistic yet anonymized datasets, offering a breakthrough solution for robust algorithm development in healthcare without compromising patient confidentiality. The sensitive nature of health data poses challenges in sharing and utilizing large datasets for ML purposes. Generative models, such as GANs and VAEs, address this issue by producing synthetic data that mirrors the statistical characteristics of real patient information.

By training on authentic healthcare datasets, generative models learn to capture the intricate patterns and relationships within the data. The synthetic datasets these models generate maintain these crucial features while eliminating personally identifiable information. This process ensures the privacy and confidentiality of individual patient records. Researchers and data scientists can then utilize these anonymized datasets to develop, test, and refine algorithms for tasks like disease prediction, treatment planning, and healthcare analytics. The ability to create realistic yet privacy-preserving synthetic data enhances the efficiency of algorithm development, fostering advancements in healthcare technology while adhering to stringent privacy regulations and ethical considerations

10.1.2 Deep Neural Network for giving suggestions

Deep Neural Networks (DNNs) are highly effective in providing intelligent and personalized suggestions across various domains, leveraging their ability to extract complex patterns and representations from large datasets. In the context of recommendation systems, DNNs have shown remarkable performance in understanding user preferences and delivering tailored suggestions. These networks are particularly powerful when dealing with intricate relationships and non-linear patterns within data.

For example, in an e-commerce setting, a DNN can analyze a user's historical purchase behavior, browsing patterns, and preferences to offer personalized product recommendations. Similarly, in content platforms, DNNs can analyze user interactions, content preferences, and engagement history to suggest articles, videos, or music that align with individual tastes.

The architecture of a DNN typically involves multiple layers of interconnected nodes, allowing the model to learn hierarchical representations of data. Through training on large datasets, DNNs can capture intricate features and relationships, making them well-suited for understanding complex user behaviors and providing accurate suggestions.

The application of DNNs for suggestion systems enhances user experience, engagement, and satisfaction by offering recommendations that align with individual preferences, ultimately contributing to the success of various online platforms and services.

In the realm of IoT-based healthcare systems, leveraging DNNs to provide personalized suggestions holds significant potential to enhance patient care and overall system efficiency [1]. DNNs can be employed to analyze a multitude of health-related data generated by IoT devices, including patient vitals, medical history, and lifestyle information. The network's ability to capture complex patterns and relationships within this diverse dataset allows it to offer tailored suggestions for various aspects of healthcare.

For instance, a DNN can analyze a patient's historical health data, medication adherence, and lifestyle choices to provide personalized suggestions for diet plans, exercise routines, or medication schedules [2]. The model can also consider real-time data from wearable devices to offer timely advice or alerts regarding potential health issues.

The architecture of a DNN enables it to learn and adapt to individual patient profiles, making the suggestions more accurate and relevant over time. This application of DNNs in IoT-based healthcare systems contributes to proactive and personalized healthcare interventions, promoting patient engagement and overall well-being. It also assists healthcare professionals in making data-driven decisions for more effective treatment strategies.

10.1.3 Fuzzy system AI for IoT-based healthcare system

Fuzzy Logic-based AI systems find practical applications in healthcare systems, providing a flexible and nuanced approach to handling uncertain and imprecise medical data. In the context of healthcare, where patient conditions and diagnoses often involve varying degrees of uncertainty, a fuzzy system can be beneficial for decision-making and diagnostics [3]. Here are a few key areas where fuzzy systems AI can be applied in healthcare:

Diagnostic Decision Support: Fuzzy systems can assist in medical diagnostics by considering ambiguous symptoms and uncertain test results. The model can integrate diverse sources of information, such as patient history, lab results, and imaging data, and provide a more comprehensive evaluation of potential conditions.

Treatment Planning: Fuzzy Logic can be used to develop AI systems that assist in creating personalized treatment plans. The model can consider factors like patient preferences, comorbidities, and potential side effects of treatments, providing a more tailored and patient-centric approach.

Health Monitoring: Fuzzy systems are suitable for handling sensor data from wearable devices and IoT devices in healthcare. These systems can assess fluctuations in vital signs and other health parameters, providing a more adaptable and context-aware health monitoring solution.

Telehealth Triage: In telehealth applications, Fuzzy Logic can be applied to assess the severity of symptoms reported by patients. The system can provide triage recommendations based on the perceived urgency, allowing healthcare professionals to prioritize cases effectively.

Medical Imaging: Fuzzy systems can enhance image analysis in medical imaging applications. For example, in radiology, Fuzzy Logic can help interpret ambiguous features in images, aiding radiologists in making more informed decisions.

The advantage of Fuzzy Logic lies in its ability to handle uncertainty, imprecision, and incomplete information, making it a valuable tool in healthcare systems where such uncertainties are inherent. Applying fuzzy systems AI in healthcare contributes to more adaptive, patient-specific, and context-aware solutions.

In an IoT-based healthcare system, the integration of a Fuzzy Logic-based AI system can bring several advantages in handling the complexity and uncertainty of health-related data generated by various IoT devices. Here are some applications of Fuzzy Logic in an IoT-based healthcare system:

Health Monitoring and Sensor Fusion: Fuzzy systems can process and fuse data from diverse IoT sensors, such as wearable devices and remote monitoring equipment. This enables a more comprehensive evaluation of a patient's health status by considering multiple, potentially conflicting, sources of information.

Patient Risk Assessment: Fuzzy Logic can be employed to assess the risk factors associated with a patient's health condition. By considering fuzzy rules that capture the uncertainty in various health parameters, the system can provide risk scores and alerts for potential health issues.

Adaptive Treatment Plans: Fuzzy systems can contribute to the development of adaptive treatment plans by considering the dynamic nature of health data from IoT devices. The AI system can adjust treatment recommendations based on real-time changes in a patient's health status, fostering personalized and responsive healthcare.

IoT Device Anomaly Detection: Fuzzy Logic can aid in the detection of anomalies in data collected from IoT devices. By establishing fuzzy rules that define normal behavior, the system can identify deviations from the expected patterns, helping in the early detection of irregularities or potential health crises.

Context-Aware Health Recommendations: Fuzzy systems can provide context-aware health recommendations by considering not only the raw sensor data but also the contextual factors affecting a patient's health. This can lead to more relevant and timely suggestions for lifestyle changes, medication adjustments, or follow-up appointments.

The incorporation of Fuzzy Logic in an IoT-based healthcare system enhances its ability to handle uncertainty, imprecision, and variability in the data, ultimately contributing to more accurate diagnostics, personalized treatment plans, and improved patient outcomes.

10.1.4 Ant Colony algorithm-based technique to find relative desirability in IoT-based healthcare system

Utilizing the Ant Colony Algorithm in an IoT-based healthcare system offers a bio-inspired technique to assess the relative desirability of various healthcare parameters. The Ant Colony Algorithm draws inspiration from the foraging behavior of ants, where individual ants deposit pheromones to communicate and collectively find the shortest path to a food source. In the context of healthcare IoT, this algorithm can be applied to optimize decision-making processes.

Here's how the Ant Colony Algorithm can be employed in an IoT-based healthcare system:

Parameter Optimization: The algorithm can optimize the weighting of different healthcare parameters based on their importance, mimicking the pheromone deposition process. For example, it may adjust the significance of patient vital signs, medication adherence, and lifestyle factors.

Dynamic Resource Allocation: In scenarios where resources are limited, such as hospital beds or medical equipment, the Ant Colony Algorithm can assist in dynamically allocating resources based on real-time demand, ensuring efficient utilization and responsiveness to changing healthcare needs.

Pathfinding for Treatment Plans: Similar to how ants find the most efficient path to a food source, the Ant Colony Algorithm can assist in determining the optimal treatment plan for patients, considering their unique health profiles, historical data, and responses to previous interventions.

Adaptive Care Plans: The algorithm can contribute to the development of adaptive care plans by adjusting the desirability of different interventions based on the evolving health conditions of patients, as observed through IoT-generated data.

IoT Device Coordination: In a network of IoT devices, such as wearable sensors and medical monitoring equipment, the Ant Colony Algorithm can help coordinate the flow of information. It may optimize the transmission of critical health data while minimizing congestion or delays.

By applying the Ant Colony Algorithm in an IoT-based healthcare system, the model can adapt and optimize its decision-making processes, ensuring a more efficient and responsive healthcare ecosystem. This bio-inspired approach aligns with the self-organizing principles observed in ant colonies and can contribute to the improvement of patient outcomes and resource utilization in healthcare settings.

10.2 THE BACKGROUND

10.2.1 IoT-based healthcare system

A smart healthcare system leverages advanced technologies and innovative solutions to enhance the efficiency, accessibility, and quality of healthcare services. Here are key components and features that characterize a smart healthcare system. They include IoT Integration, Health Information Exchange (HIE), Telehealth and Telemedicine, AI and ML, Blockchain for Security, Smart Devices and Remote Monitoring, Mobile Health (mHealth) Apps, Predictive Analytics, Robotic Assistance and Cybersecurity Measures.

The integration of IoT devices allows for real-time monitoring of patient health. Wearable devices, smart sensors, and medical IoT devices provide continuous data, enabling remote patient monitoring, preventive care, and early detection of health issues [4]. Seamless exchange of patient information among healthcare providers, hospitals, and other stakeholders ensures coordinated and comprehensive care. Interoperability standards and electronic health records (EHRs) contribute to the efficient sharing of medical data [5]. Smart healthcare systems embrace telehealth and telemedicine solutions, enabling virtual consultations, remote diagnosis, and digital communication between healthcare providers and patients. This approach enhances accessibility to medical expertise and reduces the need for physical visits. AI and ML algorithms analyze vast amounts of healthcare data to assist in diagnostics, treatment planning, and predictive analytics [6]. These technologies contribute to personalized medicine, improving accuracy and efficiency in healthcare decision-making.

Implementing blockchain technology enhances the security and integrity of healthcare data [7]. It ensures secure and transparent management of patient records, protects against unauthorized access, and facilitates secure sharing of sensitive health information. Connected medical devices, including smart wearables and home monitoring equipment, enable individuals to actively participate in their health management. These devices provide continuous data streams that can be analyzed to monitor chronic conditions and track overall wellness. Mobile applications play a vital role in empowering individuals to manage their health. Furthermore, mHealth apps provide functionalities such as medication reminders, symptom tracking, and access to health information, promoting patient engagement and adherence to treatment plans [8]. By utilizing historical data and real-time information, predictive analytics models can forecast disease trends, patient outcomes, and resource requirements. This aids healthcare providers in proactive decision-making and resource allocation.

Smart healthcare systems may incorporate robotic technologies for tasks such as surgery, rehabilitation, and patient assistance [9]. Robots equipped with AI capabilities contribute to precision and efficiency in medical procedures [10]. Given the sensitive nature of healthcare data, robust cybersecurity measures are crucial. Smart healthcare systems implement advanced security protocols to safeguard patient information, prevent data breaches, and ensure compliance with privacy regulations.

Smart healthcare system integrates cutting-edge technologies to optimize patient care, streamline workflows, and improve overall healthcare delivery. By combining IoT, AI, telehealth, and other innovations, smart healthcare systems aim to provide more accessible, personalized, and efficient healthcare services [11].

An IoT-based healthcare system represents a transformative approach to healthcare delivery, leveraging interconnected devices and technologies to enhance patient care, streamline processes, and improve overall healthcare outcomes. IoT-enabled wearable devices, such as fitness trackers, smartwatches, and health monitoring sensors, continuously collect and transmit real-time health data. These devices monitor vital signs, physical activity, sleep patterns, and other relevant metrics. IoT facilitates remote patient monitoring, allowing healthcare providers to track patients' health outside traditional clinical settings. This is particularly beneficial for managing chronic conditions, post-surgery recovery, and elderly care [12]. Monitoring devices can transmit data to healthcare professionals for timely intervention. IoT connects medical devices and equipment within healthcare facilities. This includes smart infusion pumps, connected imaging devices, and IoT-enabled diagnostic equipment. The seamless flow of data between devices improves the accuracy and efficiency of diagnostics and treatment.

Integration of IoT with health information systems and EHRs ensures a comprehensive and centralized repository of patient data. This interoperability enhances care coordination, reduces redundancies, and supports data-driven decision-making. IoT enables telehealth services, allowing patients to engage in virtual consultations with healthcare professionals. Connected cameras, microphones, and telemedicine platforms facilitate remote healthcare delivery, expanding access to medical expertise [13]. The vast amount of data generated by IoT devices is analyzed using analytics and ML algorithms. Predictive insights help in identifying trends, predicting disease outbreaks, and personalizing treatment plans based on individual health profiles.

IoT enhances the efficiency of hospital operations through smart infrastructure. This includes IoT-based asset tracking, inventory management, and environmental monitoring systems. Smart lighting, temperature control, and energy management contribute to a more sustainable and patient-friendly environment. IoT technologies assist in medication adherence and management. Smart pill dispensers, medication tracking devices, and connected prescription systems help patients adhere to prescribed medication schedules, thereby reducing medication errors [14]. Wearable devices and IoT-enabled alert systems provide immediate notifications in case of emergencies. Elderly individuals or patients with chronic conditions can have access to emergency response services through connected devices. Given the sensitive nature of healthcare data, robust security measures are implemented in IoT-based healthcare systems. Encryption, authentication, and secure data transfer protocols are essential to safeguard patient information and ensure compliance with privacy regulations. Implementing an IoT-based healthcare system offers the potential to transform healthcare delivery by fostering proactive, personalized, and data-driven approaches to patient care. However, it is crucial to address challenges related to data security, interoperability, and regulatory compliance in the deployment of IoT technologies in healthcare.

10.2.2 Advantages of Fuzzy ANN for IoT-based healthcare system

A Fuzzy Artificial Neural Network (Fuzzy ANN) brings several advantages to the healthcare system, combining the strengths of Fuzzy Logic and ANNs. Here are some key advantages:

Handling Uncertainty and Ambiguity: Fuzzy Logic is adept at handling uncertainty and ambiguity in healthcare data. By integrating Fuzzy Logic into the neural network, Fuzzy ANN can effectively manage imprecise and incomplete information, which is common in medical diagnoses and decision-making.

Linguistic Representation: Fuzzy Logic allows for linguistic representation of variables, which aligns well with the way medical experts express and interpret information. This linguistic approach enhances the interpretability of the model, making it more accessible for healthcare professionals.

Patient-Specific Modeling: Fuzzy ANN can be tailored to individual patient characteristics. This personalized modeling is valuable in healthcare, where patients exhibit diverse responses to treatments, and factors affecting health outcomes can be highly specific to each individual.

Adaptability and Learning from Experience: Artificial neural networks inherently possess the ability to learn from data. When combined with Fuzzy Logic, Fuzzy ANN can adapt and evolve based on new information and experiences, allowing it to continuously improve its performance in healthcare applications.

Integration of Qualitative and Quantitative Data: Healthcare data often includes both qualitative and quantitative information. Fuzzy ANN can seamlessly integrate both types of data, providing a comprehensive understanding of patient conditions and enabling more holistic decision-making.

Decision Support in Complex Systems: In healthcare systems with intricate and complex relationships among variables, Fuzzy ANN can offer valuable decision support. This is particularly beneficial in scenarios where multiple factors contribute to a medical outcome, and the relationships are not strictly linear.

Reducing Overfitting: Fuzzy Logic introduces a degree of flexibility into the model, which can help mitigate overfitting issues commonly observed in traditional neural networks. This flexibility allows the model to generalize well to new, unseen data.

Enhancing Interpretability: The fuzzy rule-based approach in Fuzzy ANN enhances the interpretability of the model. This is crucial in healthcare settings where clear explanations for decisions are necessary to gain trust and acceptance from healthcare professionals.

Applications in Diagnostics and Risk Assessment: Fuzzy ANN can be applied to diagnostics and risk assessment tasks in healthcare. It can provide more nuanced predictions by considering fuzzy rules, resulting in improved accuracy and reliability in identifying diseases or assessing patient risk.

Patient-Centric Approach: Fuzzy ANN's ability to account for individual patient characteristics, uncertainties, and linguistic input aligns with a patient-centric approach to healthcare. This makes it a valuable tool for creating more personalized and patient-friendly healthcare solutions.

The advantages of Fuzzy ANN in healthcare lie in its ability to handle uncertainty, interpretability, patient-specific modeling, and its potential to enhance decision-making in complex healthcare systems.

Integrating Fuzzy ANN into an IoT-based healthcare system offers several advantages, combining the strengths of Fuzzy Logic and neural networks to address the complexities of healthcare data generated by IoT devices. Here are some key advantages:

Handling Uncertainty in IoT Data: IoT devices in healthcare generate diverse and sometimes uncertain data. Fuzzy Logic within the ANN allows the system to handle uncertainties and imprecise information effectively, improving the robustness of the model in the face of varied and unpredictable data.

Adaptability to Varied Sensor Data: IoT devices produce a wide range of sensor data, and the nature of this data can vary. Fuzzy ANN can adapt to different types of sensor inputs, thereby allowing for flexibility and adaptability in processing diverse healthcare information.

Linguistic Representation of Health Parameters: Fuzzy Logic enables the use of linguistic variables, making it easier to represent and interpret health parameters. This linguistic representation is particularly useful in healthcare, where medical experts often express conditions and observations using qualitative terms.

Personalized Patient Modeling: Fuzzy ANN can be tailored to individual patient profiles by considering their unique health characteristics and responses to IoT-generated data. This personalized modeling contributes to more accurate and patient-centric healthcare interventions.

Effective Decision-Support System: In IoT-based healthcare systems, where data from multiple sensors influence decision-making, Fuzzy ANN provides an effective decision-support system. It considers the complex relationships between different variables, enhancing the system's ability to make informed decisions.

Integration of Qualitative and Quantitative Data: Fuzzy ANN seamlessly integrates both qualitative and quantitative data from IoT devices.

This integration ensures a more comprehensive analysis of patient health, accounting for both numerical measurements and subjective observations.

Robustness to Noisy IoT Data: IoT data can be prone to noise and inconsistencies. Fuzzy ANN's ability to deal with imprecision makes it more robust in the presence of noisy data, improving the reliability of healthcare predictions and recommendations.

Dynamic Adaptation to Changing Health Conditions: Fuzzy ANN's learning capabilities enable dynamic adaptation to changing health conditions. As IoT devices continuously provide real-time data, the model can adapt to new information, ensuring that healthcare interventions remain relevant and up to date.

Interpretability of Model Outputs: Fuzzy Logic enhances the interpretability of the model outputs, making it easier for healthcare professionals to understand the reasoning behind recommendations or predictions. This interpretability is crucial for gaining trust in AI-driven healthcare systems.

Enhanced Diagnostic Accuracy: Fuzzy ANN's nuanced approach to data analysis, considering fuzzy rules and linguistic variables, can contribute to improved diagnostic accuracy in IoT-based healthcare. It considers the subtleties and complexities of healthcare conditions for more precise predictions.

Incorporating Fuzzy ANN into an IoT-based healthcare system enhances its ability to handle uncertainties, personalize patient models, and provide effective decision support, ultimately contributing to more reliable and patient-centric healthcare solutions.

10.2.3 Artificial Neural Networks for automation

ANNs play a significant role in automation by mimicking the structure and function of the human brain, enabling machines to learn and make decisions independently [15]. Some of the several ways in which ANNs contribute to automation are Pattern Recognition, Machine Vision, Predictive Maintenance, Process Optimization, Autonomous Vehicles, Robotics and Automation, Natural Language Processing (NLP), Anomaly Detection, Financial Automation and Supply Chain Optimization.

ANNs excel at recognizing complex patterns and relationships within data. In automation, this capability is leveraged for tasks such as image recognition, speech recognition, and signal processing [16]. ANNs can be trained to identify patterns and make decisions based on input data. ANNs are commonly used in machine vision systems for automating visual inspection processes. They can analyze images or videos to detect defects, classify objects, or recognize specific features, enhancing the efficiency and accuracy of quality control in manufacturing. ANNs can analyze historical data from sensors and equipment to predict when machinery is likely to fail. This predictive maintenance approach helps automate maintenance schedules, reducing downtime, and extending the lifespan of equipment.

ANNs can optimize complex industrial processes by learning from historical data and making real-time adjustments. This is particularly valuable in sectors like manufacturing, where ANNs can enhance efficiency, reduce energy consumption, and improve overall process performance. In the field of transportation, ANNs are crucial for enabling autonomous vehicles. Neural networks process data from sensors, cameras, and other sources to make decisions such as steering, braking, and acceleration, allowing vehicles to navigate without human intervention. ANNs empower robots to perform tasks with a higher level of autonomy. Whether in manufacturing, logistics, or other industries, ANNs enable robots to adapt to changing environments, recognize objects, and execute tasks more efficiently. ANNs are employed in NLP applications to automate the understanding and generation of human language. This is utilized in chatbots, virtual assistants, and other systems where machines need to interpret and respond to human language.

ANNs can be trained to recognize patterns of normal behavior within a system. When deviations from the norm occur, ANNs can flag anomalies, contributing to automated monitoring and security systems. In the financial sector, ANNs are used for tasks such as fraud detection, credit scoring, and algorithmic trading [17]. These applications automate decision-making processes based on historical data and real-time market conditions. ANNs contribute to automating supply chain management by predicting demand, optimizing inventory levels, and enhancing logistics planning. This leads to more efficient and responsive supply chain operations. ANNs are a versatile tool in automation, enabling machines to learn, adapt, and make intelligent decisions across various domains. As technology continues to advance, the role of ANNs in automation is likely to expand, bringing increased efficiency and autonomy to a wide range of industries.

10.2.4 Fuzzy Logic and bio-inspired Ant Colony Algorithm-based technique to find relative desirability

The combination of Fuzzy Logic and a bio-inspired Ant Colony Algorithm forms a powerful technique for finding relative desirability in complex systems. This hybrid approach leverages the strengths of both Fuzzy Logic, which excels in handling uncertainty and imprecision, and the Ant Colony Algorithm, inspired by the foraging behavior of ants.

10.2.4.1 Fuzzy Logic in desirability assessment

Fuzzy Logic provides a framework for representing and processing uncertain or vague information. In the context of relative desirability, fuzzy sets and rules can be employed to define linguistic variables that capture the subjective nature of desirability. Membership functions define the degrees to which an element belongs to a fuzzy set, allowing for the expression of preferences and uncertainties in a more human-like manner. Fuzzy inference systems use rules to map input variables to fuzzy output sets, offering a systematic way to assess the desirability of various options.

10.2.4.2 Bio-inspired Ant Colony Algorithm

The Ant Colony Algorithm draws inspiration from the foraging behavior of ants, where the collective decision-making of individual ants leads to the discovery of optimal paths to food sources. In the context of desirability assessment, artificial ants can represent decision agents exploring the solution space. Their pheromone trails symbolize the attractiveness or desirability of different options. The algorithm involves the iterative updating of pheromone levels based on the desirability of chosen paths, allowing the system to converge toward optimal or highly desirable solutions over time.

10.2.4.3 Integration of Fuzzy Logic and Ant Colony Algorithm

Fuzzy Logic provides a linguistic framework for expressing the desirability of options, allowing for the incorporation of human-like reasoning and preferences. The Ant Colony Algorithm, on the other hand, introduces a bio-inspired approach to decision-making, thus enabling the system to dynamically adapt and explore the solution space. Fuzzy rules can guide the behavior of artificial ants, influencing their decision-making based on the fuzzy assessment of desirability. The pheromone trail updates can, in turn, be influenced by the fuzzy outputs, creating a symbiotic relationship.

This hybrid technique can be applied to various domains, such as optimization problems, resource allocation, and decision-support systems where the desirability of multiple options needs to be determined. It is particularly useful in scenarios where preferences are subjective, and the decision space is complex and dynamic. The Fuzzy Logic and Ant Colony Algorithm combination provides a robust and adaptable approach, capable of handling uncertainties, adapting to changing conditions, and converging towards optimal solutions in a human-interpretable manner. In essence, the fusion of Fuzzy Logic and the Ant Colony Algorithm offers a bio-inspired, fuzzy-guided method for finding relative desirability, contributing to more informed decision-making in complex and dynamic systems.

10.2.5 Preference-Leveled Evaluation Functions method for recommendation

The PLEF method is an approach used in recommendation systems to provide personalized suggestions based on user preferences [18]. This method employs evaluation functions that take into account various levels of preference, allowing for a nuanced and customized recommendation experience. Here is an overview of how the PLEF method works:

1. **User Preference Levels:** PLEF considers different levels of user preferences to capture the nuances of individual tastes. Users often have varying degrees of preference for different items, and PLEF takes this into account by categorizing preferences into multiple levels.

2. **Evaluation Functions:** PLEF utilizes evaluation functions that assess the desirability of items at each preference level. These functions consider various factors such as user history, explicit ratings, implicit feedback, or contextual information. Each evaluation function is designed to capture specific aspects of user preferences, allowing for a comprehensive analysis of items that align with different preference levels.

3. **Preference Modeling:** The method involves modeling user preferences by assigning weights to different features or attributes that influence the desirability of items. This could include factors like genre, content type, popularity, or recency. By incorporating these weighted features into the evaluation functions, PLEF creates a preference model that reflects the user's diverse tastes and preferences.

4. **Multi-Level Recommendation:** PLEF provides recommendations at multiple preference levels, offering a range of suggestions that cater to different aspects of a user's preferences. This multi-level approach ensures that users receive recommendations that align with their varying tastes and moods.

5. **Dynamic Adaptation:** The PLEF method may incorporate dynamic adaptation mechanisms that adjust preference levels based on user interactions and feedback. As user preferences evolve, the system can adapt and refine its recommendation strategies accordingly.

6. **Interpretability and Explainability:** PLEF aims to provide recommendations that are interpretable and explainable to users. The evaluation functions, with their emphasis on different preference levels, allow users to understand why specific recommendations are made, fostering user trust in the recommendation system.

7. **Integration of Context:** PLEF may integrate contextual information to further enhance recommendation accuracy. Contextual factors such as time, location, or user activity can influence preference levels, and the method adapts recommendations accordingly.

8. **Continuous Learning:** The PLEF method supports continuous learning by updating the preference model based on user feedback and interactions. This ensures that the recommendation system remains adaptive and reflective of users' evolving preferences over time.

PLEF can be applied in various domains, including e-commerce, content streaming, social media, and more, where personalized recommendations play a crucial role in enhancing user experience and engagement [19]. The PLEF method provides a personalized and nuanced approach to recommendation systems by considering multiple preference levels, employing evaluation functions, and adapting dynamically to user interactions. This leads to more tailored and user-centric recommendations in diverse application domains [20].

10.3 PROPOSED MODEL

Designing a model that integrates PLEFs with an Ant Colony Algorithm for finding relative desirability involves combining personalized recommendation strategies with a bio-inspired optimization approach. Here is a conceptual outline of how such a model could be structured:

User Preference Modeling: Implement PLEFs to model user preferences at different levels. Consider various factors, such as user history, explicit and implicit feedback, and contextual information, to categorize preferences into multiple levels.

Evaluation Functions: Develop evaluation functions for each preference level. These functions assess the desirability of items based on weighted features that reflect user preferences. The weights can be dynamically adjusted based on user interactions and feedback.

Ant Colony Optimization Module: Integrate an Ant Colony Algorithm module inspired by the foraging behavior of ants. Define artificial ants that represent decision agents exploring the space of available items. Pheromone levels represent the desirability of each item.

Solution Space Representation: Map the recommendation problem into a solution space where artificial ants traverse different paths, each path representing a set of recommended items. The pheromone levels on paths guide the ants toward more desirable solutions.

Pheromone Update Mechanism: Define a mechanism to update pheromone levels based on the desirability scores obtained from the PLEFs. Positive feedback from users (e.g., clicks, likes) can lead to increased pheromone levels, while negative feedback can result in decreased levels.

Dynamic Adaptation: Implement dynamic adaptation mechanisms that adjust the weights in PLEFs and the pheromone update rules based on evolving user preferences. This ensures that the model continuously adapts to changing user behavior.

Multi-Level Recommendations: Leverage the PLEFs to provide multilevel recommendations. For each preference level, the Ant Colony

Algorithm generates a set of recommended items. Users can receive suggestions that align with different aspects of their preferences.

Interpretability and Explainability: Ensure that the model provides interpretability and explainability. Users should be able to understand why certain items are recommended at different preference levels, fostering trust in the recommendation system.

Contextual Considerations: Integrate contextual information, such as time, location, or user activity, into both the PLEFs and the Ant Colony Algorithm. Context-aware recommendations enhance the relevance of suggestions in different scenarios.

Continuous Learning: Support continuous learning by incorporating feedback mechanisms. Learn from user interactions, update preference models, and adjust the pheromone levels to improve the accuracy and relevance of recommendations over time.

By combining the Preference-Leveled Evaluation Functions with the Ant Colony Algorithm, this model offers a robust, adaptive, and personalized recommendation system that takes into account varying user preferences and provides nuanced suggestions based on a bio-inspired optimization approach.

Algorithm 1: PFMDMM Based on Preference-Leveled Evaluation Functions for relatively desirable health suggestion

---

Input: Input x features of data of health selection dataset

Output: Categorizes the health Suggestion for node and Alternate Choice.

Start

1. Input x features of data of health selection dataset
2. PFMDMM to recognize the best health for the node; arranged in the correct format to feed into the ANN for classification.
3. *Training stage:* weights of the Feed Forward Neural Network were given by some arbitrary value as per Table 10.1 and is then tuned for optimal during the iterative learning procedure with the help of backpropagation algorithm.
4. *Testing stage:* the neural network is tested against a variety of test samples, to ensure whether the acquired system correctly categorizes the health to best preferred health and other health parts.
5. Categorize the health to best suggestion range and alternate opinion.

Stop

10.4 CONCLUSION

This work presents a novel BPNN and ant colony algorithm model that uses the PFMDMM data classification system to determine the optimal health of a node as given in Figure 10.1. The trial's findings demonstrated that the parameterized fuzzy measures decision-making model for best

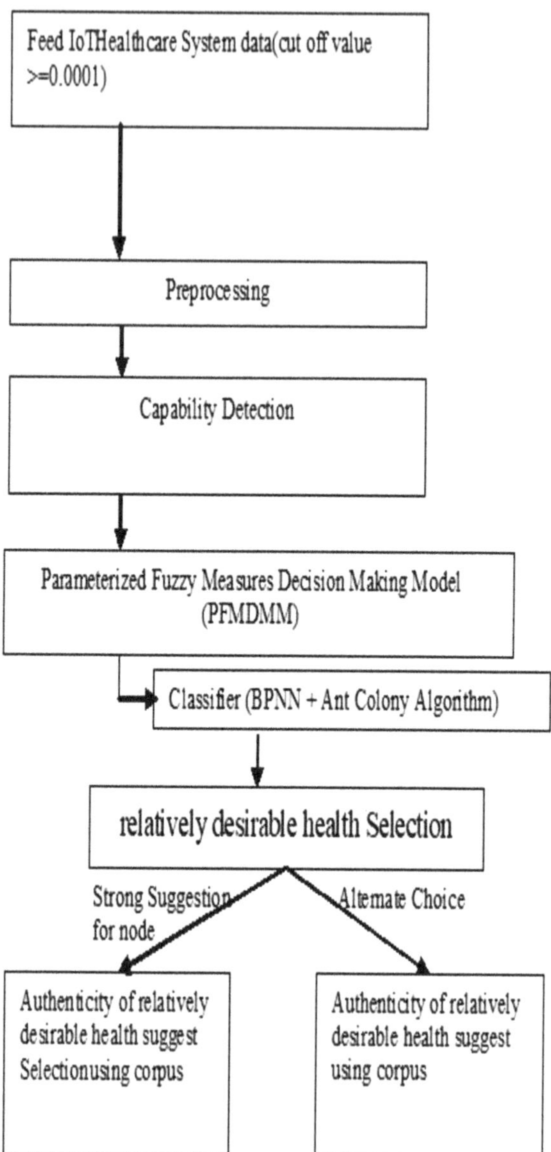

Figure 10.1 Block diagram of proposed PFMDMM based on Preference-Leveled Evaluation Functions for relatively desirable health suggestion.

Table 10.1 Examples features of data of health dataset

Time	Use x[0]	Gen x[1]	Preference-1 x[2]	Preference-2 x[3]	Preference-3 x[4]	Preference-4 x[5]	Preference-5 x[6]	Preference-6 x[7]	Preference-7 x[8]
1.45E+09	0.932833	0.003483	0.932833	10.33E-05	0.0207	0.061917	0.442633	0.12415	0.006983
1.45E+09	0.934333	0.003467	0.934333	0	0.020717	0.063817	0.444067	0.124	0.006983
1.45E+09	0.931817	0.003467	0.931817	1.67E-05	0.0207	0.062317	0.446067	0.123533	0.006983
1.45E+09	1.02205	0.003483	1.02205	1.67E-05	0.1069	0.068517	0.446583	0.123133	0.006983
1.45E+09	1.1394	0.003467	1.1394	0.000133	0.236933	0.063983	0.446533	0.12285	0.00685
1.45E+09	1.391867	0.003433	1.391867	0.000283	0.50325	0.063667	0.447033	0.1223	0.006717
1.45E+09	1.366217	0.00345	1.366217	0.000283	0.4994	0.063717	0.443267	0.12205	0.006733
—	—	—	—	—	—	—	—	—	—

health recognition exhibits encouraging performances in terms of the best health proposal for a smart node. This model is built on preference-leveled assessment functions.

To the best of our knowledge, this is the first study to employ the preference-leveled evaluation functions-based parameterized fuzzy measures decision-making model for data categorization.

The main outcomes of this research are, in particular:

- The proposed approach is capable of providing clustering decision-making within constrained time periods.
- For the method of feature extraction, this study suggests a novel application of the Parameterized Fuzzy Measures Decision Making Model Based on Preference-Leveled Evaluation Functions for health classification-based BPNN architecture.
- In relation to the test, the provided health dataset has been taken into account to assess the suggested methodology. This dataset includes strong health measurements that were gathered to build a parametric model for health classification.

Particularly, by utilizing the decision-making skills of Parameterized Fuzzy Measures Decision Making Model Based on Preference-Leveled Evaluation Functions, this study defined a novel automated health recommendation system.

REFERENCES

1. Jabri, I., Mekki, T., Rachedi, A., & Jemaa, M. B. (2019). Vehicular fog gateways selection on the internet of vehicles: A fuzzy logic with ant colony optimization based approach. Ad Hoc Networks, 91, 101879.
2. Jakšić, Z., Devi, S., Jakšić, O., & Guha, K. (2023). A comprehensive review of bio-inspired optimization algorithms including applications in microelectronics and nanophotonics. Biomimetics, 8(3), 278.
3. Sripriyanka, G., & Mahendran, A. (2022). Bio-inspired Computing Techniques for Data Security Challenges and Controls. SN Computer Science, 3(6), 427.
4. Alizadehsani, R., Roshanzamir, M., Izadi, N. H., Gravina, R., Kabir, H. D., Nahavandi, D., & Fortino, G. (2023). Swarm intelligence in internet of medical things: A review. Sensors, 23(3), 1466.
5. Azad, P., Navimipour, N. J., & Hosseinzadeh, M. (2019). A fuzzy-based method for task scheduling in the cloud environments using inverted ant colony optimisation algorithm. International Journal of Bio-Inspired Computation, 14(2), 125–137.
6. Aftab, F., Khan, A., & Zhang, Z. (2019). Bio-inspired clustering scheme for Internet of Drones application in industrial wireless sensor

network. International Journal of Distributed Sensor Networks, 15(11), 1550147719889900.

7. Salam, A., Javaid, Q., & Ahmad, M. (2021). Bio-inspired cluster-based optimal target identification using multiple unmanned aerial vehicles in smart precision agriculture. International Journal of Distributed Sensor Networks, 17(7), 15501477211034071.

8. Tariq, R., Ashraf, H., Sohail, M., & Khattak, H. A. (2022, December). Optimization Techniques for Scheduling IoT tasks in Fog-based Environments. In 2022 IEEE 19th International Conference on Smart Communities: Improving Quality of Life Using ICT, IoT and AI (HONET) (pp. 075–080). IEEE.

9. Golchay, R. (2016). From Mobile to Cloud: Using Bio-inspired Algorithms for Collaborative Application Offloading (Doctoral dissertation, Université de Lyon).

10. Ould-Yahia, Y., Banerjee, S., Bouzefrane, S., & Boucheneb, H. (2017). Exploring formal strategy framework for the security in IoT towards e-health context using computational intelligence. Internet of Things and Big Data Technologies for Next Generation Healthcare (pp. 63–90).

11. Seyyedabbasi, A., Kiani, F., Allahviranloo, T., Fernandez-Gamiz, U., & Noeiaghdam, S. (2023). Optimal data transmission and pathfinding for WSN and decentralized IoT systems using I-GWO and Ex-GWO algorithms. Alexandria Engineering Journal, 63, 339–357.

12. Husnain, G. (2021). Bio Inspired Intelligent Routing Scheme for Clustering in Vehicular Ad Hoc Networking (VANETs) (Doctoral dissertation, University of Engineering & Technology Peshawar).

13. Choudhary, A., Nizamuddin, M., & Sachan, V. K. (2020). A hybrid fuzzy-genetic algorithm for performance optimization of cyber physical wireless body area networks. International Journal of Fuzzy Systems, 22, 548–569.

14. El-Ghamry, A., Gaber, T., Mohammed, K. K., & Hassanien, A. E. (2023). Optimized and efficient image-based IoT malware detection method. Electronics, 12(3), 708.

15. Husnain, G., & Anwar, S. (2021). An intelligent cluster optimization algorithm based on Whale Optimization Algorithm for VANETs (WOACNET). PLoS One, 16(4), e0250271.

16. Benmansour, F. L., & Labraoui, N. (2021). A comprehensive review on swarm intelligence-based routing protocols in wireless multimedia sensor networks. International Journal of Wireless Information Networks, 28(2), 175–198.

17. Darvishpoor, S., Darvishpour, A., Escarcega, M., & Hassanalian, M. (2023). Nature-inspired algorithms from oceans to space: A comprehensive review of heuristic and meta-heuristic optimization algorithms and their potential applications in drones. Drones, 7(7), 427.

18. Panigrahy, S. K., & Emany, H. (2023). A survey and tutorial on network optimization for intelligent transport system using the internet of vehicles. Sensors, 23(1), 555.

19. Kotiyal, V., Singh, A., Sharma, S., Nagar, J., & Lee, C. C. (2021). ECS-NL: An enhanced cuckoo search algorithm for node localisation in wireless sensor networks. Sensors, 21(11), 3576.
20. Kord, H., & Pourgalehdari, O. (2019). ALQARM: An ant-based load and QoS aware routing mechanism for IoT. Journal of Advances in Computer Research, 10(3), 65–82.

Chapter 11

A new adaptive neural network-based fast terminal sliding mode and force/position control of time-varying constrained reconfigurable manipulators

Ruchika, Naveen Kumar, and Manju Rani

11.1 INTRODUCTION

Nature has been giving intelligent beings an appropriate breeding habitat since the beginning of life. Organisms with biological intelligence can adjust to harsh or dynamic circumstances. For example, a group of fish and birds can effectively sense the changeable situations around them and behave according to inputs that frequently have extremely basic mechanisms and little information availability. Certain creatures have group behaviors and can work together to complete tasks that would be impossible for a single person to complete on their own with little to no implicit communication. Beneficial features are those that an organism can pass on to its progeny, demonstrating high environmental adaptation. Humans possess emotion, thinking, and learning capacities because of the nerve system in the brain.

There has been a broad shift in recent years toward robots that are service-oriented, meaning they must be able to handle a variety of uncertainty and adapt to complicated dynamic situations. Researchers have been working to give robots biological intelligence to enable safe navigation and effective cooperation among the autonomous robots in changing environments. This is because biological organisms have desirable properties like adaptability, robustness, versatility, and agility. Due to the economy's rapid growth, traditional manipulators are insufficient for the flexible product system [1, 2]. So, the reconfigurable manipulators were introduced for modern standard of industries; a set of joint and interchangeable link modules with a common connecting interface is served as reconfigurable manipulators. Due to assembling and dissembling characteristics such as reconfiguration of small modules, it can be capable of completing a range of tough tasks like space exploration, military, medical/industrial, and entertainment [3, 4]. Due to highly nonlinear dynamical model, reconfigurable manipulators are difficult to control. So, controlling these types of manipulators is a marvelous task. Since many years, the control problem related to trajectory tracking of reconfigurable manipulators has admirable great diligence around the world [5–8]. However, practical

DOI: 10.1201/9781003530077-11

applications like polishing, grinding, grabbing require interaction between the reconfigurable manipulator and environment. So, the force exerted due to contact of end-effector with the environment as well as position of end-effector must be controlled simultaneously. Therefore, only position control is not sufficient for reconfigurable manipulators. For this reason, the force and position are both to be controlled simultaneously in the constrained space. As a result, various control strategies such as centralized control [9], decentralized control [10, 11], robust/adaptive control [12, 13], and sliding mode control [14, 15] have been proposed in literature. A decentralized adaptive force control method which combines the hybrid approach as well as neural network compensator was presented in [16]. A combination made up of model based and model free approach for force/position control method to interact with stiff environment is proposed in [17]. A modular manipulation technique—a control strategy based on joint torque sensing and joint by joint stabilization—has been developed by [18] for modular and reconfigurable robots. A decentralized approach, which eliminates the need to measure the joint torque and velocity, uses only position parameters to control Harmonic drive (HD)-based modular robot manipulator (MRM) presented in [19].

It should be noted that the aforementioned control systems have advanced work on tracking control without taking time-varying constraints in the surrounding environment. However, force/position control of reconfigurable manipulators has received very little scientific attention with time-varying constraints published in the literature [20–24]. For trajectory tracking control [21], Dong and Li developed an integral sliding mode control strategy to eliminate the chattering effect of modular and changeable manipulators. A new continuous time decentralized reinforcement learning for highly coupled nonlinear dynamics of time varying constrained reconfigurable manipulator is described by [24]. However, some challenges and difficulties may arise due to uncertain parameters in dynamical model. To overcome these types of difficulties in manipulator system, numerous schemes have been put forward by introducing the aforementioned scheme with compensating controllers like neural network type controller [25–27]. In the context of uncertainties and time-varying limitations, Wu et al. [28] proposed an adaptive control law for reconfigurable manipulators based on neural networks. For time-varying restricted reconfigurable manipulators, Kumar and Rani [29] introduced a hybrid force/position control strategy.

11.1.1 Scope of chapter

Unfortunately, there are very less work found in literature which control both force and position for reconfigurable manipulators with time varying constraints, although it is necessary to deal with the time-varying constraints to achieve targeted manifold. So, this brings a motivation to our work, that is, to design a controller for the reconfigurable manipulators with time-varying constraints imposed on it. However, the time-varying constrained manipulator has a more intricate and sophisticated control architecture. As a result, the controller performs worse when time-varying restrictions are ignored, leading to things like a sluggish convergence rate. Therefore, to overcome these types of problems, authors have developed a finite-time fast terminal sliding mode control (FTSMC) for reconfigurable manipulators subject to time-varying limitations. According to author's best knowledge, there has not been designed a controller which completely fulfills authors requirement of fast convergence, boundedness of disturbance terms, and error due to neural network for time-varying constrained reconfigurable manipulators.

Motivated by the aforementioned discussion, this work presents a strategy that integrates the benefits of fast terminal sliding mode control and neural network with adaptive bound. The main contribution of this approach is given as follows:

- The integration of FTSMC and radial basis function neural network (RBFNN) is used to actualize efficient position and force control of reconfigurable manipulator with time-varying constraints.
- The superiority of introducing FTSMC lies in finite time stability, that is, rapid convergence rate.
- Neural network controller RBFNN is employed to estimate the non-linear dynamical part. Also, the adaptive bound part is added to counteract the effect of collision (friction) terms and network reconstruction error.
- The stability analysis is elaborated to show the asymptotic as well as finite time stability of the system. Additionally, the effectiveness and robustness of given reconfigurable manipulator are carried out with the help of simulation results which are compared with other conventional controllers.

11.2 DYNAMICAL MODEL DESCRIPTION

To exhibit effective functioning with the proposed control framework, we are using time-varying constraint which is given as [21]:

$$\Phi(q,t) = 0 \tag{11.1}$$

The continuous and second-order differential function which is given as $\Phi(.): R^l \rightarrow R^b$ and q represents the joint's position vector. The one-dimensional reconfigurable manipulator's dynamic equation with time-varying constraint is presented as [24]:

$$E(q)\ddot{q} + C(q,\dot{q})\dot{q} + G(q) + F(\dot{q}) = U + f \tag{11.2}$$

$$f = J_\Phi^T(q)\lambda \tag{11.3}$$

The inertia matrix for the given system is represented by $E(q) \in R^{l \times l}$; the vector consists of centripetal force terms denoted as $C(q,\dot{q}) \in R^{l \times l}$. The vectors $G(q) \in R^l$ and $F(\dot{q}) \in R^l$ are used to represent gravitational force and friction force, respectively. The input torque vector of joint force exerting on the manipulator is denoted by $\tau \in R^l$. The vector $f \in R^l$ contains the constraint force terms. Also, the constrained force may be ahead, described as $f = J_\Phi^T(q)\lambda$, where $J_\Phi \in R^{h \times l}$ is Jacobian matrix which denotes the relation between joint space, Cartesian space, and $\lambda \in R^h$ is the generalized force multiplier.

Furthermore, the description of presumptions which are used in a controlled system are given as [27]:

> Assumption 1: This assumption described that the constraint is applied on a rigid and friction less surface.
>
> Assumption 2: The desired value of joint position, joint velocity, and joint acceleration is bounded, which are denoted as $q_d, \dot{q}_d, \ddot{q}_d$.
>
> Assumption 3: The Jacobian matrix for the reconfigurable manipulator should avoid any singularities, so that Rank $J_\Phi^T(q)$ = min {l, h}. After imposition of h constraints to manipulator, the system is left with only $l - h$ DOF.
>
> Assumption 4: By using upper bounding functions along some finite constants b_1 and b_2, the friction term satisfies the following inequality $\left\| F(\dot{q}) \right\| \le K_{f1} \left| \dot{q} \right| + K_{f2}$

$$q = [q_1, q_2]^T, q_1 \in R^{l-h}, \quad q_2 \in R^h \tag{11.4}$$

where $l - h = k_1$. Also, q_2 can be described as

$$q_2 = \psi(q_1, t) \text{ and } q = \left[q_1, \psi(q_1, t) \right] \tag{11.5}$$

On differentiating (11.4) w.r.t time t, concludes

$$\dot{q} = B(q_1)\dot{q}_1 + T \tag{11.6}$$

On differentiating (11.6), we obtain the following:

$$\ddot{q} = \dot{B}\dot{q}_1 + B\ddot{q}_1 + \dot{T} \tag{11.7}$$

where $B(q_1) = \left[I_{n-k}, \dfrac{\partial \psi(q_1)}{\partial q_1} \right]^T$, $T = \left[0, \dfrac{\partial \psi(q_1)}{\partial t} \right]^T$

The matrices $B(q_1)$ and $J_\Phi(q)$ satisfy the relation given by

$$B^T J_\Phi^T = 0$$

After consideration of (11.5) and (11.6), the dynamic model is transformed as:

$$\begin{aligned}
&E(q_1)B(q_1)\ddot{q}_1 + E(q_1)\dot{B}(q_1)\dot{q}_1 + C(q_1,\dot{q}_1)B(q_1)\dot{q}_1 \\
&+E(q_1)\dot{T} + C(q_1,\dot{q}_1)T + G(q_1) + F(\dot{q}_1) = U + J_\Phi^T(q_1)\lambda
\end{aligned} \tag{11.8}$$

The simplified form of the above equation is given as:

$$E_1(q_1)\ddot{q}_1 + C(q_1,\dot{q}_1)\dot{q}_1 + L(q_1,t) + G(q_1) + F(\dot{q}_1) = U + J_\Phi^T(q_1)\lambda \tag{11.9}$$

where $E(q_1)B(q_1) = E_1(q_1)$

$$E(q_1)B(\dot{q}_1) + C(q_1,\dot{q}_1)B(q_1) = C_1(q_1,\dot{q}_1) \quad L(q_1,t) = E(q_1)\dot{T} + C(q_1,\dot{q}_1)T$$

Multiplying (11.8) by $B^T(q_1)$ on both sides and in view of (11.9), we obtain

$$\bar{E}(q_1)\ddot{q} + \bar{C}(q_1,\dot{q}_1)\dot{q}_1 + \bar{L}(q_1,t) + \bar{G}(q_1) + \bar{F}(\dot{q}_1) = B^T(q_1)U \tag{11.10}$$

where $\bar{E} = B^T E_1, \bar{C} = B^T E_1, \bar{L} = B^T L, \bar{G} = B^T G, \bar{F} = B^T F$

Property 1: $\bar{E}$ is positive bounded matrix with symmetry property.
Property 2: The skew-symmetric relationship is allowed by the matrix $\left(\dot{\bar{E}} - 2\bar{C}\right)$, which is beneficial while conducting stability analysis.

11.3 FORCE/POSITION CONTROLLER DESIGN

11.3.1 Error dynamics

Positional and force errors are explained as follows:

$$e(t) = q_d(t) - q(t); \quad e_1(t) = q_{1d}(t) - q_1; \quad e_\lambda(t) = \lambda_d - \lambda \tag{11.11}$$

11.3.2 Design of a fast terminal sliding surface

This section introduces a controller which is based on fast terminal sliding surface that includes the dominance of rapid convergence and immense robustness, defined as:

$$S(t) = \dot{e}_1 + \Gamma_1 \text{signum}(e_1)^{\Lambda_1} + \Gamma_2 \text{signum}(e_1)^{\Lambda_2} = \dot{q}_{1r}(t) - \dot{q}_1(t) \tag{11.12}$$

where Γ_1, $\Gamma_2 \in R^n$, $\Lambda_1 \geq 1$ and $0 < \Lambda_2 < 1$, and $\Lambda_2 = r/s$, r, $s > 0$ belongs to set of integers and allow this inequality $r < s < 2r$.

The sliding function differentiation, expressed as a function of t, can be expressed as follows:

$$\ddot{S} = \ddot{e}_1 + \Gamma_1 e_{1r_1} + \Gamma_2 e_{1K_1} \tag{11.13}$$

$$e_{1r1} = \begin{cases} \Lambda_1 |e_1|^{\Lambda_1 - 1} \dot{e}_1, & e_1 \neq 0 \\ 0, & e_1 = 0 \end{cases}$$

and

$$e_{1k1} = \begin{cases} \Lambda_2 |e_1|^{\Lambda_2 - 1} \dot{e}_1, & e_1 \neq 0 \\ 0, & e_1 = 0 \end{cases}$$

Using the above equations, the modified dynamics in the form of sliding variable S is expressed as:

$$\bar{E}\dot{S} = -\bar{V}S - B^T U + F + v(\bar{y}) \tag{11.14}$$

where $v(\bar{y}) = \bar{E}\ddot{q}_{1r} + \bar{C}\dot{q}_{1r} + \bar{L}(q,t) + \bar{G}$ is the robot nonlinear function and the input vector is selected as $\bar{y} = \left[e_1^T, \dot{e}_1^T, q_{1d}^T, \dot{q}_{1d}^T, \ddot{q}_{1d}^T \right]^T$. In view of the RBF neural networks, we determine the dynamical part that is unknown as the approximation of $v(x)$ is given as:

$$v(\bar{y}) = W^T \Xi(\bar{y}) + \epsilon(\bar{y}) \tag{11.15}$$

where $W \in R^{N \times n3}$ symbolizes the weight matrix; a smooth basis vector (known) is expressed as $\Xi(\cdot): R^{5n3} \to R^N$, $\epsilon(\cdot): R^{5n3} \to R^{n3}$ represents the network reconstruction error. After substitution of $v(\bar{y})$ from (11.15), the robot dynamics (11.14) is rewritten as:

$$\bar{E}\dot{S}(t) = -\bar{C}S - B^T U + W^T \Xi(\bar{x}) + \epsilon(\bar{y}) + \bar{F} \tag{11.16}$$

For the force/position control, we select the following control torque input U:

$$U = U_1 + U_2 \tag{11.17}$$

where U_1 is selected for managing position and U_2 is selected for managing force:

$$U_2 = -J_{\Phi}^T\left(\lambda_d - K_{\lambda}e_{\lambda}\right) \tag{11.18}$$

After consideration of equations (11.21) and (11.22), the equation (11.20) is altered as:

$$\bar{E}S(t) = -\bar{C}S - B^T\dot{U}_1 + W^T\Xi(\bar{x}) + \bar{F} \tag{11.19}$$

To track the desired path, the torque input for position is designed as:

$$B^T U_1 = \hat{W}^T\Xi(\bar{x}) + K\operatorname{sign}(S) + K_1\operatorname{sign}(S)^{\Lambda_1} + K_2\operatorname{sign}(S)^{\Lambda_2} + \Pi \tag{11.20}$$

The network reconstruction error and effect of friction parameters are reimbursed with the help of adaptive compensator Π, which is used in given control law and detailed as follows:

On using assumption (4) and selecting $\epsilon < \epsilon_N$, we come across

$$\left\|\bar{F}(\dot{q}_1) + \epsilon(\bar{x})\right\| \leq K_{f_1} + K_{f_2}\left\|\dot{q}_1\right\| + \epsilon_N ; \rho = \begin{bmatrix} 1 & \left\|\dot{q}_1\right\| & 1 \end{bmatrix}\begin{bmatrix} K_{f_1} & K_{f_1} & \epsilon_N \end{bmatrix}^T = T^T\left(\left\|\dot{q}_1\right\|\right)\chi$$

On considering the $\dot{\delta} = -\alpha\,\delta$, $\delta(0) = $ The design constant's value is selected to be positive. We select the adaptive compensator term as:

$$\Pi = \frac{\hat{\rho}^2 S}{\hat{\rho}S + \delta} \tag{11.21}$$

where $\hat{\rho}$ stands for estimated value of bound function $\rho > 0$. The reduced dynamics (11.19) can be rewritten in considering (11.20) and (11.21):

$$\bar{E}\dot{S} = -\bar{C}S + \tilde{W}^T \Xi(\bar{x}) - K\,\mathrm{sign}(S) - K_1\,\mathrm{sign}(S)^{\Lambda_1} - K_2\,\mathrm{sign}(S)^{\Lambda_2} + \bar{F}(\dot{q}_1) + \epsilon(\bar{x}) - \frac{\hat{\rho}^2 S}{\hat{\rho}S + \delta}$$

11.4 STABILITY ANALYSIS

Considering the time-varying constraint as given in 11.1 used in dynamical model given in Section 11.2 the proposed force/position control law, if the adaptive laws taken as where Γ_W and Γ_χ are positive definite matrices, then the tracking errors asymptotically converge to zero in a finite period of time, and all the signals are bounded jointly.

$$\dot{\hat{W}} = \Gamma_W E(\bar{y})S^T; \qquad \dot{\hat{\chi}} = \Gamma_\chi T\|S\| \tag{11.22}$$

Proof: The Lyapunov function that we choose is:

$$P = \frac{1}{2}S^T \bar{E}S + \frac{1}{2}tr\left(\widetilde{W}^T \Gamma_W^{-1} \dot{\hat{W}}\right) + \frac{1}{2}tr\left(\widetilde{\chi}^T \Gamma_\chi^{-1} \dot{\hat{\chi}}\right) + \frac{\delta}{\alpha} \tag{11.23}$$

The differentiation of (11.23), gives:

$$\dot{P} = \frac{1}{2}S^T \dot{\bar{E}}S + S^T\left(-\bar{C}S - K\,\mathrm{sign}(S) - K_1\,\mathrm{sign}(S)^{\Lambda_1} - K_2\,\mathrm{sign}(S)^{\Lambda_2} - \frac{\widehat{\rho}^2 S}{\hat{\rho}\|S\| + \delta}\right)$$
$$+ S^T\left(\widetilde{W}^T \Xi(\bar{y})\right) + S^T\left(\bar{F}(\dot{q}_1) + \epsilon(\bar{y})\right) - tr\left(\widetilde{W}^T \Gamma_W^{-1} \dot{\hat{W}}\right) - tr\left(\widetilde{\chi}^T \Gamma_\chi^{-1} \dot{\hat{\chi}}\right) - \delta$$

$$\dot{P} = \frac{1}{2}S^T\left(\dot{\bar{E}} - 2\bar{C}\right)S - S^T\left(K\,\mathrm{sign}(S) + K_1\,\mathrm{sign}(S)^{\Lambda_1} + K_2\,\mathrm{sign}(S)^{\Lambda_2}\right) + S^T \widetilde{W}^T \varsigma(\bar{y})$$
$$+ S^T\left(\bar{F}(\dot{q}_1) + \epsilon(\bar{y})\right) - \frac{\widehat{\rho}^2 S^T S}{\hat{\rho}|S| + \delta} - tr\left(\widetilde{W}^T \Gamma_W^{-1} \dot{\hat{W}}\right) - tr\left(\widetilde{\chi}^T \Gamma_\chi^{-1} \dot{\hat{\chi}}\right) - \delta$$

$$\dot{P} = -S^T K\,\mathrm{sign}(S) - S^T K_1\,\mathrm{sign}(S)^{\Lambda_1} - S^T K_2\,\mathrm{sign}(S)^{\Lambda_2} + S^T \widetilde{W}^T \Xi(\bar{y}) + S^T\left(\bar{F}(\dot{q}_1) + \epsilon(\bar{y})\right)$$
$$- tr\left(\widetilde{W}^T \Xi(\bar{y})S^T\right) - tr\left(\widetilde{\chi}^T T|S|\right) - \frac{\widehat{\rho}^2 \|S\|^2}{\hat{\rho}\|S\| + \delta} - \delta$$

$$\dot{P} = -S^T K sign(S) - S^T K_1 sign(S)^{\Lambda_1} - S^T K_2 sign(S)^{\Lambda_2}$$
$$+ S^T \left(\overline{F}(\dot{q}_1) + \epsilon(\overline{y}) \right) - \widetilde{\chi^T} T |S| - \frac{(T^T \widehat{\chi})^2 \| S \|^2}{(T^T \widehat{\chi}) \|S\| + \delta} - \delta$$

$$S^T \left(\overline{F}(\dot{q}_1) + \epsilon(\overline{y}) \right) \leq \|S\| \left(\overline{F}(\dot{q}_1) + \epsilon(\overline{y}) \right) \| \leq \rho S = T^T \chi S = T^T \left(\widehat{\chi} + \widetilde{\chi} \right) \|S\|$$

Consequent to the above expression, the above equation concludes:

$$\dot{P} \leq \lambda_{\min}(K)|S| \leq 0 \tag{11.24}$$

where $\lambda_{\min}(K)$ stands for the smallest eigenvalue of the matrix K. As equations (11.22) and (11.24) show that $P > 0$ and $\dot{P}\, P \leq 0$ which concludes total stability in the Lyapunov sense.

Now, $\dot{P}\left(S(t), \widetilde{W}, \widetilde{\chi}\right) \leq 0$ implies

$$P\left(S(t), \widetilde{W}, \widetilde{\chi}\right) \leq P\left(S(0), \widetilde{W}, \widetilde{\chi}\right)$$

This demonstrates that $S(t), \widetilde{W}, \widetilde{\chi}$ are bounded. A function called $\pi(t) = \lambda_{\min}(K)|S|^2 \leq -\dot{P}$ is defined. Following the integration of the function $\pi(t)$ with respect to time, we may get the inequality

$$\int_0^t P(t)\,dt \leq P(0) - P(t) = P\left(S(0), \widetilde{W}, \widetilde{\chi}\right) - P\left(S(t), \widetilde{W}, \widetilde{\chi}\right) \tag{11.25}$$

Now $P\left(S(0), \widetilde{W}, \widetilde{\chi}\right)$ being bounded and $P\left(S(t), \widetilde{W}, \widetilde{\chi}\right)$ being decreasing and bounded function, boundedness of $S(t), \widetilde{W}$ and $\widetilde{\chi}$ is guaranteed. As S is function of e_1 and e_1 error signals are also bounded.

$$\ddot{P} \leq -2S^T K\ sign(S) \tag{11.26}$$

Now from (11.26), we have $\ddot{P} < 0$. As a result, $\dot{P}$ is uniformly continuous. Therefore, all tracking errors must go to zero in accordance with Barbalat's lemma. Also, we have

$$\lambda = A\left(C_1(q_1, \dot{q}_1)\dot{q}_1 + \overline{L}(q_1, t) + F(\dot{q}_1) + G(q_1) - U \right) \tag{11.27}$$

With $A = \left(J_\Phi^1 E_1^{-1} J_\Phi^{1T} \right)^{-1} J_\Phi^1 \left(E_1 \right)^{-1}$

Also, we obtain:

$$\lambda = A(C_1(q_1,\dot{q}_1)\dot{q}_1 + \bar{L}(q_1,t) + F(\dot{q}_1) + G(q_1) + U_1) - U_2$$

$$\left(I - K_\lambda\right)e_\lambda = AB^{+^T}\bar{E}\dot{q}_1 \tag{11.28}$$

where $B^+ = \left(B^T B\right)^{-1} B^T$. The terms A, B^T, $\bar{E}$ and $\dot{q}_1$ are bounded, suggesting that the phrase on the RHS of (11.28) is bounded. Furthermore, the convergence of e_λ toward zero is implied by choosing the gain matrix K_λ correctly. So, we can conclude that $\lambda \to \lambda_d$.

We determine the reaching time and stopping time to illustrate the convergence in finite time. The following Lyapunov function is selected for reaching time:

$$P = \frac{1}{2}S^T\bar{E}S$$

$$\dot{P} = \frac{1}{2}S^T\dot{\bar{E}}S + S^T\bar{E}\dot{S}$$

$$\dot{P} = \frac{1}{2}S^T\dot{\bar{E}}S + \left(-\bar{C}S - K\text{sign}(S) - K_1\text{sign}(S)^{\Lambda_1} - K_2\text{sign}(S)^{\Lambda_2} - \frac{\widehat{\rho^2}\,\|S\|^2}{\hat{\rho}\|S\| + \delta}\right)$$
$$+ S^T\left(\widetilde{W^T}\Xi(\bar{y})\right) + S^T\left(\bar{F}(\dot{q}_1) + \epsilon(\bar{y})\right)$$

$$\dot{P} = \frac{1}{2}S^T\left(\dot{\bar{E}} - 2\bar{C}\right)S - S^T\left(K\text{sign}(S) + K_1\text{sign}(S)^{\Lambda_1} + K_2\text{sign}(S)^{\Lambda_2}\right)$$
$$+ S^T\widetilde{W^T}\Xi(\bar{y}) + S^T\left(\bar{F}(\dot{q}_1) + \epsilon(\bar{y})\right) - \frac{\widehat{\rho^2}\,\|S\|^2}{\hat{\rho}\|S\| + \delta}$$

$$\dot{P} = -S^T\left(K\text{sign}(S) + K_1\text{sign}(S)^{\Lambda_1} + K_2\text{sign}(S)^{\Lambda_2}\right)$$
$$+ S^T\widetilde{W^T}\Xi(\bar{y}) + S^T\left(\bar{F}(\dot{q}_1) + \epsilon(\bar{y})\right) - \frac{\widehat{\rho^2}\,\|S\|^2}{\hat{\rho}\|S\| + \delta}$$

$$\dot{P} \leq -S^T\left(K\text{sign}(S) + K_1\text{sign}(S)^{\Lambda_1} + K_2\text{sign}(S)^{\Lambda_2}\right)$$
$$- \frac{\widehat{\rho^2}\,\|S\|^2}{\hat{\rho}\|S\| + \delta} + \|S\|\widetilde{W^T}\Xi(\bar{y}) + \|S\|\left(\bar{F}(\dot{q}_1) + \epsilon(\bar{y})\right)$$

The above stated function, that is, Gaussian function always lies between 0 and 1 and the above computation validates the boundedness of S, ε, F which results in

$$|S|\left|\widetilde{W}^T \Xi(\bar{y})\right| + |S|\left|\bar{F}(\dot{q}_1) + \epsilon(\bar{y})\right| \le h_1 + h_2 = h_m$$

$$\le -\wedge_{min}|S| - S^T K_1 \text{sign}(S)^{\Lambda_1} - S^T K_2 \text{sign}(S)^{\Lambda_2} + h_m$$

In this approach, the value of h is chosen so that $-\wedge_{min} > h_m$

$$\dot{P} \le -S^T K_1 \text{sign}(S)^{\Lambda_1} - S^T K_2 \text{sign}(S)^{\Lambda_2}$$

$$= -\sum_{i=1}^{n} S_i K_{1i} |S_i|^{\Lambda_1} \text{sign}(S_i) - \sum_{i=1}^{n} S_i K_{2i} |S_i|^{\Lambda_2} \text{sign}(S_i)$$

$$= -\sum_{i=1}^{n} K_{1i} |S_i|^{\Lambda_1 + 1} \text{sign}(S_i) - \sum_{i=1}^{n} K_{2i} |S_i|^{\Lambda_2 + 1} \text{sign}(S_i)$$

$$= -\sum_{i=1}^{n} K_{1i} |S_i|^{\Lambda_1 + 1} - \sum_{i=1}^{n} K_{2i} |S_i|^{\Lambda_2 + 1}$$

$$\le \eta_1 \left(\sum_{i=1}^{n} \frac{1}{2} \bar{d}_1 S_i^2 \right)^{\mu_1} - \eta_2 \left(\sum_{i=1}^{n} \frac{1}{2} \bar{d}_2 S_i^2 \right)^{\mu_2}$$

$$\le \eta_1 P^{\mu_1} - \eta_2 P^{\mu_2} \le \eta_m (P)^{\mu}$$

where $K_{min} = \min K_{1i}$, $K'_{min} = \min K_{1i}$, $\mu_{1=}(1+\Lambda_1)/2$ $\eta_1 = K_{min}\left(\dfrac{2}{\bar{d}_1}\right)^{\mu_1}$,
$\mu_{2=}(1+\Lambda_2)/2$, $\mu = $ $\min\{\mu_1, \mu_2\}$,

$\eta_2 = K_{min}\left(\dfrac{2}{\bar{d}_2}\right)^{\mu_2}$ and $\eta_m = \min\{\eta_1, \eta_2\}$

Therefore, it can be concluded from the definition of sliding surface that along the fast terminal sliding surface, the tracking error will converge to zero in a finite amount of time. Moreover it has become clear from lemma 1, $t_r \le \dfrac{P^{1-\mu}(0)}{\eta_m(1-\mu)}$ denotes the amount of timerequired for the system states to reach the sliding surface, or the settling time to attain the sliding mode.

Now, stopping time is defined as:

$$\tau_s = \text{inf} t \ge t_r : e_1(t) = 0$$

- We need to prove the existence $r \le t_s \le \infty s$. $t\tau s \le t_r$ to calculate the stopping time and the Lyapunov function for the same is described as:

$$\dot{P} = \frac{1}{2} e_1^T e_1$$

$$\dot{P} = e_1^T \dot{e}_1$$

Noting that $S(t) = 0$ at $t = t_r$.
So

$$\dot{e}_1 + \Gamma_1 \mathrm{sign}\left(e_1\right)^{\Lambda_1} + \Gamma_2 \mathrm{sign}\left(e_1\right)^{\Lambda_2} = 0$$

$$\dot{P} = -e_1 \Gamma_1 \mathrm{sign}\left(e_1\right)^{\Lambda_1} - e_1 \Gamma_2 \mathrm{sign}\left(e_1\right)^{\Lambda_2}$$

$$\dot{P} = -e_{1i} \Gamma_1 \mathrm{sign}\left(e_{1i}\right)^{\Lambda_1} - e_{1i} \Gamma_2 \mathrm{sign}\left(e_{1i}\right)^{\Lambda_2}$$

$$= -\sum_{i=1}^{n} \Gamma_{1i} \left|e_{1i}\right|^{\Lambda_1} \mathrm{sign}\left(e_{1i}\right) - \sum_{i=1}^{n} \Gamma_{2i} \left|e_{1i}\right|^{\Lambda_2} \mathrm{sign}\left(e_{1i}\right)$$

$$= -\sum_{i=1}^{n} \Gamma_{1i} \left|e_{1i}\right|^{\Lambda_1+1} - \sum_{i=1}^{n} \Gamma_{2i} \left|e_{1i}\right|^{\Lambda_2+1}$$

$$\leq n_1 \left(\sum_{i=1}^{n} \frac{1}{2} e_{i1}^2\right)^{u_1} - n_2 \left(\sum_{i=1}^{n} \frac{1}{2} e_{i1}^2\right)^{u_2} \leq n_m (P)^u$$

where $\Gamma_{1\min} = \min \Gamma_{1i}$, $\Gamma_{2\min} = \min \Gamma_{2i}$, $n_1 = \Gamma_{1\min}(2)^{u_1}$, $n_2 = \Gamma_{2\min}(2)^{u_2}$, $\Gamma_{\min} = \min\{\Gamma_{1\min}, \Gamma_{2\min}\}$, $n_m = \min\{n_1, n_2\}$, $u_1 = (1+\Lambda_1)/2$ $u_2 = (1+\Lambda_2)/2$, where $0 \leq \Lambda_2 \leq 1$ and $u_m = \min\{u_1, u_2\}$, $t_r \leq t_s + \dfrac{P^{1-\mu}\left(t_r\right)}{2^u \Gamma_{\min}(1-u)}$. The system trajectory converges to the reference trajectory within time interval t_s as shown by the results.

11.5 STUDIES USING NUMERICAL SIMULATIONS

Numerical simulation outcomes over a 2-degree of freedom (2-DOF) time-varying constrained reconfigurable robot with two distinct configurations, as illustrated in Figure 11.1, verify the efficacy of the proposed controller.

It is thought of the time-varying constraint as a type of column that can be rotated with some DOF. For configuration I (circular type), the time-varying constraint is regarded as follows: $\Phi_1(q,t) = l_1 \cos q_1 + l_2 \cos q_2 - (l_3 + l_0 \cot(\beta(t))) = 0$. where the link lengths are given as: $l_1 = 1$, $l_2 = 1$, and $\beta(t)$ is

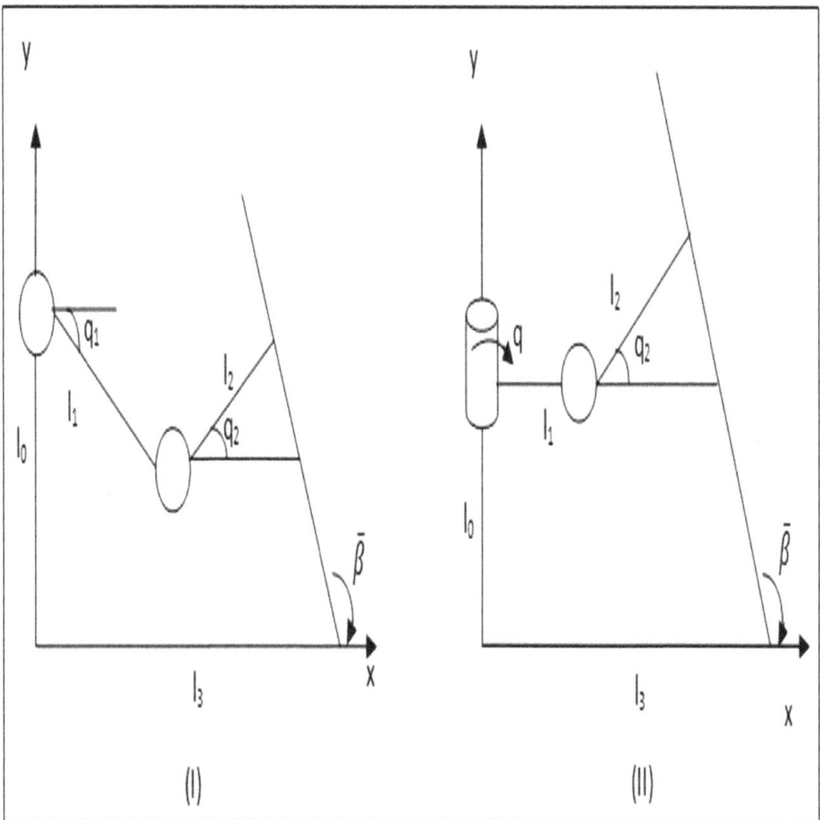

Figure 11.1 (a) Configuration I and (b) configuration II analytical charts.

the angle that is thought to exist between the *x*-axis and the time-varying restriction. $\beta(t) = 0.75\pi + 0.2\sin(t/2)$. The following are the desired position trajectory and the desired force for configuration I:

$$q_{11d} = 0.5\cos(t) + 0.2\sin(3t); q_{12d} = \arcsin\left(\frac{l_1\sin(\beta(t) - q_{11d}) - l_3\sin(\alpha(t))}{l_2}\right) + \beta(t)$$

$$\lambda_{1d} = -\frac{1}{2}m_cg\frac{l_4\cos(\beta(t))\sin(\beta(t))\cos(\beta(t) + q_2)}{l_0 - l_1\sin q_1 - l_2\sin q_2}$$

Further, another configuration with the cylindrical-type constraint, which is provided as follows, is used to verify that the control plan that has been offered is effective:

$$\Phi_2(q, t) = l_1 + l_2\cos q_2 - (l_3 + l_0\cot(\beta(t))) = 0$$

For configuration II, the expected force and position trajectory are regarded as:

$$q_{\text{II}2d} = \arcsin\left(\frac{l_1\sin(\beta(t)) - l_3\sin(\alpha(t))}{l_2}\right) + \beta(t); q_{\text{II}1d} = 0;$$

$$\lambda_{\text{II}d} = -\frac{1}{2}m_c g \frac{l_4\cos(\beta(t))\sin(\beta(t))\cos(\beta(t) + q_2)}{l_0 - l_2\sin q_2}$$

The following parameters are used for given control laws:

$\Gamma_1 = 5I_1$, $\Gamma_2 = 5I_1$, $K_1 = 10I_1$, $K_2 = 5I_1$.Twenty nodes are used to build the RBF neural network. Central locations are selected from -2 to 2 and the spread factor for the radial basis function is0.2. In the MATLAB environment, the simulation run lasted for 10 seconds. $\Gamma_v = 10I_{20}$, $\Gamma_\phi = 5I_3$ are the matrices used. To exhibit the robustness of the proposed FTSMC scheme, the simulation results are produced and compared with two existing control approaches, that is, Liu et al. [27] and Kumar and Rani [29]. For this purpose, we examine the variation in force and position errors with both configurations I and II. Figures 11.2–11.6 exemplify the simulation process of configuration I for given control law and existing controllers [27, 29] while the simulation results corresponding to Figures 11.7–11.9 encapsulates the efficiency of proposed approach for configuration II. Figures 11.2 and 11.3 show the growth of trajectory tracking errors for joint 1 and joint 2 with the existing and proposed approach for configuration I. The proficiency of constraint force is demonstrated in Figure 11.4 for configuration I. Figures 11.5 and 11.6 show the trajectory tracking performance for configuration I. It clearly exhibits that the desired tracking profile is achieved. For configuration II, the position and velocity tracking error profiles are organized in Figures 11.7 and 11.8. Figure 11.9illustratesthe progress of force tracking error. It can observed from these figures that the relation between constraint and the independent joint is a clear task in configuration I, where, in case of another configuration, this relationship is constant which manifests the robustness. Also, these figures show the fast convergence of tracking errors with the proposed control scheme, whereas there is delay in convergence of errors for some seconds with the existing approaches. These two statistics of configuration show that the proposed method fully satisfies the prerequisite need for reconfigurable manipulators. Furthermore, to illustrate the effectiveness and enhancement of given control scheme, the simulation results are demonstrated and compared with other existing approaches [27, 29].

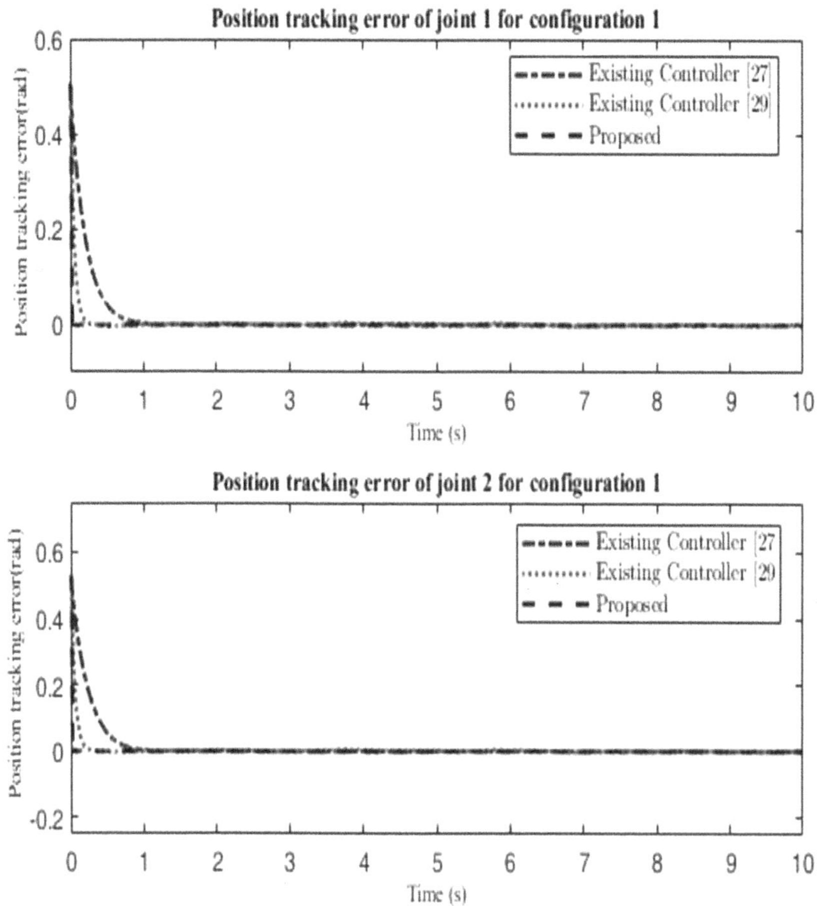

Figure 11.2 Joints 1 and 2 position tracking error for configuration-I.

From this demography, it is revealed that the desired positions and force trajectory converge to the actual one. After comparison, it can be clearly seen from the simulation results that the tracking errors of each joint and the force tracking errors rapidly converge to zero with both configurations and the tracking performance with the proposed method being satisfactory. Very importantly, it can be seen that the proposed method takes benefits of time-varying constraints appropriately and less affected by uncertainties [30].

Next to quantify the quality, efficiency, and reliability of control system measuring the quality and effectiveness of the controls scheme, we use the

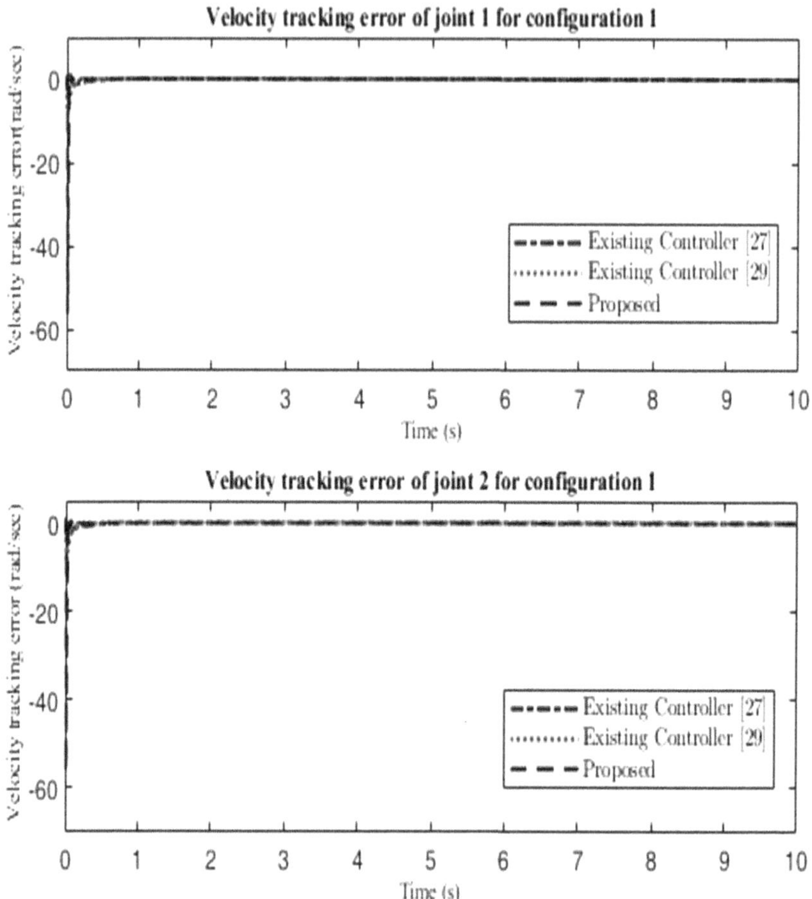

Figure 11.3 Joints 1 and 2's velocity tracking inaccuracy for configuration-I.

L^2norm performance index which calculates the root mean square (RMS) which is given as follows: $L^2 = \sqrt{\dfrac{1}{t_f - t_o} \int_{t_o}^{t_f} \|q(t)\|^2 \, dt}$. The RMS average of joint tracking errors for all configurations are calculated with the proposed as well as existing approaches as given in Table 11.1.

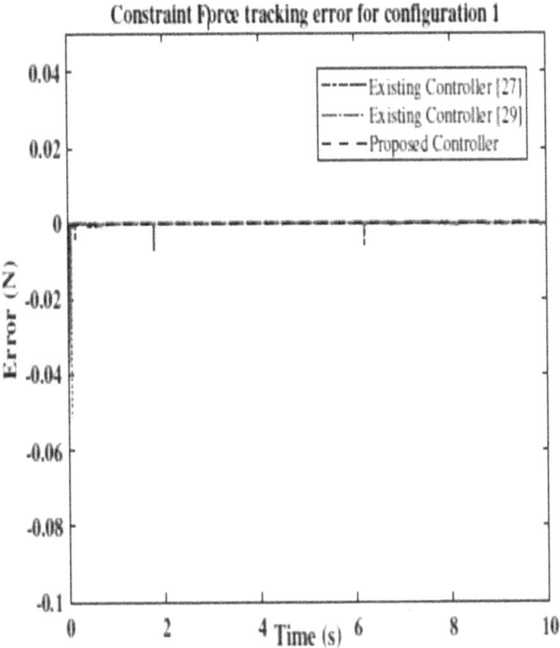

Figure 11.4 Force tracking error with respect to configuration I.

11.6 CONCLUSION

This chapter presents an FTSMC method that ensures high precision control in the face of uncertainties, external disturbances, and time-varying constraints. Firstly, the dynamical model of the systems is reduced with the imposition of time-varying kinematic constraint. Based on proposed sliding surface and reduced dynamical model, for time-varying restricted reconfigurable manipulators, a limited-time FTSMC system is presented for force/position control [31]. The proposed scheme achieves faster convergence rate due to nonlinear sliding surface and shows a high degree of resilience in the face of uncertainty and outside shocks. To approximate the robot nonlinear function, RBFNN is employed. Additionally, an adaptive compensator that is introduced to the controller compensates for the influence of friction terms and the network reconstruction error. The stability of overall system is analyzed to show the asymptotic as well as finite-time convergence of the tracking errors. The effectiveness as well as robustness of the proposed

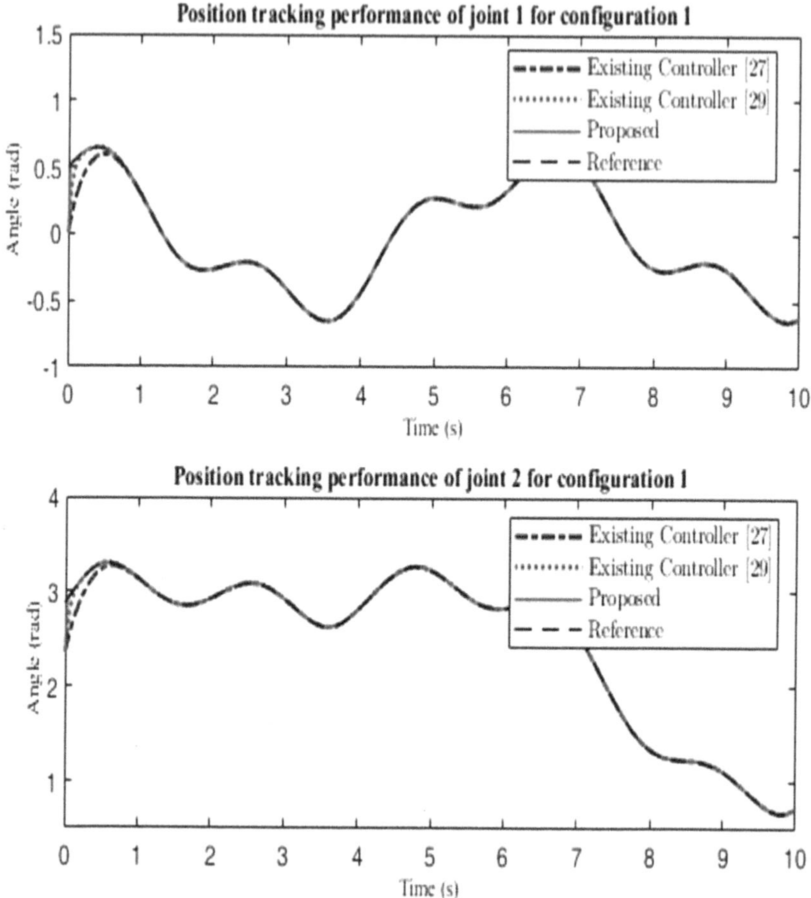

Figure 11.5 Joints 1 and 2's position tracking performance for configuration-I.

controller are shown with the help of simulation results in a comparative manner. We can infer from the entire theoretical analysis and the numerical simulation results obtained in this study that the suggested controller has improved controller efficiency and is capable of handling uncertainties and external disturbances in the system.

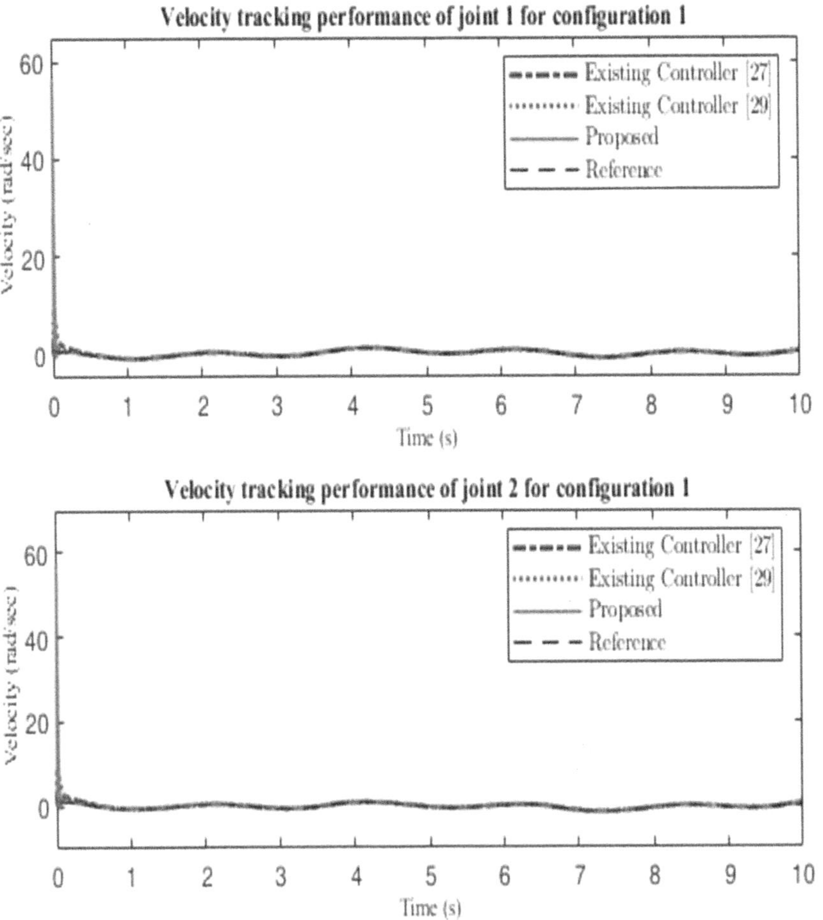

Figure 11.6 Joints 1 and 2's velocity tracking with existing and proposed approach with desired trajectory for configuration-1.

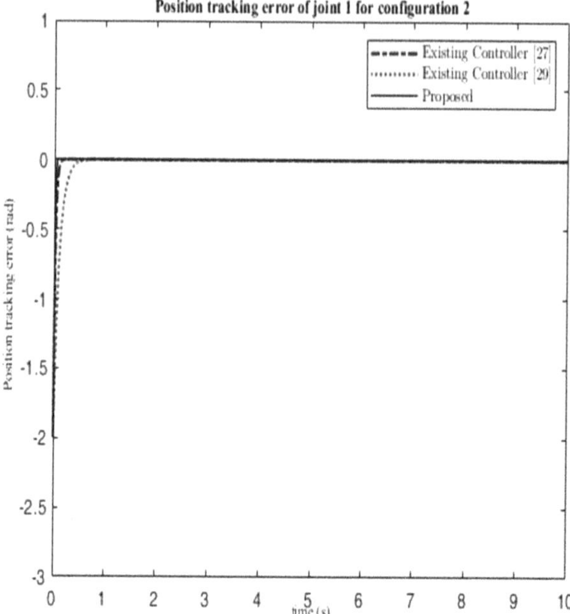

Figure 11.7 Position tracking error of joint 1 configuration-II.

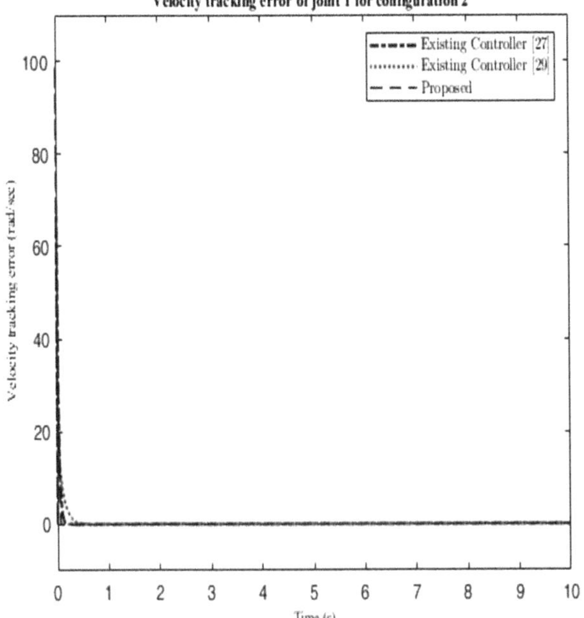

Figure 11.8 Velocity tracking error of joint 1 for configuration-II.

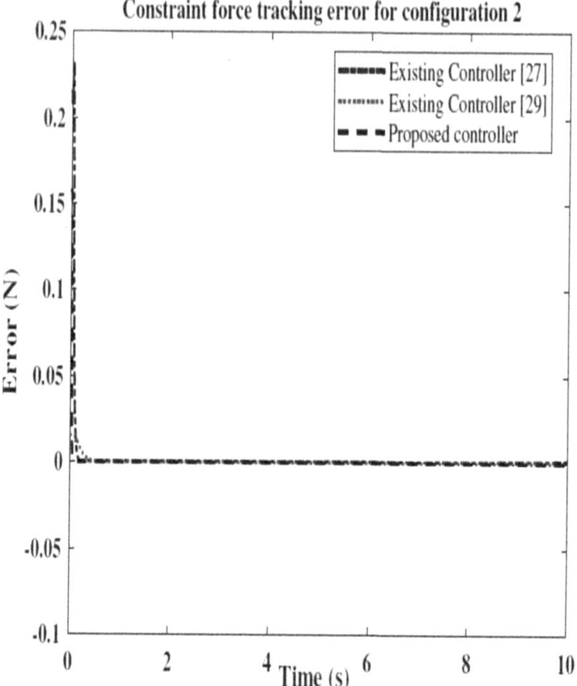

Figure 11.9 Force tracking error for configuration-II.

Table 11.1 Joint tracking errors' root mean square measure (RMS)

Controllers	L2[q₁]	L2[q₂]	L2[q₁]
Proposed Controller	0.0067	0.0696	0.0150
Existing Controller [27]	0.1784	1.2517	0.9571
Existing Controller [29]	0.0192	0.1398	0.0989

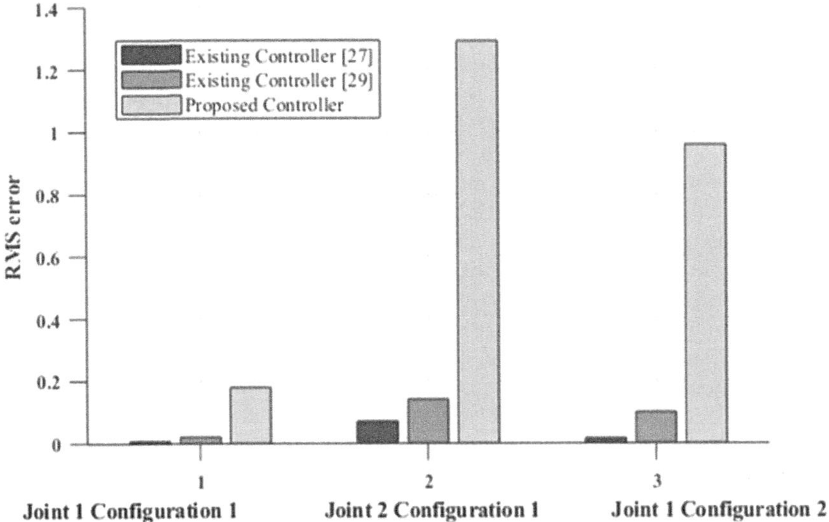

Figure 11.10 Bar diagram representing root mean square average of tracking errors.

REFERENCES

1. E. Ogas, L. Avila, G. Larregay, D. Moran, Object grasping with a robot arm using a convolutional network, International Journal of Mechatronics and Automation 7(3) (1993) 113–121.
2. A. Rodkina, On rate control of n-link manipulator robot, Mathematics and Computers in Simulation 45(3) (1998) 359–364.
3. B. Zhao, Y. Li, Local joint information based active fault tolerant control for reconfigurable manipulator, Nonlinear Dynamics 77(3) (2014) 859–876.
4. Valente, Reconfigurable industrial robots: A stochastic programming approach for designing and assembling robotic arms, Robotics and Computer-Integrated Manufacturing 41 (2016) 115–126.
5. A. Pamecha, I. Ebert-Uphoff, G.S. Chirikjian, Useful metrics for modular robot motion planning, IEEE Transactions on Robotics and Automation 13(4) (1997) 531–545.
6. C.J.J. Paredis, H.B. Benjamin, P.K. Khosla, A rapidly deployable manipulator system, Robotics and Autonomous Systems 21(3) (1996) 289–301.
7. S. Ahmad, H. Zhang, G. Liu, Distributed fault detection for modular and reconfigurable robots with joint torque sensing: A prediction error based approach, Mechatronics 23 (2013) 607–616.
8. W.W. Melek, A.A. Goldenberg, Neurofuzzy control of modular and reconfigurable robots, IEEE/ASME Transactions on Mechatronics 8(3) (2003) 381–389.

9. A. Giusti, M. Althoff, Automatic centralized controller design for modular and reconfigurable robot manipulators, IEEE/RSJ International Conference on Intelligent Robots and Systems (IROS) (2015) 3268–3275.

10. Y. Li, G. Ding, B. Zhao, Decentralized adaptive neural network sliding mode position/force control of constrained reconfigurable manipulators, Journal of Central South University 23 (2016) 2917–2925.

11. Y. Li, Z. Lu, F. Zhou, B. Dong, K. Liu, Y. Li, Adaptive sliding mode decentralized control for modular and reconfigurable robots with torque sensor, 2019 Chinese Control And Decision Conference (CCDC), Nanchang, China (2019) 5132–5137. doi: 10.1109/CCDC.2019.8832343

12. F. Zhou, Y. Li, G. Liu, Robust decentralized force/position fault-tolerant control for constrained reconfigurable manipulators without torque sensing, Nonlinear Dynamics 89 (2017) 955–969.

13. B. Ma, B. Dong, F. Zhou, Y. Li, Adaptive dynamic programming-based fault-tolerant position-force control of constrained reconfigurable manipulators, IEEE Access 8 (2020) 183286–183299.

14. T. An, Y. Qin, S. Wang, F. Zhou, K. Liu, B. Dong, Model free optimal integral sliding mode control for reconfigurable manipulators based on adaptive dynamic programming, Chinese Automation Congress (CAC) (2018) 3457–3463.

15. G. Wang, B. Dong, S. Wu, Y. Li, Sliding mode position/force control for constrained reconfigurable manipulator based on adaptive neural network, International Conference on Control, Automation and Information Sciences (ICCAIS) (2015) 96–101.

16. Y. Du, Q. Zhu, Decentralized adaptive force/position control of reconfigurable manipulator based on soft sensors, Proceedings of the Institution of Mechanical Engineers, Part I: Journal of Systems and Control Engineering 232 (2018) 381–389.

17. D.J.F. Heck, A. Saccon, N. Wouw, H. Nijmeijer, Guaranteeing stable tracking of hybrid position–force trajectories for a robot manipulator interacting with a stiff environment, Automatica 63 (2015) 235–247.

18. G. Liu, S. Abdul, A.A. Goldenberg, Distributed control of modular and reconfigurable robot with torque sensing, Robotica 26 (2008) 75–84.

19. B. Dong, K. Liu, Y. Li, Decentralized control of harmonic drive based modular robot manipulator using only position measurements: Theory and experimental verification, Journal of Intelligent & Robotic Systems 33 (2017) 3–18.

20. Y. Liu, S. Lu, S. Tong, Neural network controller design for an uncertain robot with time-varying output constraint, IEEE Transactions on Systems, Man, and Cybernetics: Systems (2016) 1–9.

21. B. Dong, Y. Li, Decentralized integral nested sliding mode control for time-varying constrained modular and reconfigurable robot, Advances in Mechanical Engineering 7(1)(2015) 1–15.

22. Y.D. Song, S. Zhou, Tracking control of uncertain nonlinear systems with deferred asymmetric time-varying full state constraints, Automatica 98 (2018) 314–322.

23. S.M. Lu, D.P. Li, Y.J. Liu, Adaptive neural network control for uncertain time-varying state constrained robotics systems, IEEE Transactions on Systems, Man, and Cybernetics: Systems (2017) 1–8.

24. B. Dong, Y. Li, Decentralized reinforcement learning robust optimal tracking control for time varying constrained reconfigurable modular robot based on ACI and Q-function, Mathematical Problems in Engineering 7(1) (2013) 1–16.
25. L. Zhu, Y. Li, Decentralized adaptive neural network control for reconfigurable manipulators, 2010 Chinese Control and Decision Conference (2010) 1760–1765.
26. K. Rani, N. Kumar, Design of intelligent optimal controller for hybrid position/force control of constrained reconfigurable manipulators, Journal of Ambient Intelligence and Humanized Computing 14 (2023) 13421–13432.
27. Y. Liu, B.Zhou, Y. Li, Adaptive neural network position/force hybrid control for constrained reconfigurable manipulators, IEEE 17th International Conference on Computational Science and Engineering (2014) 38–43.
28. Y. Wu, R. Huang, X. Li, S. Liu, Adaptive neural network control of uncertain robotic manipulators with external disturbance and time-varying output constraints, Neurocomputing 323 (2019) 108–116.
29. N. Kumar, M. Rani, A new hybrid force/position control approach for time-varying constrained reconfigurable manipulators, ISA Transactions 110 (2021) 138–147.
30. Y. Tang, Terminal sliding mode control for rigid robot, Automatica 34 (1998) 51–56.
31. J. Park, J.W. Sandberg, Universal approximation using radial-basis function networks, Neural Computing 3 (1991) 246–257.

Chapter 12

Horizons of smart data-driven technologies in solar photovoltaic glass illuminating path to smart cities

Skylines SDG 13 (Climate Action) Eliminating Environmental Snags

Bhupinder Singh, Komal Vig, and Christian Kaunert

12.1 INTRODUCTION AND BACKGROUND

The smart and sustainable cities' functional aspects, big data, and Internet of Things (IoT) integration are essential components. A data-driven approach is being progressively adopted by urban operations and planning which results in the development of data-driven urbanism (Kilkis et al., 2023). This offers a chance to use routinely gathered data to develop real-time models of smart, sustainable cities. These models make it possible to continuously monitor, comprehend, analyze, and develop cities that improve environmental health and energy efficiency (Razzaq et al., 2023). In this sense, sophisticated decision support systems are new urban intelligence functions. Still, the majority of studies that have been done thus far have been on data-driven technologies in smart cities, especially when it comes to the social and economic aspects of them. There has not been enough research done on these technologies' significant and cooperative impacts on environmental sustainability (Aljaghoub et al., 2022). This knowledge gap also encompasses eco-cities and other sustainable urban areas. With exploring the possibilities and function of data-driven smart solutions in boosting environmental sustainability within the context of smart and sustainable cities, this chapter focuses on closing this gap and defines the term "environmentally data-driven smart sustainable cities enabling SDG 13 (Climate Action) Eliminating Environmental Snags" (Izam et al., 2022).

Smart grids, smart meters, smart buildings, smart environmental monitoring, and smart urban metabolism are the main data-driven smart solutions used to improve environmental sustainability in eco-cities and smart cities alike (Li et al., 2013). As they work together and interact, these solutions have a definite synergy that produces combined impacts that outweigh the sum of their separate contributions, especially in terms of environmental advantages (Ludin et al., 2018). The concept

DOI: 10.1201/9781003530077-12

of enhancing energy efficiency, lowering pollution levels, implementing renewable energy sources, and giving real-time feedback on energy flows with high temporal and geographical resolutions are all addressed. The smart city is more environmentally sustainable: the city has a long history of environmental activities, strong environmental legislation, progressive environmental performance, adherence to strict environmental standards and ambitious goals (Weckend et al., 2016). The adoption of data-driven technological solutions in the environmental and energy sectors is a common goal in the form of strategic planning and each city makes data-driven decisions that are specific to the environmental issues that it faces. Big data is the answer, but every city develops its own set of questions based on its own attributes, such as goals, rules, plans, initiatives, and priorities (Omer, 2009).

12.1.1 Data-driven technology: Overview and relevance

Data-driven decision-making involves the process of arriving at decisions by analyzing and interpreting information obtained from digital sources (Memon et al., 2020). The utilization of a data-driven approach offers several advantages to organizations, notably the following as shown in Figure 12.1.

Enhanced decision-making accuracy: By employing a more informed and reliable procedure, errors are minimized, leading to more precise decision-making.

Facilitation of agile strategy development: Data-driven tools contribute to increased productivity and efficiency, thereby reducing the time required for decision-making and promoting the development of more agile strategies.

Contribution to energy transition: The integration of technologies such as artificial intelligence (AI), blockchain, and big data not only optimizes processes but also facilitates the creation of more sustainable products, thus driving the energy transition.

The protection of the environment has become a top issue in the modern world, pushing us to look for creative solutions. Data science in particular has shown to be a vital ally in this endeavor (Shastri & Singh, 2022). With the increasing urgency of environmental issues, data science has established itself as a force to be reckoned with, capable of analyzing large datasets and providing insights into complex ecological dynamics. This newly acquired knowledge serves as the cornerstone for developing successful environmental protection programs (Wu et al., 2021). It can find hidden patterns and relationships by using data science to enable a thorough investigation of environmental data. This capacity is useful for identifying pollution sources, forecasting natural disasters, and protecting biodiversity. Effective resource management is essential to sustainable practices, and data science plays a critical role in this regard by predicting waste production, water use, and

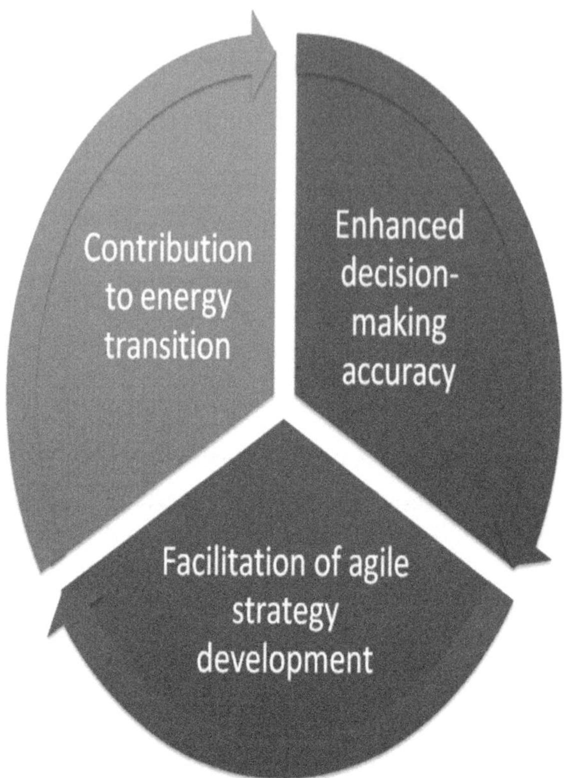

Figure 12.1 Advantages: Utilization of a data-driven.

energy demand (Mahmoudi et al., 2021). By enabling efficient resource allocation, this forecasting contributes to preventing resource depletion. Data science also helps with conservation efforts by predicting animal populations and detecting ecosystems that are at risk, which enables focused preservation efforts (Omer, 2008a). The data-driven technologies use predictive modeling methods to foresee the consequences of climate change, offering important insights into the condition of the environment in the future.

12.1.2 Significance of SDG 13 (Climate Action)

The cities have major negative environmental consequences due to their concentration of economic activity, intense resource usage, and heavy reliance on non-renewable energy sources (Zurita et al., 2018). The cities are encountering increasing difficulties in their pursuit of more ecologically sustainable arrangements, particularly in light of the present period of unparalleled urbanization and increased worldwide unpredictability (Njoh et al., 2019). This growing difficulty suggests that city administrations, especially in developed countries with superior technical and ecological capabilities, face substantial challenges arising from problems related to urbanization (Buchholz & Brandenburg, 2018). The increased energy use, pollution, deteriorating environmental conditions, poor planning and scheduling techniques, inadequate systems for making decisions, and social and economic inequality are some of these problems as depicted in Figure 12.2. In particular, when cities develop, a number of issues arise that threaten the environmental sustainability of cities by putting tremendous strain on urban systems and raising the need for energy resources and services (Shukla et al., 2018).

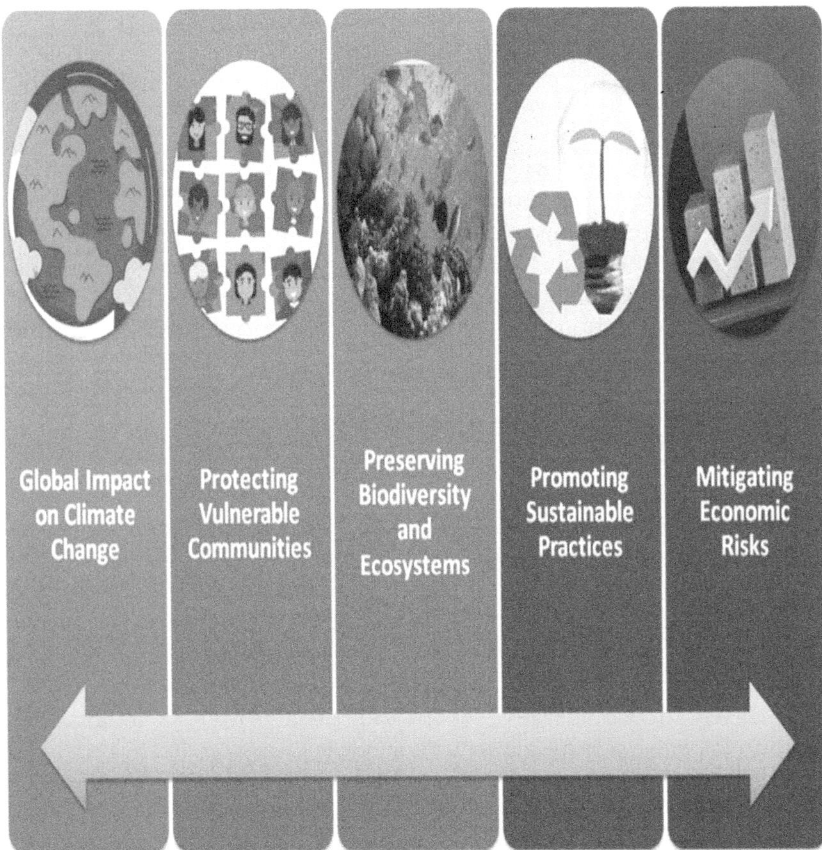

Figure 12.2 SDG 13: Significance.

The Sustainable Development Goal 13 (SDG 13) emphasizes the need for coordinated action to address the problems caused by climate change and its effects (Narayanan et al., 2019). All countries on all continents are currently experiencing the negative effects of climate change, which disrupts national economies and negatively damages people. The effects are being felt now and it is projected that in the future there will be greater financial and human costs (Rahman, 2017). The people are already experiencing major impacts such as changes in weather patterns, increasing sea levels, and more frequent and extreme weather events. The main cause of climate change is human activity, which produces greenhouse gas (GHG) emissions (Singh, 2019). These emissions are rising and have reached previously unheard-of amounts. In the absence of preventive action, it is predicted that the average surface temperature would rise globally throughout the course of the twenty-first century, possibly rising by more than 3°C in some areas (Hong et al., 2016).

12.1.3 Objectives of the chapter

This chapter has the following objectives as shown in Figure 12.3:

- explore the data-driven technology
- assess the current state of solar PV glass technology
- investigate the impact of smart data-driven technologies on efficiency and sustainability
- assimilation of data-driven to combat environmental problems
- propose solutions for eliminating environmental problems through innovative technologies

12.1.4 Structure of the chapter

This chapter comprehensively explores various dimensions of the applications of Smart Data Driven Technologies in Solar Photovoltaic Glass Illuminating Path to Smart Cities: Lensing SDG 13 (Climate Action) Eliminating Environmental Problems. Section 12.2 elaborates the Solar Photovoltaic Glass and its Role in Sustainable Energy Generation. Section 12.3 lays down the Applications of Smart Cities and Sustainable Urban Planning. Section 12.4 explores Smart Data-Driven Technologies. Section 12.5 travels the Comparative Analysis of Data-Driven Technologies and Traditional Solar PV Systems. Section 12.6 discusses the Evaluation of Environmental Benefits and Energy Output. Section 12.7 examines the Impact of Smart Technologies on the Efficiency of Solar PV Glass. Finally, Section 12.8 concludes the Chapter with Future Scope. The complete structure is designed in Figure 12.4.

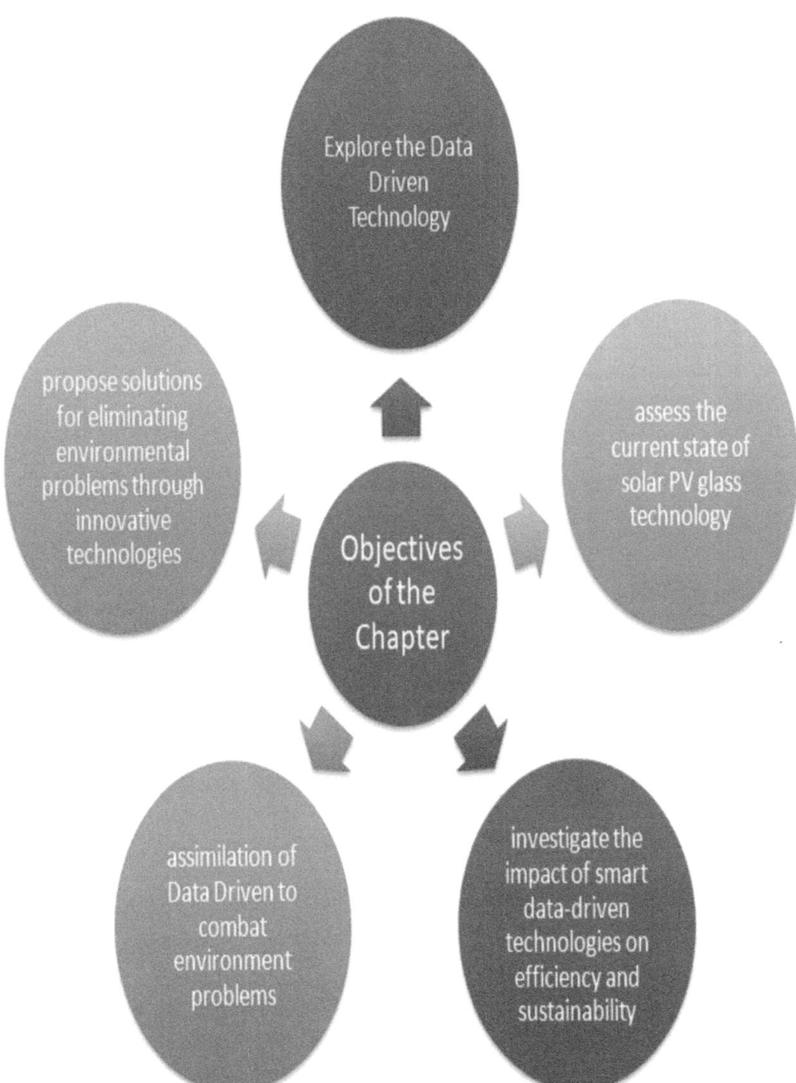

Figure 12.3 Objectives of the chapter.

12.2 SOLAR PHOTOVOLTAIC GLASS AND ITS ROLE IN SUSTAINABLE ENERGY GENERATION

Solar technologies are expected to experience a substantial surge in growth over the coming decade. Glasses are assuming a crucial role as transparent

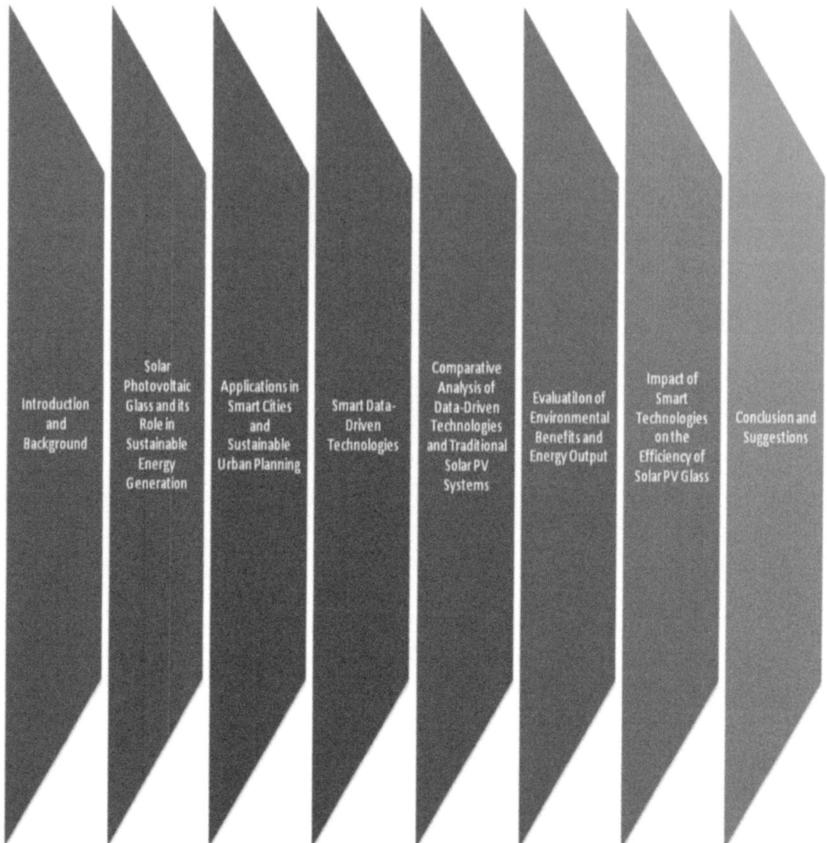

Figure 12.4 Structure of the chapter.

materials in both photovoltaic (PV) cells and concentrating solar power systems (Fereshtehpour et al., 2021). It is characterized by a brief energy payback time and environmental compatibility as glasses prove to be suitable materials for sustainable energy concepts. As the need for clean and renewable energy sources grows worldwide, solar PV glass is essential to the advancement of sustainable energy generation (Dhaouadi et al., 2021). This particular glass serves as an essential part of PV systems, efficiently absorbing solar energy and transforming it into electrical power to support sustainable power production (Herman, 2018). There are recent solar applications that simply highlight the significant enhancement made in energy efficiency for solar conversion systems achieved through the surface structuring and coating of glasses (Morelli et al., 2022). Because solar energy provides a clean, abundant and secure source for future renewable technologies, solar cells are seen as very potential answers to the world's energy problems. An attractive alternative for raising the energy conversion efficiency of single-gap solar cells is the intermediate band solar cell. In this area, methods like highly mismatched semiconductor alloys and quantum dot (QD) nanostructures are being investigated to increase energy conversion efficiency (Herrschel & Dierwechter, 2018).

Although the world's need for energy is growing, reserves of conventional energy sources like fossil fuels are progressively running out. The usage of fossil fuels results in the release of pollutants and GHGs like CO_2, which worsen air quality and pose a threat to public health (Cattaneo et al., 2015). So, based on current patterns in economic growth, the world's energy needs are projected to reach 28 TW in 2050 and 46 TW in 2100. This energy requirement should be mostly satisfied by renewable, ecologically beneficial sources. To build future renewable and sustainable technologies it is imperative that new energy sources be explored (Bellinkx, 2020).

It identifies encapsulated glass-to-glass PV modules and solar photocatalytic glass surfaces as integral components of green architecture (Omer, 2008b), seamlessly integrating renewable power generation with the reduction of air pollutants in urban environments (Hoornweg et al., 2011). There are emerging solar technologies for power generation such as transparent PV modules, solar chimneys, and thermoelectric systems that have the potential to become prominent areas of application for solar glass in the future.

12.3 SMART DATA-DRIVEN TECHNOLOGIES IN SOLAR PHOTOVOLTAIC GLASS: APPLICATIONS IN SMART CITIES AND SUSTAINABLE URBAN PLANNING

Solar PV glass is primarily used as a transparent component for PV cells which are often incorporated into solar panels. With the use of solar PV glass, the panels convert sunlight into electrical power; these cells provide a clean, sustainable energy source (Belli et al., 2020). PV glass stands out for its unique qualities such as its durability and transparency which make it an essential component in the construction of solar modules (Campisi et al., 2021). Smart cities represent urban areas that utilize data and technology to enhance the well-being of their residents. They emphasize sustainability, efficiency, and connectivity as these cities strive to optimize resource utilization, improve public services and cultivate a more enjoyable living environment (Seuwou et al., 2020). The integration of PV glass plays a crucial role in achieving these goals because traditionally, solar power generation involves expansive solar arrays located on the outskirts of cities. However, PV glass introduces innovative solar panels seamlessly incorporated into the design of buildings, infrastructure and urban landscapes (Fay, 2012). This integration allows structures to generate their own clean energy, contributing to a more sustainable environment while aesthetically blending with the surrounding architectural context.

The major factor in the advancement of sustainable urban planning and smart cities is solar PV glass. Its uses support the objectives of building greener, more efficient, and livable cities by improving the visual appeal and general resilience of urban settings in addition to optimizing energy usage (Dhakal & Chevalier, 2017). There are multifarious applications in Smart Cities and Sustainable Urban Planning concerning Smart Data-Driven Technologies in Solar Photovoltaic Glass. They are listed below, as shown in Figure 12.5.

Monitoring Energy in Real Time: Solar PV glass is equipped with intelligent, data-driven technologies that enable real-time energy production monitoring (LeBlanc et al., 2009). With this capability, energy consumption

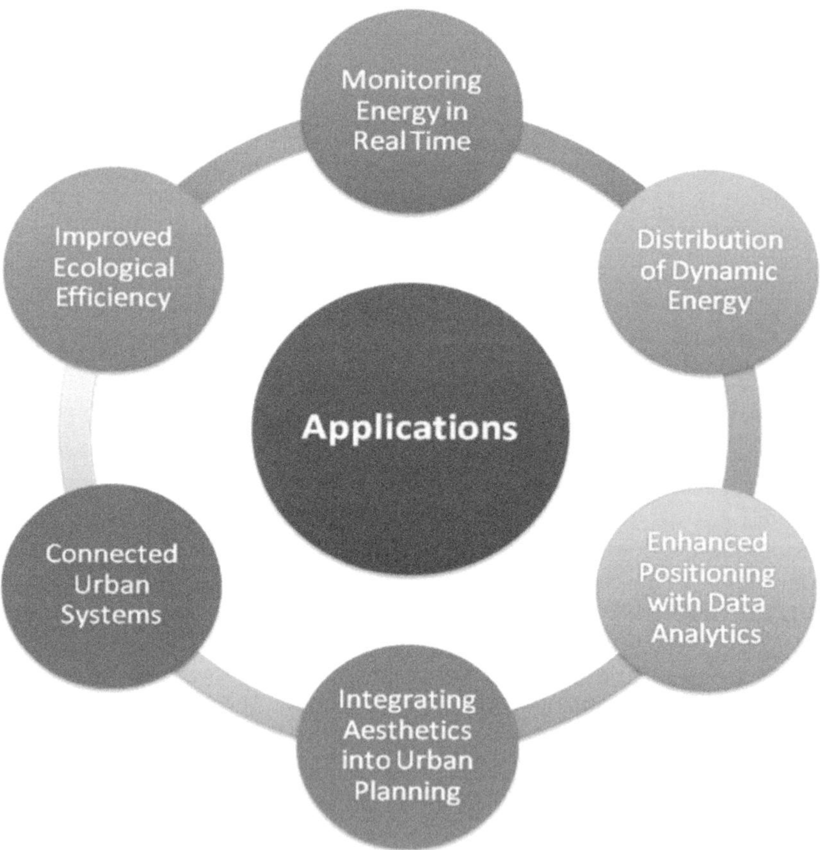

Figure 12.5 Applications in smart cities and sustainable urban planning.

in smart cities may be optimally managed and optimized in response to variations in demand and environmental factors.

Distribution of Dynamic Energy: The integration of data-driven technology with solar PV glass enables dynamic modifications to the distribution of energy. As the demands of various urban sectors change, smart cities may strategically distribute energy resources to suit those needs which makes the energy infrastructure more responsive and adaptable (Bhushan et al., 2020).

Enhanced Positioning with Data Analytics: Urban planners are assisted in placing solar panels as efficiently as possible by data analytics tools connected to smart PV glass. The accurate environmental data analysis allows solar panels to be strategically positioned for optimal efficiency which increases the total efficacy of renewable energy production (Afshan et al., 2022).

Integrating Aesthetics into Urban Planning: Solar panels are smoothly integrated into urban infrastructure and architectural designs by the use of smart PV glass (Nilsson et al., 2018). This adds to the allure of smart cities by encouraging sustainable energy solutions and improving the architectural design of public areas and buildings.

Connected Urban Systems: The greater goal of networked urban systems is aligned with the uses of smart data-driven technology in solar PV glass. Through this integration, a comprehensive and effective urban environment is promoted by fostering cooperation between various smart city components, including energy management, transportation, trash disposal and public services.

Improved Ecological Efficiency in City Development: Because smart PV glass offers a clean and renewable energy source, it is essential to the advancement of sustainable urban development (Bak et al., 2017). By incorporating data-driven technology, solar energy systems may efficiently reduce carbon footprints which is in line with the objectives of developing resilient and ecologically conscious metropolitan environments.

12.4 SMART DATA-DRIVEN TECHNOLOGIES: ROLE OF DATA ANALYTICS AND SENSORS IN SOLAR ENERGY SYSTEMS

Climate change and advancing toward a sustainable future now heavily depend on renewable energy. Alternative energy sources such as biomass, solar, wind or geothermal energy have far smaller environmental effect and may be used in a variety of settings and situations (Bhattacharya et al., 2021). They also offer diversity and plenty. Solar PV energy is the most constant performer. It is a clean, easily accessible resource that is expected to account for the increase in renewable energy over the coming years (Singh, 2022a). The importance of data analytics in optimizing solar energy generation and efficiency is rapidly expanding as a result of the global effort to fully realize its potential.

By analyzing and interpreting large-scale statistics, data analytics applied to solar energy allows businesses to optimize electricity generation (Parasher et al., 2019). Real-time solar component monitoring, proactive maintenance, accurate energy forecasts, problem detection, energy consumption analysis and cost savings are the means by which this objective is accomplished. The goal of using data analytics in the solar industry is to increase energy output and operational effectiveness (Ejaz & Anpalagan, 2019). Solar plants may use this program to precisely predict their energy output and optimize their operations for maximum efficiency. By employing sophisticated analytical algorithms these systems are able to predict factors such as meteorological conditions, energy storage requirements, and power generating requirements, which in turn creates a more adaptable and efficient power grid (Wiek et al., 2011). For solar energy companies, data analytics offers a wide range of benefits, including financial savings, productivity, accuracy, automation and environmental sustainability (Riffat et al., 2016). These businesses may improve operational procedures and performance via the use of advanced analytical techniques, which lowers costs and boosts profitability (Bamwesigye & Hlavackova, 2019). The future generations will have a more sustainable and eco-friendlier future because of the adoption of data-driven initiatives in the solar energy sector. To maximize the benefits of data-driven solar energy and create a cleaner and more optimistic future it is vital that we embrace its potential (Singh, 2023).

12.5 EVALUATION OF ENVIRONMENTAL BENEFITS AND ENERGY OUTPUT: SOLAR PHOTOVOLTAIC GLASS ILLUMINATING PATH TO SMART CITIES

The important step in illuminating the way to smart cities is the study of solar PV glass's energy output and environmental advantages (Singh, 2022b). The incorporation of solar PV glass appears as a transformational option as we seek sustainable living while navigating the challenges of urban growth. The main advantage for the environment is the decrease in GHG emissions (Singh & Kaunert, 2024). Solar PV glass, in contrast to conventional energy sources, uses sunshine to create electricity, reducing the need for fossil fuels and the carbon impact that comes with producing power conventionally. This shift is in line with the international agreement to mitigate climate change and lessen its effects on the environment (Mariotti et al., 2020).

The way solar PV glass produces energy is a major factor in how smart cities are developing. These glass installations aid in the production of clean, renewable energy by absorbing and converting solar radiation into electrical power (Rashid Khan et al., 2021). This is complemented by developments in energy storage technology which provide a steady power supply even in low-sun seasons (Mirjalili et al., 2023). The incorporation of solar PV glass into urban architecture is an example of a forward-thinking approach to energy generation, which is why efficiency and sustainability are given top priority in smart cities (Sirin et al., 2023). The improved aesthetics of metropolitan areas must be taken into consideration when assessing the advantages to the environment. The smooth integration of solar PV glass into building designs and infrastructure not only contributes to a cleaner environment but also harmonizes with the cityscape (Hunt et al., 2009). These works' striking aesthetics not only demonstrate a dedication to sustainability but also provide a benchmark for contemporary urban planning.

Thoroughly assessing the viability of sustainable technology from an economic standpoint is essential in the development of smart cities. Long-term cost reductions are possible with solar PV glass because of its capacity to produce energy and lessen reliance on conventional power systems (Li et al., 2023). The initial costs of these technologies are frequently covered by lower energy costs and, occasionally, by excess energy produced that may be returned to the grid (Fassbender et al., 2022). An encouraging route to the development of smart cities is shown by the assessment of solar PV glass's energy production and environmental advantages. Apart from being an eco-friendly decision, the use of solar PV glass in cityscapes signifies a flexible and sustainable strategy for supplying contemporary cities with the energy they require (Ragheb et al., 2016).

12.6 IMPACT OF SMART TECHNOLOGIES ON THE EFFICIENCY OF SOLAR PV GLASS

Solar power has become a very sustainable renewable energy source in the quest to find the ideal balance between economic growth, environmental sustainability and energy security. To meet the worldwide challenge of balancing these factors there are significant research and development expenditures have been made in the solar energy sector which has led to continuous innovation (Ambati et al., 2022). The 1960s saw the advent of solar technology and the first conception of solar PV as a future idea. Since then, there have been significant changes. Significant advancements made in technology have resulted in cost-effectiveness, enhanced dependability and expandability, making solar energy a more desirable option for renewable energy use (Dimitriou & Karagkouni, 2022). The industry has grown at an exponential rate as a result of this change. A number of exciting technologies are either in the process of being developed or have already been released into the market; they are expected to have a gradual influence on the growth of the solar energy industry. There are many new goods and concepts that are being developed including bigger, stronger, and more effective solar panels (Maghrabie et al., 2021).

A significant development is the Internet of Energy (IoE), a subdomain that focuses on power supply and generation inside the larger IoT architecture. IoE represents a move toward a more sustainable power infrastructure

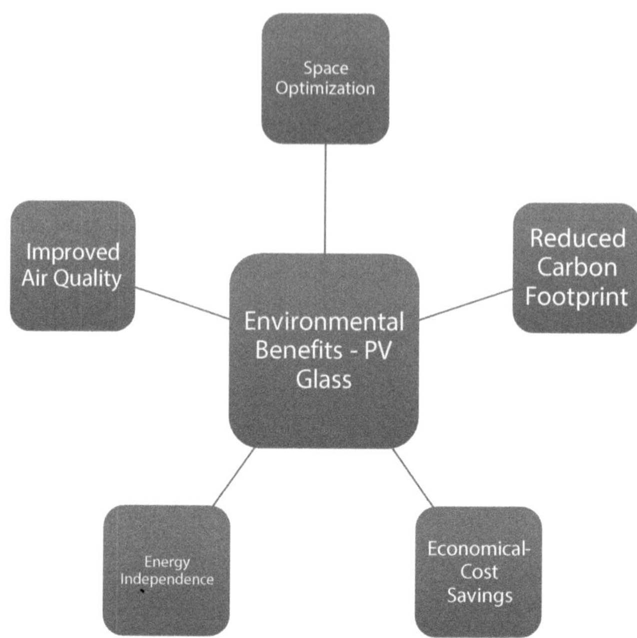

Figure 12.6 Benefits of solar photovoltaic glass illuminating path to smart cities: Skylines SDG 13.

where energy is used at the time and place of creation and is strongly aligned with the idea of energy decentralization (Jayachandran et al., 2022). The new technology platforms must be managed with a high level of automation due to this paradigm change. AI is crucial when combined with other new developments in technology. With the help of this technical layer, solar energy providers are able to make data-driven decisions in real time and do predictive maintenance. It is anticipated that the solar energy landscape would see increased efficiency and concurrently strengthened reliability (Boronuosi et al., 2023). Real benefits which are given in Figure 12.6 are elaborated as below.

12.6.1 Environmental impact

Environmental effects are modifications to the built environment or the natural world brought about by a particular activity that has an adverse effect on the air, soil, water, wildlife, aquatic life and the inhabitants of an ecosystem. The action's possible short- or long-term repercussions, pollution, contamination or harm are all considered environmental impacts (Fthenakis & Moskowitz, 1995). These effects frequently have a direct bearing on problems pertaining to general quality of life and public health. The reductions in fine particulate matter and carbon monoxide levels are examples of stricter laws that have been successful in lowering pollution levels (Ahmad et al., 2024).

PV technologies have evolved that offer more affordable production techniques than classic silicon crystalline-based modules even though silicon crystalline PV modules are still widely utilized worldwide. These consist of cadmium telluride, copper indium selenide (CIS), and amorphous silicon. To remain compatible with these new or improved technologies, new standards and testing procedures are being created (Omer, 2008b). The continuous depletion of fossil fuels has fueled the rise in popularity of renewable energy in recent decades, highlighting the growing significance of electrical energy. Within this framework, the PV sector has been expanding steadily and quickly. The PV systems have the potential to greatly increase the amount of power produced worldwide (Khalid et al., 2023).

12.6.2 Reduction of carbon footprint

Solar PVs facilitate the production of clean, sustainable energy as they play a critical role in lowering carbon emissions. Due to the fact that the usage of traded solar cells and modules lowers emissions of CO_2 and other local pollutants, the worldwide trade of solar PV goods has resulted in significant reductions in emissions (Khan et al., 2024). Solar PV systems provide environmental benefits including reduced emissions of carbon and GHGs and the preservation of precious water resources. In contrast to fossil fuel energy resources, PV systems have a far smaller carbon footprint. This compares to emissions from burning oil by orders of magnitude (Gervais et al., 2021). The cost-effectiveness of solar PV has increased in comparison to non-renewable energy sources due to the declining cost of solar energy, which increases the technology's potential for wider adoption and further lowers carbon emissions. Solar PV is essential in combating climate change as it offers a sustainable and clean substitute for fossil fuels which eventually lowers carbon emissions (Fthenakis & Moskowitz, 2000).

12.6.3 Assessment of potential for mitigating climate change

Climate change stands as a significant obstacle to worldwide food production and security as exerting detrimental impacts on agricultural output in various ways. There are adverse effects observed at present including diminished crop yields, a decline in the nutritional quality of crucial cereals, and decreased productivity in livestock (Ferrari & Cirisano, 2020). When compared to other energy sources, solar PV systems have a considerable positive environmental impact which supports sustainable development and the ecological benefits of human activity. To fulfill the increasing electricity demand and maintain environmental quality, technical advancements are essential (Alkhalidi et al., 2019). These advancements reduce carbon emissions by switching from carbon-emitting fossil fuels to renewable energy sources. As large amounts of land are needed for the construction of these facilities globally, it presents a number of ecological difficulties. It offers information on a number of environmental concerns such as those pertaining to land use, human health, animals, vegetation and the environment's overall influence (Bakhiyi et al., 2014). When compared to a thermal power plant, the land area needed for a PV system is numerically speaking, either less than or equivalent to that needed to generate one kilowatt-hour of power.

12.6.4 Economic viability: Analysis of economic feasibility of implementing smart data-driven technologies

The important factor that has to be carefully considered to determine if solar PV glass is feasible is its economic viability (Tawalbeh et al., 2021). The initial cost of producing and installing solar PV glass which includes costs for the integration of PV cells and the creation of specialized glass materials is a crucial factor. As, economic viability is also largely dependent on how durable and efficient the technology is longer lifespans and more efficiency translate into better returns on investment over the course of the system's lifetime (Zhang et al., 2023). The important factors to take into account include solar PV glass's potential energy output and its capacity to produce electricity effectively in a variety of environmental settings (Fraser & Girtan, 2023). The current energy costs and government incentives or subsidies for renewable energy technology also have an impact on economic viability and can affect the payback period and total financial returns.

12.7 CONCLUSION AND FUTURE SCOPE

Data analytics within the concept of solar energy encompasses the utilization of methods for collecting, analyzing and interpreting data to derive meaningful insights, recognize patterns and inform decisions based on data. This practice is typically employed across various facets of solar energy including the assessment of system performance, energy generation, financial analysis, optimization of maintenance and integration into the grid. The process involves harnessing data from diverse sources such as solar panels, sensors, weather stations, smart meters, and operational systems. The monitoring of solar panel performance becomes more accurate and effective when sophisticated analytics are used. The real-time analysis is made feasible by the integration of sensors and data analytics allowing for the early identification of errors or inefficiencies. Each solar panel's temperature, irradiance, voltage and current are measured via sensors. The evaluation of ideal panel performance is made possible by these data measures. Any discrepancies can be used to identify the problem and take proactive measures to solve them before it becomes more serious. Utilizing modern analytics makes it possible to monitor solar panels more efficiently, which maximizes production of the solar panels.

REFERENCES

Afshan, S., Ozturk, I., & Yaqoob, T. (2022). Facilitating renewable energy transition, ecological innovations and stringent environmental policies to improve ecological sustainability: Evidence from MM-QR method. *Renewable Energy*, *196*, 151–160.

Ahmad, A., Prakash, O., Kausher, R., Kumar, G., Pandey, S., & Hasnain, S. M. (2024). Parabolic trough solar collectors: A sustainable and efficient energy source. *Materials Science for Energy Technologies*, 7, 99–106.

Aljaghoub, H., Abumadi, F., AlMallahi, M. N., Obaideen, K., & Alami, A. H. (2022). Solar PV cleaning techniques contribute to Sustainable Development Goals (SDGs) using Multi-criteria decision-making (MCDM): Assessment and review. *International Journal of Thermofluids*, *16*, 100233.

Alkhalidi, A., Khawaja, M. K., Kelany, A., & Ghaffar, A. (2019). Investigation of repurposed material utilization for environmental protection and reduction of overheat power losses in PV panels. *International Journal of Photoenergy*, *2019*.

Ambati, M. S. K., Dalapati, G. K., Lawaniya, R., Samanta, A., Kumar, A., & Chakrabortty, S. (2022). Photovoltaic/catalysis integration toward a 100% renewable energy infrastructure. In *Sulfide and Selenide Based Materials for Emerging Applications* (pp. 553–582). Elsevier.

Bak, C., Bhattacharya, A., Edenhofer, O., & Knopf, B. (2017). Towards a comprehensive approach to climate policy, sustainable infrastructure, and finance. *Economics*, *11*(1), 20170033.

Bakhiyi, B., Labrèche, F., & Zayed, J. (2014). The photovoltaic industry on the path to a sustainable future—Environmental and occupational health issues. *Environment International*, *73*, 224–234.

Bamwesigye, D., & Hlavackova, P. (2019). Analysis of sustainable transport for smart cities. *Sustainability*, *11*(7), 2140.

Belli, L., Cilfone, A., Davoli, L., Ferrari, G., Adorni, P., Di Nocera, F., & Bertolotti, E. (2020). IoT-enabled smart sustainable cities: Challenges and approaches. *Smart Cities*, *3*(3), 1039–1071.

Bellinkx, V. (2020). *The role of sustainable development law in energy transition* (Doctoral dissertation, University of Antwerp).

Bhattacharya, A., Ivanyna, M., Oman, W., & Stern, N. (2021). *Climate action to unlock the inclusive growth story of the 21st Century* IMF Working Papers 2021/147, International Monetary Fund.

Bhushan, B., Khamparia, A., Sagayam, K. M., Sharma, S. K., Ahad, M. A., & Debnath, N. C. (2020). Blockchain for smart cities: A review of architectures, integration trends and future research directions. *Sustainable Cities and Society*, *61*, 102360.

Boronuosi, F., Aghababaei, S., Azad, S., Ameli, M. T., & Nazari-Heris, M. (2023). Building-Integrated Photovoltaic (BIPV) and Its Application, Design, and Policy and Strategies. In *Natural Energy, Lighting, and Ventilation in Sustainable Buildings* (pp. 91–109). Cham: Springer Nature Switzerland.

Buchholz, P., & Brandenburg, T. (2018). Demand, supply, and price trends for mineral raw materials relevant to the renewable energy transition wind energy, solar photovoltaic energy, and energy storage. *Chemie Ingenieur Technik*, *90*(1–2), 141–153.

Campisi, T., Severino, A., Al-Rashid, M. A., & Pau, G. (2021). The development of the smart cities in the connected and autonomous vehicles (CAVs) era: From mobility patterns to scaling in cities. *Infrastructures*, *6*(7), 100.

Cattaneo, G., Faes, A., Li, H. Y., Galliano, F., Gragert, M., Yao, Y., & Perret-Aebi, L. E. (2015). Lamination process and encapsulation materials for glass–glass PV module design. *Photovoltaics International*, *27*, 1–8.

Dhakal, K. P., & Chevalier, L. R. (2017). Managing urban stormwater for urban sustainability: Barriers and policy solutions for green infrastructure application. *Journal of Environmental Management*, *203*, 171–181.

Dhaouadi, R., Al-Othman, A., Aidan, A. A., Tawalbeh, M., & Zannerni, R. (2021). A characterization study for the properties of dust particles collected on photovoltaic (PV) panels in Sharjah, United Arab Emirates. *Renewable Energy*, *171*, 133–140.

Dimitriou, D., & Karagkouni, A. (2022). Due Diligence of Transport Infrastructure Operators Sustainability: A Circular Economy Driven Approach. *Frontiers in Sustainability*, *3*, 916038.

Ejaz, W., & Anpalagan, A. (2019). *Internet of things for smart cities: Technologies, big data and security* (pp. 1–15). Berlin/Heidelberg, Germany: Springer International Publishing.

Fassbender, E., Ludwig, F., Hild, A., Auer, T., & Hemmerle, C. (2022). Designing transformation: Negotiating solar and green strategies for the sustainable densification of urban neighbourhoods. *Sustainability*, *14*(6), 3438.

Fay, M. (2012). *Inclusive green growth: The pathway to sustainable development*. World Bank Publications, Washington, DC.

Fereshtehpour, M., Sabbaghian, R. J., Farrokhi, A., Jovein, E. B., & Sarindizaj, E. E. (2021). Evaluation of factors governing the use of floating solar system: A study on Iran's important water infrastructures. *Renewable Energy*, *171*, 1171–1187.

Ferrari, M., & Cirisano, F. (2020). High transmittance and highly amphiphobic coatings for environmental protection of solar panels. *Advances in Colloid and Interface Science*, *286*, 102309.

Fraser, R., & Girtan, M. (2023). A Selective Review of Ceramic, Glass and Glass–Ceramic Protective Coatings: General Properties and Specific Characteristics for Solar Cell Applications. *Materials*, *16*(11), 3906.

Fthenakis, V. M., & Moskowitz, P. D. (1995). Thin-film photovoltaic cells: Health and environmental issues in their manufacture use and disposal. *Progress in Photovoltaics: Research and Applications*, *3*(5), 295–306.

Fthenakis, V. M., & Moskowitz, P. D. (2000). Photovoltaics: Environmental, health and safety issues and perspectives. *Progress in Photovoltaics: Research and Applications*, *8*(1), 27–38.

Gervais, E., Herceg, S., Nold, S., & Weiß, K. A. (2021). Sustainability strategies for PV: Framework, status and needs. *EPJ Photovoltaics*, *12*, 5.

Herman, K. (2018). The role of technology in the global environment. In *Global Environmental Politics: Concepts, Theories and Case Studies* (pp. 155–177). Routledge.

Herrschel, T., & Dierwechter, Y. (2018). *Smart transitions in city regionalism: Territory, politics and the quest for competitiveness and sustainability.* Routledge.

Hong, J., Chen, W., Qi, C., Ye, L., & Xu, C. (2016). Life cycle assessment of multicrystalline silicon photovoltaic cell production in China. *Solar Energy*, *133*, 283–293.

Hoornweg, D., Sugar, L., & Trejos Gómez, C. L. (2011). Cities and greenhouse gas emissions: Moving forward. *Environment and Urbanization*, *23*(1), 207–227.

Hunt, D. V., Jefferson, I., Gaterell, M. R., & Rogers, C. D. (2009). Planning for sustainable utility infrastructure. *Proceedings of the Institution of Civil Engineers-Urban Design and Planning*, *162*(4), 187–201.

Izam, N. S. M. N., Itam, Z., Sing, W. L., & Syamsir, A. (2022). Sustainable development perspectives of solar energy technologies with focus on solar Photovoltaic—A review. *Energies*, *15*(8), 2790.

Jayachandran, M., Gatla, R. K., Rao, K. P., Rao, G. S., Mohammed, S., Milyani, A. H., & Geetha, S. (2022). Challenges in achieving sustainable development goal 7: Affordable and clean energy in light of nascent technologies. *Sustainable Energy Technologies and Assessments*, *53*, 102692.

Khalid, H. M., Rafique, Z., Muyeen, S. M., Raqeeb, A., Said, Z., Saidur, R., & Sopian, K. (2023). Dust accumulation and aggregation on PV panels: An integrated survey on impacts, mathematical models, cleaning mechanisms, and possible sustainable solution. *Solar Energy*, *251*, 261–285.

Khan, S., Sudhakar, K., Hazwan Yusof, M., & Sundaram, S. (2024). Review of building integrated photovoltaics system for electric vehicle charging. *The Chemical Record*, *24*(3), 1–29.

Kilkis, S., Krajacic, G., Duic, N., & Rosen, M. A. (2023). Sustainable development of energy, water and environment systems in the critical decade for climate action. *Energy Conversion and Management*, *296*, 117644.

LeBlanc, R. J., Matthews, P., & Richard, R. P. (Eds.). (2009). *Global atlas of excreta, wastewater sludge, and biosolids management: Moving forward the sustainable and welcome uses of a global resource.* Un-habitat.

Li, S., Ma, T., & Wang, D. (2023). Photovoltaic pavement and solar road: A review and perspectives. *Sustainable Energy Technologies and Assessments*, *55*, 102933.

Li, D. H., Yang, L., & Lam, J. C. (2013). Zero energy buildings and sustainable development implications–A review. *Energy*, *54*, 1–10.

Ludin, N. A., Mustafa, N. I., Hanafiah, M. M., Ibrahim, M. A., Teridi, M. A. M., Sepeai, S., & Sopian, K. (2018). Prospects of life cycle assessment of renewable energy from solar photovoltaic technologies: A review. *Renewable and Sustainable Energy Reviews*, *96*, 11–28.

Maghrabie, H. M., Elsaid, K., Sayed, E. T., Abdelkareem, M. A., Wilberforce, T., & Olabi, A. G. (2021). Building-integrated photovoltaic/thermal (BIPVT) systems: Applications and challenges. *Sustainable Energy Technologies and Assessments, 45*, 101151.

Mahmoudi, S., Huda, N., & Behnia, M. (2021). Critical assessment of renewable energy waste generation in OECD countries: Decommissioned PV panels. *Resources, Conservation and Recycling, 164*, 105145.

Mariotti, N., Bonomo, M., Fagiolari, L., Barbero, N., Gerbaldi, C., Bella, F., & Barolo, C. (2020). Recent advances in eco-friendly and cost-effective materials towards sustainable dye-sensitized solar cells. *Green Chemistry, 22*(21), 7168–7218.

Memon, S., Katsura, T., Radwan, A., Zhang, S., Serageldin, A. A., Abo-Zahhad, E. M., & Kiani, A. (2020). Modern eminence and concise critique of solar thermal energy and vacuum insulation technologies for sustainable low-carbon infrastructure. *International Journal of Solar Thermal Vacuum Engineering, 1*(1), 52–71.

Mirjalili, S. M. A., Aslani, A., & Zahedi, R. (2023). Towards sustainable commercial-office buildings: Harnessing the power of solar panels, electric vehicles, and smart charging for enhanced energy efficiency and environmental responsibility. *Case Studies in Thermal Engineering, 52*, 103696.

Morelli, G., Magazzino, C., Gurrieri, A. R., Pozzi, C., & Mele, M. (2022). Designing smart energy systems in an industry 4.0 paradigm towards sustainable environment. *Sustainability, 14*(6), 3315.

Narayanan, A., Mets, K., Strobbe, M., & Develder, C. (2019). Feasibility of 100% renewable energy-based electricity production for cities with storage and flexibility. *Renewable Energy, 134*, 698–709.

Nilsson, M., Chisholm, E., Griggs, D., Howden-Chapman, P., McCollum, D., Messerli, P., & Stafford-Smith, M. (2018). Mapping interactions between the sustainable development goals: Lessons learned and ways forward. *Sustainability science, 13*, 1489–1503.

Njoh, A. J., Etta, S., Ngyah-Etchutambe, I. B., Enomah, L. E., Tabrey, H. T., & Essia, U. (2019). Opportunities and challenges to rural renewable energy projects in Africa: Lessons from the Esaghem Village, Cameroon solar electrification project. *Renewable Energy, 131*, 1013–1021.

Omer, A. M. (2008a). Energy, environment and sustainable development. *Renewable and Sustainable Energy Reviews, 12*(9), 2265–2300.

Omer, A. M. (2008b). Green energies and the environment. *Renewable and Sustainable Energy Reviews, 12*(7), 1789–1821.

Omer, A. M. (2009). Energy use and environmental impacts: A general review. *Journal of Renewable and Sustainable Energy, 1*(5), 1–29.

Parasher, Y., Singh, P., & Kaur, G. (2019). Green Smart Town Planning. *Green and Smart Technologies for Smart Cities* (pp. 19–41).

Ragheb, A., El-Shimy, H., & Ragheb, G. (2016). Green architecture: A concept of sustainability. *Procedia-Social and Behavioral Sciences, 216*, 778–787.

Rahman, M. W., Mahmud, M. S., Ahmed, R., Rahman, M. S., & Arif, M. Z. (2017, April). Solar lanes and floating solar PV: New possibilities for source of energy generation in Bangladesh. In *2017 Innovations in Power and Advanced Computing Technologies (i-PACT)* (pp. 1–6). IEEE.

Rashid Khan, H. U., Awan, U., Zaman, K., Nassani, A. A., Haffar, M., & Abro, M. M. Q. (2021). Assessing hybrid solar-wind potential for industrial decarbonization strategies: Global shift to green development. *Energies*, *14*(22), 7620.

Razzaq, I., Amjad, M., Qamar, A., Asim, M., Ishfaq, K., Razzaq, A., & Mawra, K. (2023). Reduction in energy consumption and CO_2 emissions by retrofitting an existing building to a net zero energy building for the implementation of SDGs 7 and 13. *Frontiers in Environmental Science*, *10*, 1028793.

Riffat, S., Powell, R., & Aydin, D. (2016). Future cities and environmental sustainability. *Future cities and Environment*, *2*(1), 1–23.

Seuwou, P., Banissi, E., & Ubakanma, G. (2020). The future of mobility with connected and autonomous vehicles in smart cities. *Digital Twin Technologies and Smart Cities*, 37–52.

Shastri, A., & Singh, B. (2022). Demystifying data justice: Legal response to India's privacy and security standards: Challenges in cloud computing. *ECS Transactions*, *107*(1), 179.

Shukla, A. K., Sudhakar, K., Baredar, P., & Mamat, R. (2018). Solar PV and BIPV system: Barrier, challenges and policy recommendation in India. *Renewable and Sustainable Energy Reviews*, *82*, 3314–3322.

Singh, B. (2019). Affordability of Medicines, Public Health and TRIPS Regime: A Comparative Analysis. *Indian Journal of Health and Medical Law*, *2*(1), 1–7.

Singh, B. (2022a). Relevance of Agriculture-Nutrition Linkage for Human Healthcare: A Conceptual Legal Framework of Implication and Pathways. *Justice and Law Bulletin*, *1*(1), 44–49.

Singh, B. (2022b). Understanding legal frameworks concerning transgender healthcare in the age of dynamism. *Electronic Journal of Social and Strategic Studies*, *3*, 56–65.

Singh, B. (2023). Blockchain technology in renovating healthcare: Legal and future perspectives. In *Revolutionizing Healthcare Through Artificial Intelligence and Internet of Things Applications* (pp. 177–186). IGI Global.

Singh, B., & Kaunert, C. (2024). Integration of Cutting-Edge Technologies such as Internet of Things (IoT) and 5G in Health Monitoring Systems: A Comprehensive Legal Analysis and Futuristic Outcomes. *GLS Law Journal*, *6*(1), 13–20.

Sirin, C., Goggins, J., & Hajdukiewicz, M. (2023). A review on building-integrated photovoltaic/thermal systems for green buildings. *Applied Thermal Engineering*, 120607.

Tawalbeh, M., Al-Othman, A., Kafiah, F., Abdelsalam, E., Almomani, F., & Alkasrawi, M. (2021). Environmental impacts of solar photovoltaic systems: A critical review of recent progress and future outlook. *Science of the Total Environment*, *759*, 143528.

Weckend, S., Wade, A., & Heath, G. A. (2016). *End of life management: Solar photovoltaic panels* (No. NREL/TP-6A20-73852). Golden, CO: National Renewable Energy Lab (NREL).

Wiek, A., Withycombe, L., Redman, C., & Mills, S. B. (2011). Moving forward on competence in sustainability research and problem solving. *Environment*, *53*(2), 3–13.

Wu, W., & Skye, H. M. (2021). Residential net-zero energy buildings: Review and perspective. *Renewable and Sustainable Energy Reviews*, *142*, 110859.

Zhang, H., Yu, Z., Zhu, C., Yang, R., Yan, B., & Jiang, G. (2023). Green or not? Environmental challenges from photovoltaic technology. *Environmental Pollution, 320*, 121066.

Zurita, A., Castillejo-Cuberos, A., García, M., Mata-Torres, C., Simsek, Y., García, R., & Escobar, R. A. (2018). State of the art and future prospects for solar PV development in Chile. *Renewable and Sustainable Energy Reviews, 92*, 701–727.

Chapter 13

Data-driven robotics

A performance analysis of wearable sensor-based wound assessment system for sensor data collection

Monica Bhutani, Monica Gupta, Tarun Jain, Tej Prakash Sharma, Akansha Solanki and Aavaig Malhotra

13.1 INTRODUCTION

An injury brought on by a decline in the patient's skin's structure and function is referred to as a wound. Wounds are often identified as cuts, scratches, and punctured skin (Feneley et al., 2015; Fujita et al., 2018). They can happen because of an accident, surgery, stitches, and sutures. Wounds that damage the outer skin of a human are usually severe, but they still need to be cleaned and treated well (Hawkes et al., 2017; Indujha et al., 2021). One may be required to visit a doctor after getting proper first aid for painful and infected wounds. With deep wounds, one should immediately see a doctor as it is impossible to close the wound without an expert attending (Feng et al., 2009; Khalil et al., 2019). Not attending to the wounds properly can lead to it becoming severe and causing infection, which may be fatal for a human (Greer et al., 2019; Haque et al., 2020; Lou et al., 2020).

13.1.1 Types of wounds

Wounds can be classified as under:

Acute wound: It is a new wound that has not yet moved through various stages of wound healing in a sequential manner. It often happens due to incision or trauma and heals timely without usually causing complications. They are generally treated with disinfectants and bandages (Hirose, 1985; Mukherjee et al., 2017).

Chronic wound: The wounds of this type fails to progress through the different phases of healing and shows no significant improvement over 30 days in a timely and orderly manner (Godfrey et al., 2008; Ohura et al., 2019).

Some of the factors that contribute to the chronicity of a wound may include the following –

- Pressure and trauma.
- Increased infection and bacterial load.

DOI: 10.1201/9781003530077-13

- Diseases like diabetes slow down the process of wound healing (Komori et al., 2019).
- Improper treatment and irregular tracking of the progress of wound healing.
- Aberrant or senescent cells.

Due to the fact that manual wound examination relies on individual analysis, which frequently produces inconsistent results, it is a challenging task (Sattar et al., 2019). Therefore, wound imaging can be used because of its potential to offer a standardized point of view (Hartmann et al., 2021). When discussing the analysis of chronic wounds, we can divide the methods into contact and non-contact methods (King et al., 2016; Kukreja et al., 2020).

13.1.2 Contact-based chronic wound analysis methods

- Manual planimetry using rulers
- Transparency tracing
- Color dye injection

In the past, these contact-based techniques were frequently employed as they were inconvenient for medical personnel and caused excruciating agony for patients (Banerjee et al., 2018; Chu and Patterson, 2018). These treatments lacked precision and accuracy because to the unevenness of the damage and the shape of the wound. As technology is always evolving, there are more and more applications for non-contact wound analysis (Townsend, 2000; Wada et al., 2010). Advances made in data processing have resulted in the expansion of digital imaging methods for wound assessment (Katzschmann et al., 2018; Periyasamy et al., 2012). Optical coherence, digital camera imaging, thermal imaging, laser Doppler imaging, confocal microscopy, and hyperspectral imaging are a few methods for visualizing wounds. Optical coherence tomography (OCT), ultrasonic imaging, near-infrared (NIR) reflectance spectroscopic imaging are other cutting-edge imaging methods (Kontoudis et al., 2019; Wang et al., 2015).

The challenges faced during wound treatment and assessment have led to practitioners adopting a holistic and systematic approach (Birglen, 2010). The approach should involve wound assessment at the initial and ongoing phases of the treatment. It has several purposes: providing baseline information against which progress can be monitored, suitable dressing can be selected, and correct measures can be undertaken (Corpin et al., 2019; Herlihy, 2006). Poor assessment of the wound can be fatal for the patient. Therefore, the best possible assessment method should be used (Hirayasu et al., 2018). Without the proper diagnosis, the course of treatment could be compromised, delaying recovery and leading to significant consequences. Therefore, correct wound assessment is required for adequate treatment of the wounds and should be an integral part of wound care practice (Kota et al., 2001; Galloway et al., 2016). For many reasons, wound assessment and pre- and post-wound treatment care can be difficult for nurses (Xu et al., 2021).

Some of the reasons are as follows:

- Even professionals have difficulty with wound healing because it is a complex process, especially in chronic wounds where the typical healing trajectory is not followed.
- Moisture, strain and, temperature, pH of the wounded area play a vital role in its healing.
- Any fluctuations in the parameters mentioned delay the healing process or may even lead to complexities.
- People with health issues like diabetes, cancer, etc., often take more than the usual time required for healing from an injury.

This work aims to design a system to solve the issue of timely wound assessment by the patients and concerned doctors/nurses. Timely analysis of various physical parameters around the wounded area can significantly help give the required attention to the injured area.

13.1.3 Novel contributions of this work

- The proposed system uses a Wi-Fi connection to connect with the phone of a patient/doctor and visually present data in the form of graphs that depict the variation of temperature/humidity vs. time.
- The module stores data in the EPROM so that even if Wi-Fi is disconnected, the previous data remains intact and appends with the graph as soon as the Wi-Fi connection is re-established.

13.2 LITERATURE SURVEY

In the field of wound assessment, traditional methods have relied heavily on visual inspection and manual measurements. However, these methods often suffer from inter-observer variability and subjective interpretation, leading to inconsistencies and limitations in documenting wounds (Yang et al., 2009). This has sparked a growing interest in developing advanced wound assessment systems that can be attached to the outer skin of the human body, allowing for continuous tracking of the healing process (Zhang et al., 2019).

To address the challenges faced by current wound assessment systems, we conducted a thorough study of multiple wound assessment tools (Xu et al., 2021). The analysis is summarized in Table 13.1. Our goal was to identify the most effective and feasible means of assessing wounds and providing accurate results. Through our research, we encountered various approaches

Table 13.1 Related research work

Reference	Journal Name and Year	Novelty
Tang et al., 2021	Advanced Healthcare Materials, 2021	To discuss the creation of intelligent wound dressings, their integration with wearable sensors and medication delivery systems, and their use in anything from wound monitoring to timely therapeutic application.
Sattar et al., 2019	IEEE Access, 2019	Developing an Internet of Things (IoT)-based intelligent Wound Assessment System, monitoring internal and external wound variables in real time, and assessing the wound's condition using measured factor values.
Zhang et al., 2021	Microbial Biotechnology, 2021	To create an adaptable, integrated sensing device that can use a specially made smartphone app to detect the local temperature of wounds in real time.
Pang et al., 2020	Advanced Science, 2020	To construct a smart, flexible gadget that can release antibiotics on demand after being UV-triggered to diagnose infections at an early stage. Data from the sensor is transmitted over Bluetooth to a smartphone.
Xu et al., 2021	Advanced Functional Materials, 2021	To develop an NFC-based device that senses the psychological aspects of a wound and electrically delivers an antibacterial medication when necessary. The device's battery life is decreased by the smartphone and its miniature circuitry.

proposed in literature, each offering unique insights into improving wound assessment methodologies.

By examining these diverse approaches, we gained valuable knowledge about the existing limitations and potential solutions in this field (Agurto et al., 2015; Kashef et al., 2020). We discovered innovative technologies that utilize sensors and imaging techniques to capture real-time data about wounds. These advancements offer objective measurements, reducing the reliance on subjective observations (Luo et al., 2004).

Furthermore, some researchers explored the use of artificial intelligence algorithms to analyze wound images and provide automated assessments. By leveraging machine learning capabilities, these systems can not only accurately assess wound status but also predict future healing outcomes (Lee et al., 2008). Such technology could revolutionize wound care by enabling healthcare professionals to make informed decisions based on reliable and consistent data (Ansary et al., 2023; Mydhili et al., 2011).

To summarize, the limitations and inconsistencies associated with traditional wound assessment methods have prompted a surge of interest in creating advanced systems capable of continuously monitoring wound healing (Majumder et al., 2019; Piazza et al., 2019). Our comprehensive review of literature revealed promising approaches that incorporate cutting-edge technologies such as sensors, imaging techniques, and artificial intelligence algorithms (Nosato et al., 2012; Tolley et al., 2014). It is through continued exploration and implementation of these innovations that we can enhance wound assessment practices and improve patient outcomes (Subramaniam et al., 2022; Shin et al., 2016).

13.3 PROPOSED METHODOLOGY

We have created a thorough method in our suggested system to keep an eye on the state of wounds, including burns, pressure ulcers, diabetic foot ulcers, surgical wounds, and other kinds of wounds that need to be checked on frequently. We use a variety of sensors, including strain, pressure, temperature, humidity, and strain data collection capabilities, to enable precise tracking and assessment.

By gathering these diverse metrics, we can gain valuable insights into the wound's condition and make informed decisions regarding its treatment and healing process. Temperature sensing allows us to gauge if there are any signs of infection or inflammation in the affected area. High temperatures could indicate an ongoing infection, while lower temperatures might suggest poor blood circulation. Monitoring temperature fluctuations over time provide crucial information about the wound's progress (Hawkes et al., 2017).

Humidity sensing provides us with vital information about the moisture levels around the wound site. Excessive moisture can delay the healing process by promoting bacterial growth, while excessively dry conditions can impede the formation of new tissue. By closely monitoring humidity, we can create an optimal environment for the wound to heal efficiently.

Strain sensing is another critical aspect to consider when assessing wound condition. Strain sensors help us measure any tension or deformation occurring at the wound site (Martinengo et al., 2019). This information helps identify if excessive movement or mechanical stress is hindering proper healing. By detecting strain patterns, healthcare professionals can adopt treatment plans accordingly to reduce further damage or aid in the recovery process (Ministry of Health, Labor and Welfare, 2007).

Pressure sensing is also incorporated within our system to evaluate the amount of force applied to the wound during activities such as sitting or lying down. Continuous high pressure on certain areas of the body may lead to pressure ulcers, which are commonly known as bedsores. By monitoring pressure levels regularly, caregivers can take necessary measures to prevent the development of these ulcers and alleviate discomfort for patients who spend extended periods in immobile positions (Bernard et al., 2009).

It is important to note that this proposed system aims not only to collect data but also to translate it into actionable insights (Aracri et al., 2021; Cheng et al., 2019). By using advanced algorithms and machine learning techniques, we can analyze the gathered information and provide healthcare professionals with comprehensive reports and recommendations for effective wound management (Choi et al., 2020).

In addition, the integration of temperature, humidity, strain, and pressure sensors within our monitoring system offers a holistic approach to track the condition of various types of wounds (Cacas-Bocanegra et al., 2020). By leveraging these diverse data points, we empower healthcare providers to make informed decisions regarding treatment plans, thereby enhancing patient care and improving healing outcomes (Choi et al., 2020). The sensors used in the system are:

- SHT20 sensor is used for sensing the temperature and humidity of the wounded area.
- DF350 strain gauge sensor is used to sense the strain on the wounded area.
- RP-S40-ST FSR is used to sense the pressure of the wounded area.

We collect data from the different sensors and send it to the microcontroller. After that, the data is sent to the ThingsBoard platform via a Wi-Fi connection between the module and mobile phone. The data transmitted from the module is visualized on the custom app as live graphs. The data is then stored in the EPROM of the module as well as in a database with the timestamp to keep a history of the variation in the physical parameters of wound assessment.

13.3.1 System description

The block diagram of Wound Assessment System is shown in Figure 13.1 and the detailed description is as follows:

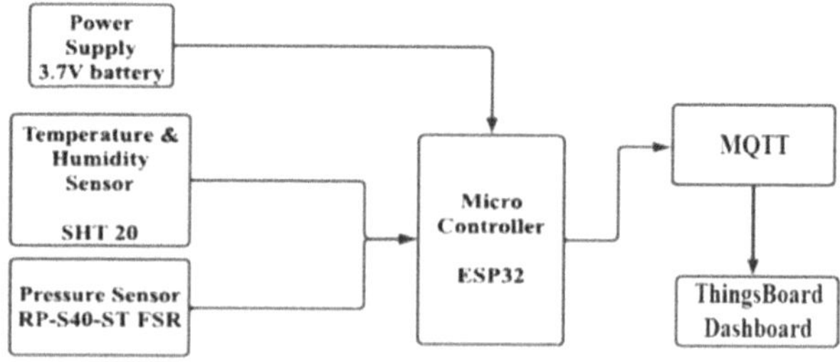

Figure 13.1 Block diagram of Wound Assessment System.

13.3.2 Hardware implementation

We use the SHT20 sensor for sensing the temperature and humidity, the DF350 strain gauge sensor for feeling the strain, and the RP-S4-ST FSR sensor for sensing the pressure on the affected area. These sensors send all the physical quantities to ESP32-C3 MINI-1.

We are using Eagle software for PCB Designing. Using PCB Designing, we will interface all the sensors with a microcontroller on a PCB board that will help to reduce the complexity & size of the system. It will help to make the interfacing easier.

13.3.3 Software implementation

Code for the microcontroller is written in Arduino IDE, in which we used a Wi-Fi manager to integrate Wi-Fi and MQTT server in the microcontroller. First, the microcontroller starts the Wi-Fi network and checks whether the sensors are connected. It also checks whether the sender and receiver connections are established via a Wi-Fi network. Once the connection is established, the microcontroller tries to read data from sensors. It sends all the data to the MQTT broker, which sends it to the ThingsBoard server for visualization.

Here is an elaboration on the code for the microcontroller written in Arduino IDE, integrating Wi-Fi and MQTT for sensor data transmission to ThingsBoard:

Libraries:

> Wi-Fi: Includes functions for connecting to and managing Wi-Fi networks.
> PubSubClient: Provides libraries for MQTT communication.
> Sensor-specific libraries: Include libraries for specific sensors you are using.

Code Structure:
Setup:

Initialize serial communication for debugging.
Define Wi-Fi credentials (SSID and password).
Define MQTT broker details (address, port).
Define ThingsBoard server details (address, access token).
Initialize sensor objects and configure pins.

Connect to Wi-Fi network:
WiFi.begin (ssid, password);
Check connection status and print info.

Connect to MQTT broker:
PubSubClient client (wifiClient);
client.setServer (brokerAddress, brokerPort);
client.connect (clientId);
Check connection status and print info.

Loop:
Read sensor data from each sensor object.
Prepare a message string containing data and timestamps.
Publish the message to the ThingsBoard topic:
client.publish (thingsboardTopic, message);
Check for errors and print info.
Handle any incoming MQTT messages (if applicable).
Delay to control data transmission rate.
Code Snippet:

```
  }

  Serial.println("Connected to the WiFi network");


  // Connect to MQTT broker

  client.setServer(brokerAddress, brokerPort);

  if (client.connect(clientId)) {

  Serial.println("Connected to MQTT broker");

  } else {

  Serial.print("MQTT connection failed, rc=");

  Serial.print(client.state());

  Serial.println(" retrying in 5 seconds");

  delay(5000);

  }

  }


  void loop() {

  // Read sensor data and prepare message

  // Publish message to ThingsBoard

  client.publish(thingsboardTopic, message);

  // Handle incoming messages

  client.loop();

  delay(1000); // Adjust as needed

  }
```

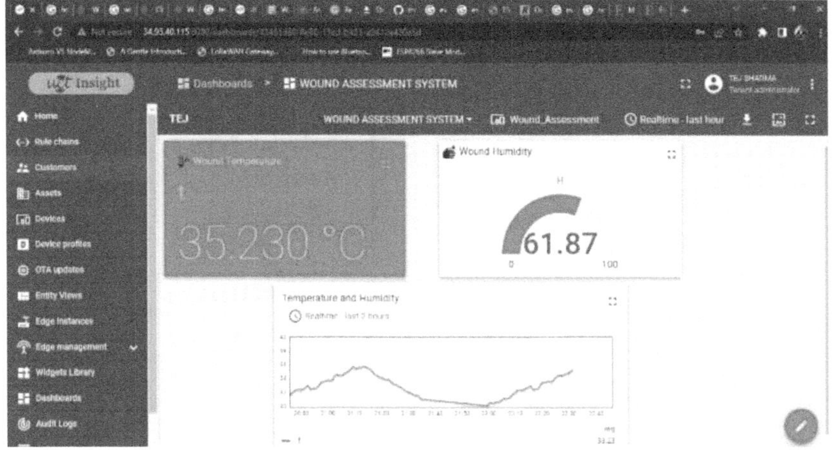

Figure 13.2 ThingsBoard Dashboard running on the Web.

ThingsBoard is an open-source platform for managing devices, processing, storing, and displaying data related to the Internet of Things. It offers the capacity to provide and oversee assets and devices on the network. ThingsBoard uses drag-and-drop widgets that can be placed on the screen and configured according to the user's requirements. The screenshot of ThingsBoard Dashboard is shown in Figure 13.2.

Using this platform, we create a dashboard with the required widgets to show data on the output screen. When a Wi-Fi connection is established between Esp32 and a mobile phone, the data is sent to the ThingsBoard server, visualized in graphs. This dashboard is tested for virtual data but not for real wound-based data.

Figure 13.3 illustrates its working. Figure 13.4 depicts the circuit design of the proposed system, while Figure 13.5 reveals the schematic diagram of the proposed design.

13.4 RESULTS

The goal of the suggested work and methodology is to present a design for a distributed wound assessment system that is portable and simple to use. It offers a design for a prospective Wound Assessment System that considers and improves upon existing wound assessment tools to produce a brand-new assessment tool. The results are summarized in Table 13.2. The microcontrollers of our proposed module, which is based on the Esp32 microcontroller, were compared to an Arduino-based Wound Assessment System. It was discovered that our proposed

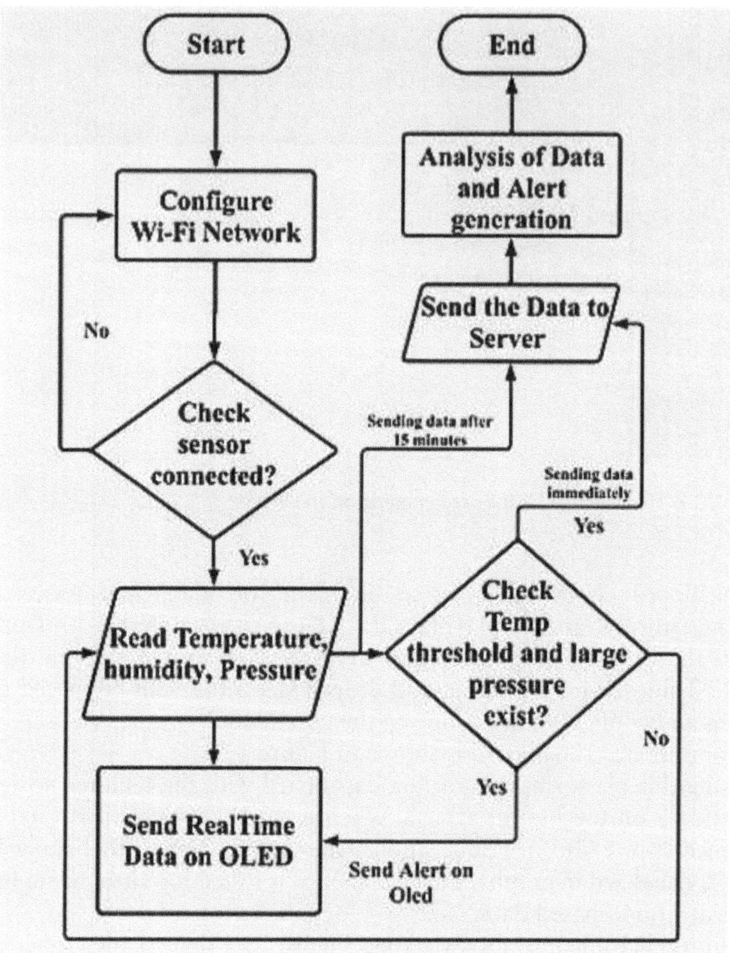

Figure 13.3 Flowchart depicting temperature and humidity sensing through SHT20 sensor and sending data to the microcontroller and then to ThingsBoard App via Wi-Fi.

module uses current wireless technologies, weighs less, and has a significantly smaller form factor.

13.5 CONCLUSION

In conclusion, our research underscores the pivotal role of data-driven robotics in advancing the field of wound assessment, as elucidated in the book

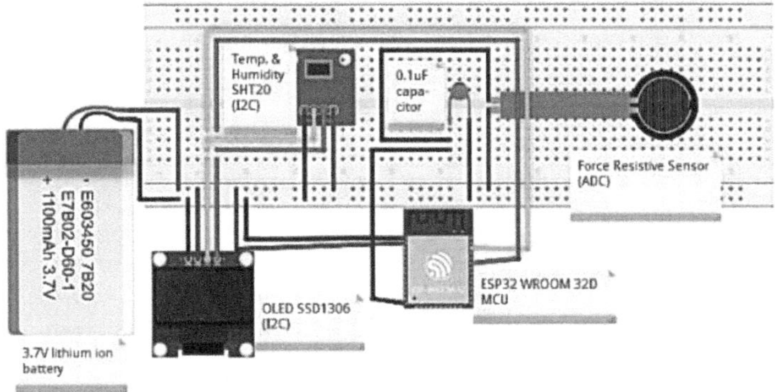

Figure 13.4 Circuit diagram of Wound Assessment System using Wi-Fi and Esp32.

chapter on "Data-Driven Robotics: A Performance Analysis of Wearable Sensor-based Wound Assessment System for Sensor Data Collection." Evaluating current Wound Assessment Systems and tools reveals insights that resonate with the core principles of leveraging technology for precise and consistent evaluations.

Traditional subjective visual inspection procedures have inherent limitations, leading to consistent and accurate reporting. Through our evaluation-based research technique, we aimed to determine the optimal requirements for an efficient wound assessment tool, aligning seamlessly with the objectives outlined in the book chapter.

The selection process scrutinized how well these tools align with the criteria for practical wound assessment, considering the number of physical factors and the overall effectiveness in providing accurate evaluations. By emphasizing data-driven approaches and integrating wearable sensor-based systems, our research aligns with the overarching theme of the book chapter, highlighting the transformative impact of technology on healthcare practices.

The outcomes of this evaluation not only contribute to understanding the efficacy of existing wound assessment instruments but also advocate for the integration of cutting-edge technologies to drive advancements in healthcare. The book chapter serves as a fitting context, emphasizing the symbiotic relationship between data-driven robotics and wearable sensor-based Wound Assessment Systems. This alignment reinforces the trajectory of technology-driven innovations in healthcare, promising enhanced patient care and accelerated healing processes.

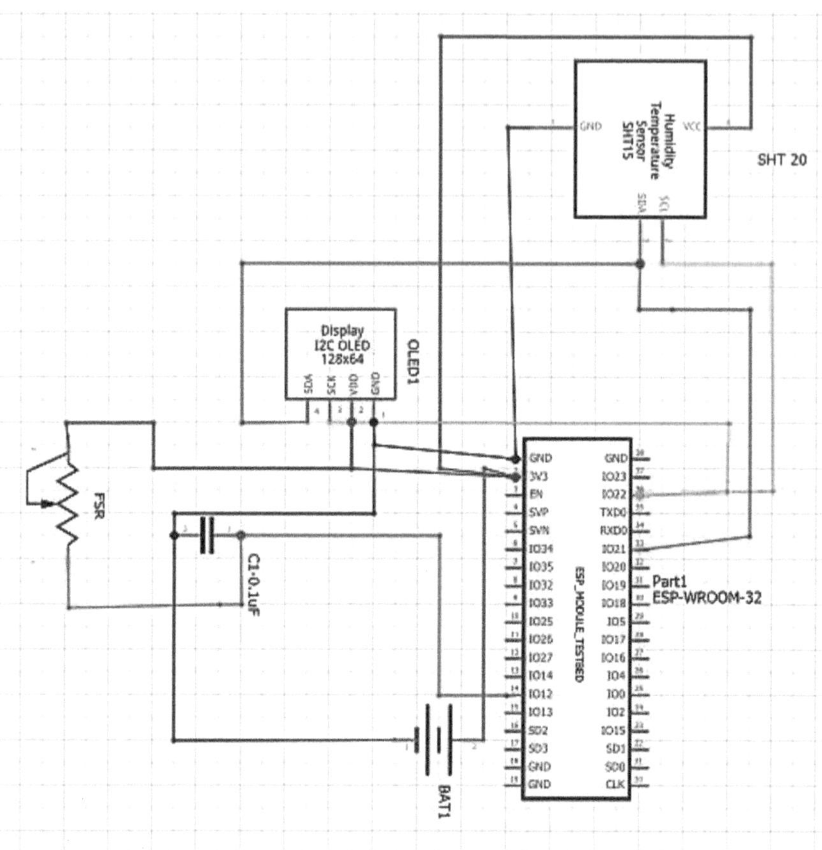

Figure 13.5 Schematic diagram depicting the various connections in the Wound Assessment System.

Table 13.2 Dimensions, weight, and connectivity comparison of microcontrollers of Arduino Uno-based and Esp32-based Wound Assessment System

Parameters	Arduino Uno	ESP32 (Proposed System)
Size (mm)	53.4 mm × 68.6 mm	28 mm × 25.5 mm
Weight (gm)	25gm	2.5 gm
Connectivity	Does not have inbuilt wireless connectivity modules	It has integrated Bluetooth and Wi-Fi modules

13.6 FUTURE SCOPE

In considering future directions for research and development, our study on wearable sensor-based wound assessment within the context of "Data-Driven Robotics: A Performance Analysis of Wearable Sensor-based Wound Assessment System for Sensor Data Collection" suggests several promising avenues.

Primarily, combining cutting-edge machine learning algorithms with artificial intelligence (AI) methods has a lot of promise. Incorporating these technologies into wearable sensor-based systems can enhance the system's ability to analyze and interpret data, leading to more precise and insightful wound assessments. Future work should explore the optimal ways to leverage AI for real-time data processing, thereby improving the system's efficiency and reliability in assessing diverse wound types.

Secondly, exploring additional sensor modalities and their integration into the assessment system could broaden its scope and capabilities. Investigating sensors that capture data beyond conventional parameters, such as incorporating thermal imaging or three-dimensional mapping, could offer a more comprehensive understanding of wound characteristics. This expansion aligns with the book chapter's theme of utilizing diverse sensor data for a nuanced analysis of wound conditions.

Furthermore, future research should focus on refining the interoperability of wearable sensor-based systems with existing healthcare infrastructure. Streamlining data exchange and integration with electronic health records can facilitate seamless communication between the assessment system and healthcare professionals, fostering a more integrated and patient-centric approach.

Lastly, user-centered design considerations and usability testing should be integral to future work. Ensuring that the wearable sensor-based system is user-friendly, comfortable, and quickly adopted by healthcare professionals and patients is essential for its successful implementation. This aligns with the broader narrative of the book chapter, emphasizing the practical application and usability of data-driven robotics in real-world healthcare settings.

In summary, future work should delve into integrating advanced technologies, exploring additional sensor modalities, refining interoperability, and prioritizing user-centered design considerations.

REFERENCES

Agurto, C., Barriga, S., Burge, M., Soliz, P. (2015). Characterization of diabetic peripheral neuropathy in infrared video sequences using independent component analysis. In 2015 IEEE 25th International Workshop on Machine Learning for Signal Processing (MLSP), Boston, MA, USA. https://doi.org/10.1109/MLSP.2015.7324362

Ansary, S.I., Deb, S., Deb, A.K. (2023). Design and development of an adaptive robotic gripper. Journal of Intelligent & Robotic Systems, 109, 13. https://doi.org/10.1007/s10846-023-01948-6

Aracri, S., Giorgio-Serchi, F., Suaria, G., Sayed, M.E., Nemitz, M.P., Mahon, S., Stokes, A.A. (2021). Soft robots for ocean exploration and offshore operations: A perspective. Soft Robotics, 8(6), 625–639. DOI:10.1089/soro.2020.0011

Banerjee, H., Tse, Z.T H., & Ren, H. (2018). Soft robotics with compliance and adaptation for biomedical applications and forthcoming challenges. International Journal of Robotics and Automation, 33(1), 69–80. https://doi.org/10.2316/Journal.206.2018.1.206-4981

Bernard, T., D'Elia, C., Kabadi, R., Wong, N. (2009). An early detection system for foot ulceration in diabetic patients. In 2009 IEEE 35th Annual Northeast Bioengineering Conference, Cambridge, MA, pp. 1–2. https://doi.org/10.1109/NEBC.2009.4967797

Birglen, L. (2010). From flapping wings to underactuated fingers and beyond: A broad look to self-adaptive mechanisms. Mechanical Sciences, 1(1), 5–12.

Breger, J.C., Yoon, C.K., Xiao, R., Kwag, H.R., Wang, M.O., Fisher, J.P., Nguyen, T.D., Gracias, D.H. (2015). Self-Folding Thermo-Magnetically Responsive Soft Microgrippers. ACS Applied Materials & Interfaces, 7(5), 3398–3405. https://doi.org/10.1021/am508621s

Casas-Bocanegra, D., Gomez-Vargas, D., Pinto-Bernal, M.J., Maldonado, J., Munera, M., Villa-Moreno, A., Stoelen, M.F., Belpaeme, T., Cifuentes, C.A. (2020). An open-source social robot based on compliant soft robotics for therapy with children with ASD. Actuators, 9(91). https://doi.org/10.3390/act9030091

Cheng, M., Fan, S., Jiang, L. (2019). Design of a highly compliant underactuated prosthetic hand. In 2019 IEEE International Conference on Robotics and Biomimetics (ROBIO), Piscataway, NJ, pp. 2833–2838.

Choi, J., Ahn, S.H., Kim, C., Park, J.H., Song, H.Y., Cho, K.J. (2020). Design of continuum robot with variable stiffness for gastrointestinal stenting using conformability factor. IEEE Transactions on Medical Robotics and Bionics, 2, 529–532.

Chu, C.Y., Patterson, R.M. (2018). Soft robotic devices for hand rehabilitation and assistance: A narrative review. Journal of NeuroEngineering and Rehabilitation, 15(9).

Corpin, R.R.A., Guingab, H.A.R., Manalo, A.N.P., Sampana, M.L.B., Abello, A.N.A., dela Cruz, A.R., Roxas, E.A., Suarez, C.G., Serrano, K.K.D. (2019). Prediction of diabetic peripheral neuropathy (DPN) using plantar pressure analysis and learning model. In 2019 IEEE 11th International Conference on Humanoid, Nanotechnology, Information Technology, Communication and Control, Environment, and Management (HNICEM), pp. 1–6. https://doi.org/10.1109/hnicem48295.2019.9072889

Feneley, R.C.L., Hopley, I.B., Wells, P.N.T. (2015). Wound assessment system performance analysis. Journal of Medical Engineering & Technology, 39(10), 459.

Feng, Y., Schlosser, F.J., Sumpio, B.E. (2009). The Semmes Weinstein monofilament examination as a screening tool for diabetic peripheral neuropathy, Journal of Vascular Surgery, 50(3), 675–682.

Fujita, M., Tadakuma, K., Komatsu, H., Takane, E., Nomura, A., Ichimura, T., Konyo, M., Tadokoro, S. (2018). Jamming layered membrane gripper mechanism for grasping differently shaped-objects without excessive pushing force for search and rescue missions. Advanced Robotics, 32(11), 590–604.

Galloway, K.C., Becker, K.P., Phillips, B., Kirby, J., Licht, S., Tchernov, D., Wood, R.J., & Gruber, D.F. (2016). Soft robotic grippers for biological sampling on deep reefs. Soft Robotics, 3(1), 23–33. https://doi.org/10.1089/soro.2015.0019

Godfrey, R., Conway, D., Meagher, D., Ólaighin, G. (2008). Direct measurement of human movement by accelerometry. Medical Engineering Physics, 30(10), 1364–1386. https://doi.org/10.1016/j.medengphy.2008.09.005

Greer, J.D., Morimoto, T.K., Okamura, A.M., Hawkes, E.W. (2019). A soft, steerable continuum robot that grows via tip extension. Soft Robotics, 6, 95–108.

Haque, F., Reaz, M.B.I., Chowdhury, M.E.H., Hashim, F.H., Arsad, N., Ali, S.H.M. (2020). Diabetic sensorimotor polyneuropathy severity classification using adaptive neuro-fuzzy inference system. IEEE Access, 9, 7618–7631. https://doi.org/10.1109/access.2020.3048742

Hartmann, F., Baumgartner, M., Kaltenbrunner, M. (2021). Becoming sustainable, the new frontier in soft robotics. Advanced Materials, 33, 2004413. https://doi.org/10.1002/adma.202004413

Hawkes, E.W., Blumenschein, L.H., Greer, J.D., Okamura, A.M. (2017). A soft robot that navigates its environment through growth. Science Robotics, 2(8). https://doi.org/10.1126/scirobotics.aan3028

Herlihy, D.V. (2006). Bicycle: The History. Yale University Press, New Haven, CT.

Hirayasu, K., Sasaki, H., Kishimoto, S., Kurisu, S., Noda, K., Ogawa, K., Tanaka, H., Sakakibara, Y., Matsuno, S., Furuta, H., Arita, M., Nara, K., Nanjo, K. (2018). Difference in normal limit values of nerve conduction parameters between Westerners and Japanese people might need to be considered when diagnosing diabetic polyneuropathy using a Point-of-Care Sural Nerve Conduction Device (NCstat®/DPNCheckTM). Journal of Diabetes Investigation, 9(5), 1173–1181.

Hirose, S. (1985). Connected differential mechanism and its applications. In International Conference on Advanced Robotics, 319–325.

Indujha, M.S., Jaiganesh, A., Poornima, N., Jebasharon, J. (2021). Automated wound assessment system for foot ulcer patients with diabetes. i-manager's Journal on Pattern Recognition, 8(1), 12–18.

Kashef, S.R., Amini, S., Akbarzadeh, A (2020). Robotic hand: A review on linkage-driven finger mechanisms of prosthetic hands and evaluation of the performance criteria. Mechanism and Machine Theory, 145, 103677. https://doi.org/10.1016/j.mechmachtheory.2019.103677

Katzschmann, R.K., DelPreto, J., MacCurdy, R., Rus, D. (2018). Exploration of underwater life with an acoustically controlled soft robotic fish. Science Robotics, 3(16), eaar3449. https://doi.org/10.1126/scirobotics.aar3449

Khalil, M.E., Ghazal, M., Burns, C., El-Baz, A. (2019). Chronic wound healing assessment system based on different features modalities and non-negative matrix factorization (nmf) feature reduction. IEEE Access, 7, 80110–80121.

King, R.C., Villeneuve, E., White, R.J., Sherratt, R.S., Holderbaum, W., Harwin, W.S. (2016). Application of data fusion techniques and technologies for wearable health monitoring. Medical Engineering & Physics, 42, 1–12. https://doi.org/10.1016/j.medengphy.2016.12.011

Komori, H. et al. (2019). Screening System for Diabetes Peripheral Neuropathy using Foot Plantar images on different hardness floor. IEEE.

Komori, H., Watanabe, K., Tsuichihara, S., Takemura, H., Imai, M., Haraguchi, M., Chou, S. (2019). Screening system for diabetes peripheral neuropathy using foot plantar images on different hardness floor. *2019 IEEE International Conference on Systems, Man and Cybernetics (SMC)*, Bari, Italy, 2019, pp. 2614–2619. doi: 10.1109/SMC.2019.8914204

Kontoudis, G.P., Liarokapis, M., Vamvoudakis, K.G. (2019). An adaptive, humanlike robot hand with selective interdigitation: Towards robust grasping and dexterous, in-hand manipulation. In 2019 IEEE-RAS 19th International Conference on Humanoid Robots (Humanoids), Toronto, ON, Canada, pp. 251–258. IEEE.https://doi.org/10.1109/Humanoids43949.2019.9035037

Kota, S., Joo, J., Li, Z., Rodgers, S.M., Sniegowski, J. (2001). Design of compliant mechanisms: Applications to MEMS. Analog Integrated Circuits and Signal Processing, 29(7), 7.

Kukreja, G.S., Alok, A., Reddy, A.K., Nersisson, R. (2020). IoT based foot neuropathy analysis and remote monitoring of foot pressure and temperature. 2020 5th International conference on computing, communication and security (ICCCS), Patna, India, pp. 1–6, doi: 10.1109/ICCCS49678.2020.9277004.

Lee, H., Park, K., Lee, B., Choi, J.S., Elmasri, R. (2008). Issues in data fusion in healthcare monitoring. In PETRA '08: Proceedings of the 1st international conference on PErvasive Technologies Related to Assistive Environments, pp. 1–8. https://doi.org/10.1145/1389586.1389590

Lou, D., Qian, P., Pei, X., Dong, S., Li, S. Tan, W., Ma, L. (2020). Flexible wound healing system for pro-regeneration, temperature monitoring and infection early warning. Biosensors and Bioelectronics, 162, 112275. https://doi.org/10.1016/j.bios.2020.112275

Luo, M., Mei, T., Wang, X., Yu, Y. (2004). Grasp characteristics of an underactuated robot hand. In IEEE International Conference on Robotics and Automation, 2004. Proceedings. ICRA '04., New Orleans, LA, 3, pp. 2236–2241.IEEE. https://doi.org/10.1109/ROBOT.2004.1307394

Majumder, S., Mondal, T., Deen, M.J. (2019). A simple, low-cost and efficient gait analyzer for wearable healthcare applications. IEEE Sensors Journal, 19(6), 2320–2329.

Martinengo, L., Olsson, M., Bajpai, R., Soljak, M., Upton, Z., Schmidtchen, A., Car, J., Järbrink, K. (2019). Prevalence of chronic wounds in the general population: Systematic review and meta-analysis of observational studies. Annals of Epidemiology, 29, 8–15.

Ministry of Health, Labor and Welfare Statics Association. (2007). Kokumin Eisei no Doukou. Ministry of Health, Labor and Welfare, 54, Tokyo.

Mukherjee, R., Tewary, S., Routray, A. (2017). Diagnostic and prognostic utility of non-invasive multimodal imaging in chronic wound monitoring: A systematic review. Journal of Medical Systems, 41, 1–17.

Mydhili, S.K., Naseera, M.F., Kumar, R.R. (2011). An IoT based Foot Healthcare System for Diabetic Patients and a Futuristic Approach for transforming

Sensor Data into real-time Medical Advice. In Proceedings of the Advancement in Electronics & Communication Engineering.

Nosato, H., Sakanashi, H., Takahashi, E., Murakawa, M. (2012). A content-based image retrieval method for optical colonoscopy images based on image recognition techniques. Proceedings of SPIE, 9414, 1–9.

Ohura, N., Ryota, M., Sakisaka, M., Terabe, Y., Morishige, Y., Atsushi, U., Okoshi, T., Iizaka S., Akihiko, T. (2019). Convolutional neural networks for wound detection: The role of artificial intelligence in wound care. Journal of Wound Care, 28(10), S13–S24. https://doi.org/10.12968/jowc.2019.28.Sup10.S13

Pang, Q., Lou, D., Li, S., Wang, G., Qiao, B., Dong, S., Ma, L., Gao, C., Wu, Z. (2020). Smart, flexible electronics-integrated wound dressing for real-time monitoring and on-demand treatment of infected wounds. Advanced Science, 7(6), 1902673. https://doi.org/10.1002/advs.201902673

Periyasamy, R., Anand, S., Ammini, A.C. (2012). Investigation of Shore meter in assessing foot plantar hardness in patients with diabetes mellitus—a pilot study. International Journal of Diabetes in Developing Countries, 32(3), 169–175.

Piazza, C., Grioli, G., Catalano, M.G., Bicchi, A. (2019). A century of robotic hands. Annual Review of Control, Robotics, and Autonomous Systems, 2(1), 1–32. https://doi.org/10.1146/annurev-control-060117-105003

Sattar, H., Bajwa, I.S., Ul Amin, R., Sarwar, N., Jamil, N., Abbas Malik, M.G., Mahmood, A., Shafi, U. (2019). An IoT-based intelligent wound monitoring system. IEEE Access, 7, 144500–144515.

Shin, H.C., Roberts, K., Lu, L., Fushman, D.D., Yao, J., Summers, R.M. (2016). Learning to read chest X-rays: Recurrent neural cascade model for automated image annotation, computer vision and pattern recognition. 2016 IEEE Conference on Computer Vision and Pattern Recognition (CVPR), Las Vegas, NV, pp. 2497–2506. doi: 10.1109/CVPR.2016.274

Subramaniam, S., Jumder, S., Faisal, A., Deen, M.J. (2022). Insole-based system for health monitoring: : Current solutions and research challenges. Sensors, 22(2), 438. https://doi.org/10.3390/s22020438

Tang, N., Zheng, Y., Cui, D., Haick, H. (2021). Multifunctional dressing for wound diagnosis and rehabilitation. Advanced Healthcare Materials, 10, 2101292. https://doi.org/10.1002/adhm.202101292

Tolley, M.T., Shepherd, R.F., Mosadegh, B., Galloway, K.C., Wehner, M., Karpelson, M., Wood, R.J., Whitesides, G.M. (2014). A resilient, untethered soft robotics. Soft Robotics, 1, 213–223.

Townsend, W. (2000). The BarrettHand grasper—programmably flexible part handling and assembly. Industrial Robot, 27(3), 181–188. https://doi.org/10.1108/01439910010371597

Wada, K., Ikeda, Y., Inoue, K., & Uehara, R. (2010). Development and preliminary evaluation of a caregiver's manual for robot therapy using the therapeutic seal robot Paro. In 19th International Symposium in Robot and Human Interactive Communication, pp. 533–538.

Wang, L., Pedersen, P.C., Strong, D.M., Tulu, B., Agu, E., Ignotz, R. (2015. Smartphone-based wound assessment system for patients with diabetes. IEEE Transactions on Biomedical Engineering, 62(2), 477–488.

Xu, G., Lu, Y., Cheng, C., Li, X., Xu, J., Liu, Z., Liu, J., Liu, G., Shi, Z., Chen, Z., Zhang, F., Jia, Y., Xu, D., Yuan, W., Cui, Z., Low, S.S., Liu, Q. (2021). Battery-free and wireless smart wound dressing for wound infection monitoring and electrically controlled on-demand drug delivery. Advanced Functional Materials, 31(26), 2100852. https://doi.org/10.1002/adfm.202100852

Yang, D., Zhao, J., Gu, Y., Wang, X., Li, N., Jiang, L., Liu, H., Huang, H., Zhao, D. (2009). An anthropomorphic robot hand developed based on underactuated mechanism and controlled by EMG signals. Journal of Bionic Engineering, 6(3), 255–263. https://doi.org/10.1016/S1672-6529(08)60119-5

Zhang, Y., Lin, B., Huang, R., Lin, Z., Li, Y., Li, J., Li, X. (2021). Flexible integrated sensing platform for monitoring wound temperature and predicting infection, Microb. Biotechnology, 14(4), 1566–1579.

Zhang, Z., Dequidt, J., Back, J., Liu, H., Duriez, C. (2019). Motion Control of Cable-Driven Continuum Catheter Robot Through Contacts. IEEE Robotics and Automation Letters, 4, 1852–1859.

Chapter 14

Leveraging data-driven solutions for energy efficiency and pollution reduction in smart sustainable cities

Bhupinder Singh and Christian Kaunert

14.1 INTRODUCTION AND BACKGROUND

The synergy between these solutions amplifies their environmental impact, creating effects greater than the sum of their individual contributions. Smart grid technologies offer a plethora of benefits, optimizing energy use, managing consumption, and integrating alternative energy sources (Petrovic & Kocic, 2020). The enabling of bidirectional energy flows and real-time data exchange between suppliers and consumers, smart grids enhance visibility, service reliability, cost control, peak load management, and energy production sharing (Bibri & Krogstie, 2020a). Integration of big data and the Internet of Things (IoT) is essential to the operation of smart and sustainable cities (Bibri, 2020a). As a result, data-driven urbanism is now extremely responsive to urban operations and planning, allowing real-time models for smart, sustainable cities to be created using routinely observed data (Myeong & Shahzad, 2021). This method makes it easier to continuously monitor, comprehend, analyze, and develop cities to improve their environmental health and energy efficiency (Bibri, 2020b).

Using new urban intelligence functions is a more sophisticated way to assist decisions (Kaluarachchi, 2022). Although a lot of research has been done on data-driven technologies in smart cities, especially in the areas of economics and society, there is still a clear knowledge vacuum on the underlying, significant, and cooperative impacts on environmental sustainability. In the context of sustainable cities, such as eco-cities this disparity is also apparent (Bibri & Krogstie, 2020b). Thus, this chapter establishes the idea of "environmentally data-driven smart sustainable cities" and investigates the role and potential of data-driven smart solutions in improving environmental sustainability in both smart cities and sustainable cities (Stubinger & Schneider, 2020). To provide insight into these new urban phenomena, Stockholm and Barcelona which are regarded as the most technologically and environmentally advanced cities in Europe, respectively are compared and examined using a qualitative research approach, namely a descriptive/illustrative case study (Bibri, 2023). The results show that important data-driven smart solutions used to improve environmental sustainability in both eco-cities and smart cities are smart grids, smart meters, smart buildings, smart environmental monitoring, and smart urban metabolism (Olaniyi et al., 2023).

These solutions clearly work in concert, producing combined effects that are greater than the sum of their separate effects (Zhang et al., 2021). This is especially true when it comes to improving energy efficiency, reducing environmental pollution, implementing renewable energy, and providing real-time feedback on energy flows with high temporal and spatial resolutions (Mousavi et al., 2023). When it comes to putting best practices for environmental sustainability into effect, Stockholm leads Barcelona because of its long history of environmental activities, strong environmental legislation, progressive performance, high standards, and ambitious ambitions (Mohapatra et al., 2022). However, the ways in which applicable data-driven technological solutions are being implemented in the energy and environmental sectors vary across the two cities (Ahmad et al., 2022). The study comes to the conclusion that local governments do not have a single agenda for strategic planning and that each city has its own specific goals, policies, strategies, routes, and aspirations that are reflected in its data-driven choices (Nica et al., 2020).

14.1.1 Urbanization and its impact

Urbanization's effects on urban areas are too important to ignore as they continue to change cityscapes around the world (Wu et al., 2022). Cities face several difficulties as a result of population growth and people moving from rural to urban areas. The stress that urbanization puts on the current infrastructure which includes public services, housing, and transportation networks is the main effect. The demand for these resources rises as more people move into cities, frequently exceeding the capacity of the existing infrastructure (Lazaroiu & Harrison, 2021). The natural ecosystems and green areas are encroached upon by expanding cities to accommodate the rising population, which leads to deforestation and a loss of biodiversity (Sarkar et al., 2020). The increase in industrial activity and vehicular traffic in metropolitan areas exacerbates air and water pollution which lowers the standard of living for those living in cities (Ahsan et al., 2023). These environmental issues highlight the need for creative methods of urban planning that promote sustainability and the welfare of inhabitants and the environment (Anthopoulos & Kazantzi, 2022).

14.1.2 Need for smart sustainable cities

An urban environment that is seamlessly connected by digital means and makes use of analytics, technology, and data to improve the quality of life for its citizens is known as a cutting-edge smart city (Verma et al., 2023). It uses a range of digital technology and sensors to provide inhabitants with quick and easy access to services (Nigro et al., 2021). Smart cities prioritize sustainable development and innovation to improve the quality of life for their residents by increasing the efficiency of services like waste management, electricity, and transportation (Gray & Kovacova, 2021). These cities also use technology to make their urban environments more eco-friendly and efficient in an effort to reduce carbon emissions and advance sustainability (Mishra & Singh, 2023a). Smart cities are based on the use of data-driven technologies like big data, artificial intelligence (AI), and the IoT, which are all utilized to create more efficient ways of doing things for resource management and enhance services for the city's inhabitants (Burke & Zvarikova, 2021).

14.1.3 Objectives of the chapter

This chapter aims to achieve the following objectives to

- sightsee data-driven approaches and analyze energy efficiency and pollution reduction strategies.
- explore the significance, and necessity of sustainability and underscore the utilization of data-driven technologies in crafting urban environments that are both efficient and innovative, drawing examples from various global contexts.
- comprehensive examination of smart cities, encompassing their prerequisites, potential elements, and existing research challenges, while also identifying opportunities for further investigation.

14.1.4 Significance of the study

The demand for urban infrastructure increases as population growth and traffic density increase in metropolitan regions (Lu et al., 2021). Over time, these infrastructures may become greener and more efficient systems by investing in creative and effective sustainable methods (Antony & Sunder, 2023). For a number of reasons, including improved quality of life, reduced environmental impact, and long-term economic benefits, sustainable urban development is crucial (Sarker, 2022). Cities may reap significant rewards from investing in sustainable projects that will promote economic expansion and job development while reducing energy use, waste production, and pollution (Grillone et al., 2020). Also, better living circumstances are a result of sustainable development which creates chances for social inclusion among various societal groups (Fang et al., 2021).These actions have a major positive environmental impact by reducing energy use and garbage generation (Xia et al., 2021). For example, cities may help battle climate change by reducing their reliance on non-renewable energy sources by implementing strategies that help mitigate carbon dioxide emissions (Bachmann et al., 2022). Improving public transportation systems also lowers living expenses for locals, lessens noise pollution and traffic jams, and enhances air quality (Andronie et al., 2021).

The financial benefits that sustainable urban development may provide are yet another strong argument in favor of giving it top priority (Ali et al., 2020). Cities that support sustainable projects attract businesses and create jobs by making themselves more attractive to them (Chapman, 2021). Businesses may provide higher pay and boost the local economy by using energy efficiency methods, which reduce operating costs and increase profitability (Sarker et al., 2020). Cutting back on municipal garbage generation decreases the price of necessities, freeing up money for vital industries like housing, healthcare, and education (Myeong & Shahzad, 2021).

14.2 SMART SUSTAINABLE CITIES: A CONCEPTUAL FRAMEWORK

The notion of the smart city has surfaced as a compelling approach to tackle the pressing issues brought about by worldwide urbanization and to face the principal roadblocks to attaining sustainable development (Shahat Osman & Elragal, 2021). While sustainability is frequently linked to environmental concerns, modifications to the built environment also affect the social dimensions (Kostepen et al., 2020). A dynamic idea, social sustainability emphasizes the development of infrastructure to meet societal demands and concerns by fusing the design of the physical environment with the design of the social sphere (Zhou et al., 2021). While the main goal of smart cities is to increase productivity via creative uses of digital data and technology, a social sustainability approach highlights the vital relationships that exist between people and their environment (Bouramdane, 2023).

The effective integration of human, digital, and physical systems within the built environment is what defines a smart city, which strives to provide its citizens with a successful, sustainable, and inclusive future (Seyrfar et al., 2021). A city's level of intelligence is not just determined by the technology it uses; rather, it is also determined by how well it integrates technology to improve the way the city functions overall, both within its many systems and as a unit. Better communication with municipal inhabitants and more efficient government systems are part of this integration (Singh, 2023a). A city may envisage a more engaging future and follow a new, more productive direction by leveraging its current foundations through the process of becoming smarter (Singh, 2024). Cities aiming to be both smart and sustainable need to combine the best aspects of both sustainable and smart city development in a way that goes beyond focusing on each one alone (Majeed et al. 2021). The goal of this combination is to produce a cooperative synergy that outperforms the sum effect of its intelligent and sustainable parts. For the purpose of establishing transparency, promoting participation, and creating additional value, smart ideas, and sustainability both rely on digital information and data (Tripathi & De, 2021).

The infrastructure that reinforces these initiatives technologically has to be openly accessible. The process of tracking local progress toward the Sustainable Development Goals (SDGs) provides individuals with information on the unique conditions surrounding their sustainable development (Yigitcanlar et al., 2021). This enables them to be informed, lobby their governments for specific reforms, or launch independent efforts to help achieve their objectives (Chen et al., 2022). For this aim, local data and projects work better than national-level measurements (Hui et al., 2023). As a result, creative projects and ideas frequently originate from the activities of one person or a small group of people, acting as models for other cities (Grimaldi et al., 2021). Local changes are more noticeable and quicker than those at the federal level. With the increasing digitization of society comes a greater consciousness of sustainability, which in turn drives a greater need for practical technical solutions in the context of open government, service delivery, and local enterprises (Singh & Kaunert, 2024).

14.2.1 Energy efficiency in smart sustainable cities—Data-driven decision-making

Cities' notable negative environmental consequences are highlighted by the concentration of economic activity, resource-intensive practices, and high non-renewable energy use in urban areas (Singh, 2023b). Cities today contain more than 50% of the world's population, and by 2050, the UN projects that this number will increase to 70% (Sharma & Singh, 2022). These cities also account for over 75% of greenhouse gas emissions and 70% of the world's energy consumption. Cities are facing more and more difficulties in making the shift to more environmentally friendly practices as a result of the world's rising unpredictability and tremendous urbanization (Dharshini & Akshayakumar, 2020). Thus, in technologically and environmentally developed countries, this phenomenon presents formidable obstacles for city governments. These challenges include those associated with urban growth, such as increased energy consumption, environmental degradation, ineffective infrastructure management, insufficient planning strategies, subpar decision-making systems, and socioeconomic disparities (Iddianozie & Palmes, 2020).

The cities need to use smart techniques and creative solutions to handle the complex issues of climate change. These approaches must be made possible by cutting-edge technology and backed by scientific understanding (Deb & Schlueter, 2021). This is necessary for planning, observing, comprehending, and evaluating cities to improve, maximize, and preserve their environmental sustainability. Advanced information and communication technology (ICT) is emphasized in the UN's 2030 Agenda as a way to improve infrastructure, safeguard the environment, and accelerate human growth (Wey & Peng, 2021). The United Nations, in their research "Big Data and the 2030 Agenda for Sustainable Development" examined the possibilities of the smart city concept in line with SDG 7 which calls for universal access to modern, cheap, sustainable, and dependable energy (Chen & Han, 2021).

14.2.2 Role of data in energy management

Urban systems are strained by the multitude of issues brought about by urban expansion, which raises the demand for energy resources and services. Energy is a major urban area that contributes most to global greenhouse gas emissions, making it a major cause of climate change (Albuquerque et al. 2021). The importance of addressing climate change and its effects is emphasized in UN Sustainable Development Goal (SDG) 13. Modern cities are essential to strategic sustainable development and are at the forefront of creating and implementing new technologies that facilitate the shift toward sustainability in the face of urbanization, despite these obstacles (Aranda et al., 2023).

14.2.3 Data-driven decision-making

Issues causing concern are pollution, depletion of non-renewable resources, overuse of renewable resources, and damage to ecosystems (Ajiboye et al., 2022). To detect and identify trends in energy production and consumption and to enable more efficient solutions to alleviate the multifaceted consequences of energy use, advanced computational data analytics methodologies are essential. Everyone agrees that cutting-edge, data-driven technological solutions have the power to combat climate change and improve energy efficiency (Petrovic, 2021). By making energy usage and greenhouse gas emissions visible through processes, goods and services, advanced ICT plays a crucial role in increasing environmental sustainability and creates the groundwork for the concept of smart energy. This vision aims to minimize dependency on fossil fuels, boost the use of renewable and local energy sources made possible by new technology, and create extremely energy-efficient systems (Relich, 2023).

14.2.3.1 Real-time monitoring and control

The development of smart meters, smart grids, green buildings, and solar photovoltaic panels has gained significance in the pursuit of sustainability. In order for a city to efficiently manage its urban infrastructures, systems and services, modern information and communication technology (ICT) plays a critical role as a distributed infrastructure (Cai et al., 2022). This technology functions as a digital nervous system. By improving the energy systems of cities, modern technologies offer a chance to have a good environmental influence (Wei et al., 2021). Together with strong computing infrastructures, recent developments in the IoT and big data analytics present prospects to create practical solutions for pollution reduction and energy efficiency, especially in the context of smart sustainable cities. This global urbanization paradigm is seen to be essential for transitioning to sustainability in an increasingly urbanized world (Anthony Jnr et al., 2020a).

14.2.3.2 Smart grids and energy distribution

Cities are getting more and more integrated with computing, datafication and instrumentation to coordinate their systems and domains. Data is therefore being used to manage, govern, and regulate several facets of city life, allowing for real-time planning, comprehension, analysis, and monitoring (Pandiyan et al., 2023). This change is redefining the possibilities for continuous reflection in the short term to make cities smarter and more sustainable in the long run. It is also changing the way cities may be designed across multiple time periods. Decentralized renewable energy sources and localized manufacturing facilities serve as the main energy suppliers in modern, environmentally friendly energy models (Petrovic, 2021). Two-way cars and smart houses with various renewable energy configurations are also becoming major players in contemporary power networks. These networks must be very flexible and include intelligent subsystems for effective energy distribution and storage to handle variations in energy supply and demand (Fafoutellis et al., 2020). Smart grids offer comprehensive support for the design, prototyping, implementation and testing of the energy systems for developing grid-connected inverters, vehicle-to-grid (V2G) systems, primary control systems for wind, solar or batteries, secondary control systems for microgrid management or tertiary control systems that interface with the main power grid (Anthony Jnr et al., 2020b).

14.2.3.3 Intelligent power infrastructure

Large volumes of data have been produced by sensors as a result of the exponential expansion of networked devices in urban settings, thus highlighting the importance of big data analytics for processing, administration, analysis, and storage. Big data and the IoT are seen to provide the foundation for creating smart, sustainable cities (Sarmas et al., 2022). Big data technologies are becoming indispensable for smart cities, especially when it comes to enhancing their performance in environmental sustainability. The same is true for sustainable cities, where data-driven urbanism is influencing how urban processes are organized and function. These cities are becoming both smart and sustainable (Gonçalves et al., 2020).

14.3 POLLUTION REDUCTION THROUGH DATA-DRIVEN APPROACHES

Human health is greatly influenced by the quality of the air people breathe, which is why both developed and developing countries throughout the world are concerned about it (Belli et al., 2020). Devices that indicate the real-time occupancy status of public parking lots and locations for car and bicycle rentals, optimizing traffic flow and the utilization of spaces and transportation resources (Pan & Zhang, 2020). The precise real-time geo-location of public transport vehicles to estimate and monitor schedules and routes accurately (Su, 2020). Devices gauge the fill levels of garbage containers in real time to optimize waste collection (Townsend, 2021).

The real-time measurement of pollution levels, including CO_2, ozone, and water quality, helps to alert the population and enhance public policies addressing environmental concerns (Ding & Liu, 2020). In the framework of smart sustainable cities, data-driven intelligent solutions present a substantial opportunity to improve environmental sustainability (Nelson & Negurița, 2020). Notably, technologies like smart buildings, smart grids, sophisticated metering infrastructure, smart tools and appliances for the home, smart environmental control and monitoring, and so on, work in concert to produce cumulative effects that outweigh the sum of their separate effects on the environment (Miao et al., 2021). The integration of alternative energy sources into power generation, transmission, and distribution networks is made easier by smart grid technologies, which also optimize energy use, manage energy consumption, conserve energy, and lower costs (Su et al., 2021). They provide real-time data on use and energy price and provide bidirectional energy flows and information sharing between suppliers and customers. Benefits include peak load shifting, grid capacity management, real-time visibility, service dependability, control over expenses and power consumption, and energy generation and sharing (consumer). Comparably, energy consumption has been shown to be effectively decreased by smart building technology (Shafiullah et al., 2022). Devices that provide real-time traffic data assist drivers in planning routes effectively and aid in urban development decisions, such as formulating urbanization policies and determining road layouts and expansions (Singh, 2023c).

14.4 DATA ANALYTICS, IOT, AND ARTIFICIAL INTELLIGENCE IN SMART CITIES

The emergence of IoT technology ushers in a new era of abundant data in the age of linked urban landscapes inside smart cities. The significant amount of data generated by IoT devices is a difficulty as well as an opportunity for the ongoing development of smart cities (Gracias et al., 2023). In this context, AI and data analytics show themselves to be powerful instruments that can help sort through this abundance of data, make wise decisions, and improve municipal administration effectiveness (Popescu, 2022). Beyond the boundaries of siloed systems, countries all around the world are adopting a data-driven governance paradigm in the dynamic field of urban development (Ifaei et al., 2022). During this transformative journey, city leaders are faced with important issues that they must answer to acquire the trust of residents and stakeholders and to explain the basis of their decision-making (Danish & Senjyu, 2023).

The assessment entails determining if the data at hand offers enough information to support well-informed decisions and takes into account variables including the seriousness of the issue, the effect on citizens, and the long-term advantages. This necessitates a thorough comprehension of how initiatives would improve the quality of life for individuals, meet the demands of the community, and promote sustainability and general well-being (Wu et al., 2022). Setting up key performance indicators (KPIs) and routinely assessing the efficacy of interventions necessitates a cooperative effort that takes into account social and ethical issues in addition to technological ones (Wei & Jiang, 2021). To make sure that data-driven decisions are in line with community needs and values, city officials must work in conjunction with data scientists, urban planners, community representatives, and other stakeholders (Hashem et al., 2023).

Urban systems today function as separate silos, concentrating on certain areas such as garbage, water, transit, and property taxes (Mishra & Singh, 2023b). The alert systems for potential dangers such as floods, fires, storms, and hurricanes, aim to implement appropriate prevention and response mechanisms, including preemptive evacuations and relief services (Lv & Shang, 2023).

14.5 POLICY AND GOVERNANCE FRAMEWORKS: GOVERNMENT INITIATIVES AND REGULATIONS

To make data a crucial component of governance, however, issues with data availability, access, security, and privacy must be resolved (Shang & Lv, 2023). Governments place a strong emphasis on public involvement (C2G) to improve the inclusivity and transparency of city planning, realizing the importance of real-time data for decision-making. An important step toward responsive government is the online integration of citizen-centric services, which enables urban local authorities to customize services in response to community requirements. "Data is the new oil" is an analogy that highlights the growing significance of data (Kutty et al., 2022).

The achieving of sustainability goals via the use of digital technology is not without its potential and obstacles. Although intelligent solutions can speed up unsustainable development patterns and improve efficiency as well as save resources and promote a circular economy, they also have disadvantages, including excessive energy consumption. It takes a complex analysis to determine how smart applications will affect the environment overall (Hodgkins, 2020). First-order effects include the direct energy and resource consumption of digital infrastructures, hardware and software; second-order effects are potential savings from efficiency improvements and systemic effects pose a risk of neutralization. Understanding social and environmental implications in the context of AI and Automated Decision-Making (ADM) systems can be difficult as these effects are mediated by psychological, economic, and social dynamics, which can have significant repercussions (Singh, 2023b). Automated decision-making procedures have a regulating effect on people by modifying behavior, influencing decisions, and guaranteeing the accomplishment of predetermined objectives. As a result, it makes sense to give automated decision-making systems a regulatory role (Ma et al., 2022). The significance of decision criteria and objectives in automated decision-making for achieving sustainability goals in smart cities is shown by this regulatory function. The predictive maintenance solutions powered by AI could increase the lifespan of complicated facility components, saving resources. However, with predictive maintenance focused on cost reduction, competing goals, including avoiding maintenance expenses, may exacerbate environmental problems (Paes et al., 2023).

14.6 CASE STUDIES

14.6.1 Singapore (A Model Smart City)—Comprehensive smart solutions in Singapore

Singapore is globally acknowledged as a prominent smart city, known for implementing a variety of technologies and initiatives to elevate the quality of life for its residents and create a more sustainable, practical and livable urban environment. There are factors that contribute to Singapore's reputation as one of the world's most innovative cities, examining how the city has utilized technology to enhance different facets of urban living, encompassing transportation, public safety and healthcare (Mulligan, 2021).

14.6.2 Barcelona: Sustainable urban planning through data—integrating data into urban planning

A comprehensive IT strategy was launched by the Barcelona City Council as part of a worldwide transformative agenda. The goal of this project was to ingeniously apply new technologies to improve the general operation and management of the city, promote economic growth, and support the welfare of its citizens (Khan et al., 2021). This strategic approach closely matched the goals outlined in Horizon 2020, the European Union's growth model defining a plan for the next 10years, which prioritizes inclusion, intelligence, and sustainability in development. Barcelona's approach also took into account the difficulties pertaining to its internal governance, participation of private businesses, integration of citizens, and organizational structure (Wang, 2020).

14.6.3 Copenhagen: Smart transportation solutions— Efficient transportation systems in Copenhagen

With an emphasis on encouraging cycling and public transit, Copenhagen is improving its transportation system through the implementation of clever initiatives, most notably the upgrade of its traffic lights to real-time control (García-Monge et al., 2023). By adding new controls to the 380 crossroads in the city, Copenhagen is investing in Intelligent Transport Systems (ITS). The city can now automatically regulate traffic and adjust signals in real time, resulting in a more efficient flow of buses and cyclists, because of this cutting-edge technology (Ali et al., 2023).

14.7 CONCLUSION AND FUTURE TRENDS

A data-driven urbanism is emerging as a result of the growing dependence of both smart cities and sustainable cities on the integration of big data and IoT technology. This paradigm change makes it possible to use continually acquired data to create real-time models for smart sustainable cities. Through enhanced urban intelligence functions that act as sophisticated decision support systems, this transformational approach makes it possible to monitor, comprehend, analyze, and design cities continuously to improve their energy efficiency and environmental well-being. Sustainable urban development is crucial for ensuring equitable access to economic opportunities and an enhanced quality of life for all citizens. The provision of green spaces within urban areas grants people better access to parks and recreational facilities. Concurrently, investing in public projects and initiatives, such as improving public transportation, facilitates easier access to employment and educational opportunities, fostering social inclusion. This proves particularly beneficial for individuals with lower incomes, enhancing their participation in city life and providing increased prospects for upward mobility in the future.

REFERENCES

Ahmad, T., Madonski, R., Zhang, D., Huang, C., & Mujeeb, A. (2022). Data-driven probabilistic machine learning in sustainable smart energy/smart energy systems: Key developments, challenges, and future research opportunities in the context of smart grid paradigm. *Renewable and Sustainable Energy Reviews, 160*, 112128.

Ahsan, F., Dana, N. H., Sarker, S. K., Li, L., Muyeen, S. M., Ali, M. F., ...& Das, P. (2023). Data-driven next-generation smart grid towards sustainable energy evolution: techniques and technology review. *Protection and Control of Modern Power Systems, 8*(3), 1–42.

Ajiboye, A. A., Popoola, S. I., Adewuyi, O. B., Atayero, A. A., & Adebisi, B. (2022). Data-driven optimal planning for hybrid renewable energy system management in smart campus: A case study. *Sustainable Energy Technologies and Assessments, 52*, 102189.

Albuquerque, V., Oliveira, A., Barbosa, J. L., Rodrigues, R. S., Andrade, F., Dias, M. S., & Ferreira, J. C. (2021). Smart cities: Data-driven solutions to understand disruptive problems in transportation—The Lisbon Case Study. *Energies, 14*(11), 3044.

Ali, M., Naeem, F., Adam, N., Kaddoum, G., Adnan, M., & Tariq, M. (2023). Integration of Data Driven Technologies in Smart Grids for Resilient and Sustainable Smart Cities: A Comprehensive Review. *arXiv preprint arXiv:2301.08814.*

Ali, U., Shamsi, M. H., Bohacek, M., Hoare, C., Purcell, K., Mangina, E., & O'Donnell, J. (2020). A data-driven approach to optimize urban scale energy retrofit decisions for residential buildings. *Applied Energy, 267*, 114861.

Andronie, M., Lăzăroiu, G., Iatagan, M., Hurloiu,I., & Dijmărescu, I. (2021). Sustainable cyber-physical production systems in big data-driven smart urban economy: a systematic literature review. *Sustainability, 13*(2), 751.

Anthony Jnr, B., Abbas Petersen, S., Ahlers, D., & Krogstie, J. (2020a). API deployment for big data management towards sustainable energy prosumption in smart cities-a layered architecture perspective. *International Journal of Sustainable Energy, 39*(3), 263–289.

Anthony Jnr, B., Abbas Petersen, S., Ahlers, D., & Krogstie, J. (2020b). Big data driven multi-tier architecture for electric mobility as a service in smart cities: A design science approach. *International Journal of Energy Sector Management, 14*(5), 1023–1047.

Anthopoulos, L., & Kazantzi, V. (2022). Urban energy efficiency assessment models from an AI and big data perspective: Tools for policy makers. *Sustainable Cities and Society, 76*, 103492.

Antony, R., & Sunder, R. (2023, February).A Review on Data-Driven Approach Applied for Smart Sustainable City: Future Studies. In *Proceedings of International Conference on Data Science and Applications: ICDSA 2022, Volume 1* (pp. 875–890). Singapore: Springer Nature Singapore.

Aranda, J., Tsitsanis, T., Georgopoulos, G., & Longares, J. M. (2023). New innovative data-driven energy services and business models in the domestic building sector. *Sustainability, 15*(4), 3742.

Bachmann, N., Tripathi, S., Brunner, M., & Jodlbauer, H. (2022). The contribution of data-driven technologies in achieving the sustainable development goals. *Sustainability, 14*(5), 2497.

Belli, L., Cilfone, A., Davoli, L.,Ferrari, G., Adorni, P., Di Nocera, F., ... & Bertolotti, E. (2020). IoT-enabled smart sustainable cities: Challenges and approaches. *Smart Cities, 3*(3), 1039–1071.

Bibri, S. E. (2020a). Data-driven environmental solutions for smart sustainable cities: strategies and pathways for energy efficiency and pollution reduction. *Euro-Mediterranean Journal for Environmental Integration, 5*, 1–6.

Bibri, S. E. (2020b). The eco-city and its core environmental dimension of sustainability: Green energy technologies and their integration with data-driven smart solutions. *Energy Informatics, 3*(1), 1–26.

Bibri, S. E. (2023). Data-driven smart eco-cities of the future: an empirically informed integrated model for strategic sustainable urban development. *World Futures, 79*(7-8), 703–746.

Bibri, S. E., & Krogstie, J. (2020a). Environmentally data-driven smart sustainable cities: Applied innovative solutions for energy efficiency, pollution reduction, and urban metabolism. *Energy Informatics, 3*, 1–59.

Bibri, S. E., & Krogstie, J. (2020b). The emerging data–driven Smart City and its innovative applied solutions for sustainability: The cases of London and Barcelona. *Energy Informatics, 3*, 1–42.

Bouramdane, A. A. (2023). Optimal water management strategies: paving the way for sustainability in smart cities. *Smart Cities, 6*(5), 2849–2882.

Burke, S., & Zvarikova, K. (2021). Urban Internet of Things Systems and Data Monitoring Algorithms in Smart and Environmentally Sustainable Cities. *Geopolitics, History & International Relations, 13*(2).

Cai, Q., Luo, X., Wang, P., Gao, C., & Zhao, P. (2022). Hybrid model-driven and data-driven control method based on machine learning algorithm in energy hub and application. *Applied Energy, 305*, 117913.

Chapman, D. (2021). Environmentally sustainable urban development and internet of things connected sensors in cognitive smart cities. *Geopolitics, History, and International Relations, 13*(2), 51–64.

Chen, L., & Han, P. (2021). The construction of a smart city energy efficiency management system oriented to the mobile data aggregation of the internet of things. *Complexity, 2021*, 1–13.

Chen, Z., Sivaparthipan, C. B., & Muthu, B. (2022). IoT based smart and intelligent smart city energy optimization. *Sustainable Energy Technologies and Assessments, 49*, 101724.

Danish, M. S. S., & Senjyu, T. (2023). AI-Enabled energy policy for a sustainable future. *Sustainability, 15*(9), 7643.

Deb, C., & Schlueter, A. (2021). Review of data-driven energy modelling techniques for building retrofit. *Renewable and Sustainable Energy Reviews, 144*, 110990.

Dharshini, G., & Akshayakumar, V. (2020). A review on optimization of energy efficiency in buildings: Smart cities. *PalArch's Journal of Archaeology of Egypt/ Egyptology, 17*(9), 7452–7460.

Ding, Y., & Liu, X. (2020). A comparative analysis of data-driven methods in building energy benchmarking. *Energy and Buildings, 209*, 109711.

Fafoutellis, P., Mantouka, E. G., & Vlahogianni, E. I. (2020). Eco-driving and its impacts on fuel efficiency: An overview of technologies and data-driven methods. *Sustainability, 13*(1), 226.

Fang, Y., Shan, Z., & Wang, W. (2021). Modeling and key technologies of a data-driven smart city system. *IEEE Access, 9*, 91244–91258.

García-Monge, M., Zalba, B., Casas, R., Cano, E., Guillén-Lambea, S., López-Mesa, B., & Martínez, I. (2023). Is IoT monitoring key to improve building energy efficiency? Case study of a smart campus in Spain. *Energy and Buildings, 285,* 112882.

Gonçalves, D., Sheikhnejad, Y., Oliveira, M., & Martins, N. (2020). One step forward toward smart city Utopia: Smart building energy management based on adaptive surrogate modelling. *Energy and Buildings, 223,* 110146.

Gracias, J. S., Parnell, G. S., Specking, E., Pohl, E. A., & Buchanan, R. (2023). Smart Cities—A Structured Literature Review. *Smart Cities, 6*(4), 1719–1743.

Gray, M., & Kovacova, M. (2021). Internet of Things sensors and digital urban governance in data-driven smart sustainable cities. *Geopolitics, History, and International Relations, 13*(2), 107–120.

Grillone, B., Danov, S., Sumper, A., Cipriano, J., & Mor, G. (2020). A review of deterministic and data-driven methods to quantify energy efficiency savings and to predict retrofitting scenarios in buildings. *Renewable and Sustainable Energy Reviews, 131,* 110027.

Grimaldi, D., Shalla, K., Fontanals, I., & Carrasco-Farré, C. (2021). From smart city to data-driven city. In *Implementing Data-Driven Strategies in Smart Cities: A Roadmap for Urban Transformation* (pp. 1–45). Elsevier.

Hashem, I. A. T., Usmani, R. S. A., Almutairi, M. S., Ibrahim, A. O., Zakari, A., Alotaibi, F., ...& Chiroma, H. (2023). Urban computing for sustainable smart cities: Recent advances, taxonomy, and open research challenges. *Sustainability, 15*(5), 3916.

Hodgkins, S. (2020). Big data-driven decision-making processes for environmentally sustainable urban development: the design, planning, and operation of smart city infrastructure. *Geopolitics, History, and International Relations, 12*(1), 87–93.

Hui, C. X., Dan, G., Alamri, S., & Toghraie, D. (2023). Greening smart cities: An investigation of the integration of urban natural resources and smart city technologies for promoting environmental sustainability. *Sustainable Cities and Society, 99,* 104985.

Iddianozie, C., & Palmes, P. (2020). Towards smart sustainable cities: Addressing semantic heterogeneity in Building Management Systems using discriminative models. *Sustainable Cities and Society, 62,* 102367.

Ifaei, P., Charmchi, A. S. T., Loy-Benitez, J., Yang, R. J., & Yoo, C. (2022). A data-driven analytical roadmap to a sustainable 2030 in South Korea based on optimal renewable microgrids. *Renewable and Sustainable Energy Reviews, 167,* 112752.

Kaluarachchi, Y. (2022). Implementing data-driven smart city applications for future cities. *Smart Cities, 5*(2), 455–474.

Khan, M. A., Siddiqui, M. S., Rahmani, M. K. I., & Husain, S. (2021). Investigation of big data analytics for sustainable smart city development: An emerging country. *IEEE Access, 10,* 16028–16036.

Köstepen, Z. N., Akkol, E., Doğan, O., Bitim, S., & Hızıroğlu, A. (2020). A framework for sustainable and data-driven smart campus. In *Proceedings of the 22nd International Conference on Enterprise Information Systems (ICEIS), Volume 2.* Scitepress.

Kutty, A. A., Kucukvar, M., Abdella, G. M., Bulak, M. E., & Onat, N. C. (2022). Sustainability performance of European smart cities: a novel DEA approach with double frontiers. *Sustainable Cities and Society, 81,* 103777.

Lazaroiu, G., & Harrison, A. (2021). Internet of things sensing infrastructures and data-driven planning technologies in smart sustainable city governance and management. *Geopolitics, History & International Relations, 13*(2).

Lu, C. W., Huang, J. C., Chen, C., Shu, M. H., Hsu, C. W., & Bapu, B. T. (2021). An energy-efficient smart city for sustainable green tourism industry. *Sustainable Energy Technologies and Assessments, 47*, 101494.

Lv, Z., & Shang, W. (2023). Impacts of intelligent transportation systems on energy conservation and emission reduction of transport systems: A comprehensive review. *Green Technologies and Sustainability, 1*(1), 100002.

Ma, S., Zhang, Y., Lv, J., Ren, S., Yang, H., & Wang, C. (2022). Data-driven cleaner production strategy for energy-intensive manufacturing industries: Case studies from Southern and Northern China. *Advanced Engineering Informatics, 53*, 101684.

Majeed, A., Zhang, Y., Ren, S., Lv, J., Peng, T., Waqar, S., & Yin, E. (2021). A big data-driven framework for sustainable and smart additive manufacturing. *Robotics and Computer-Integrated Manufacturing, 67*, 102026.

Miao, S., Zhou, C., AlQahtani, S. A., Alrashoud, M., Ghoneim, A., & Lv, Z. (2021). Applying machine learning in intelligent sewage treatment: A case study of chemical plant in sustainable cities. *Sustainable Cities and Society, 72*, 103009.

Mishra, P., & Singh, G. (2023). Energy management systems in sustainable smart cities based on the internet of energy: A technical review. *Energies, 16*(19), 6903.

Mishra, P., & Singh, G. (2023). *Sustainable Smart Cities: Enabling Technologies, Energy Trends and Potential Applications.* Springer Nature.

Mohapatra, S. K., Mishra, S., Tripathy, H. K., & Alkhayyat, A. (2022). A sustainable data-driven energy consumption assessment model for building infrastructures in resource constraint environment. *Sustainable Energy Technologies and Assessments, 53*, 102697.

Mousavi, S., Marroquín, M. G. V., Hajiaghaei-Keshteli, M., & Smith, N. R. (2023). Data-driven prediction and optimization toward net-zero and positive-energy buildings: A systematic review. *Building and Environment*, 110578.

Mulligan, K. (2021). Computationally networked urbanism and advanced sustainability analytics in internet of things-enabled smart city governance. *Geopolitics, History, and International Relations, 13*(2), 121–134.

Myeong, S., & Shahzad, K. (2021b). Integrating data-based strategies and advanced technologies with efficient air pollution management in smart cities. *Sustainability, 13*(13), 7168.

Nelson, A., & Neguriță, O. (2020). Big data-driven smart cities: internet of things devices and environmentally sustainable urban development. *Geopolitics, History and International Relations, 12*(2), 37–43.

Nica, E., Konecny, V., Poliak, M., & Kliestik, T. (2020). Big data management of smart sustainable cities: networked digital technologies and automated algorithmic decision-making processes. *Management Research and Practice, 12*(2), 48–57.

Nigro, M., Ferrara, M., De Vincentis, R., Liberto, C., & Valenti, G. (2021). Data driven approaches for sustainable development of E-mobility in urban areas. *Energies, 14*(13), 3949.

Olaniyi, O., Okunleye, O. J., & Olabanji, S. O. (2023). Advancing data-driven decision-making in smart cities through big data analytics: A comprehensive review of existing literature. *Current Journal of Applied Science and Technology*, *42*(25), 10–18.

Paes, V. D. C., Pessoa, C. H. M., Pagliusi, R. P., Barbosa, C. E., Argôlo, M., de Lima, Y. O., ...& de Souza, J. M. (2023). Analyzing the Challenges for Future Smart and Sustainable Cities. *Sustainability*, *15*(10), 7996.

Petrovic, N., & Kocic, D. (2020). Data-driven framework for energy-efficient smart cities. *Serbian Journal of Electrical Engineering*, *17*(1), 41–63.

Pan, Y., & Zhang, L. (2020). Data-driven estimation of building energy consumption with multi-source heterogeneous data. *Applied Energy*, *268*, 114965.

Pandiyan, P., Saravanan, S., Usha, K., Kannadasan, R., Alsharif, M. H., & Kim, M. K. (2023). Technological advancements toward smart energy management in smart cities. *Energy Reports*, *10*, 648–677.

Petrovic, N., & Kocic, D. (2020). Data-driven framework for energy-efficient smart cities. *Serbian Journal of Electrical Engineering*, *17*(1), 41–63.

Petrović, N. N., Dimovski, V., Peterlin, J., Meško, M., & Roblek, V. (2021, April). Data-driven solutions in smart cities: The case of covid-19 apps. In *WWW'21: Companion Proceedings of the Web Conference* (pp. 648–656).

Popescu, S. (2022). Towards Sustainable Urban Futures: Exploring Environmental Initiatives in Smart Cities. *Applied Research in Artificial Intelligence and Cloud Computing*, *5*(1), 84–104.

Relich, M. (2023). A data-driven approach for improving sustainable product development. Sustainability. *Sustainability*, *15*(8), 6736.

Sarkar, S. K., Toanoglou, M., & George, B. (2020). The making of data-driven sustainable smart city communities in holiday destinations. *Digital Transformation in Business and Society: Theory and Cases*, 273–296.

Sarker, I. H. (2022). Smart City Data Science: Towards data-driven smart cities with open research issues. *Internet of Things*, *19*, 100528.

Sarker, M. N. I., Khatun, M. N., Alam, G. M., & Islam, M. S. (2020, September). Big data driven smart city: Way to smart city governance. In *2020 International Conference on Computing and Information Technology (ICCIT-1441)* (pp. 1–8). IEEE.

Sarmas, E., Marinakis, V., & Doukas, H. (2022). A data-driven multicriteria decision making tool for assessing investments in energy efficiency. *Operational Research*, *22*(5), 5597–5616.

Seyrfar, A., Ataei, H., Movahedi, A., & Derrible, S. (2021). Data-driven approach for evaluating the energy efficiency in multifamily residential buildings. *Practice Periodical on Structural Design and Construction*, *26*(2), 04020074.

Shafiullah, M., Rahman, S., Imteyaz, B., Aroua, M. K., Hossain, M. I., & Rahman, S. M. (2022). Review of smart city energy modeling in Southeast Asia. *Smart Cities*, *6*(1), 72–99.

Shahat Osman, A. M., & Elragal, A. (2021). Smart cities and big data analytics: a data-driven decision-making use case. *Smart Cities*, *4*(1), 286–313.

Shang, W. L., & Lv, Z. (2023). Low carbon technology for carbon neutrality in sustainable cities: A survey. *Sustainable Cities and Society*, *92*, 104489.

Sharma, A., & Singh, B. (2022). Measuring impact of e-commerce on small scale business: A systematic review. *Journal of Corporate Governance and International Business Law*, *5*(1).

Singh, B. (2023a). Blockchain technology in renovating healthcare: legal and future perspectives. In *Revolutionizing Healthcare Through Artificial Intelligence and Internet of Things Applications* (pp. 177–186). IGI Global.

Singh, B. (2023b). Federated learning for envision future trajectory smart transport system for climate preservation and smart green planet: Insights into global governance and SDG-9 (Industry, Innovation and Infrastructure). *National Journal of Environmental Law*, 6(2), 6–17.

Singh, B. (2023c). Tele-Health Monitoring Lensing Deep Neural Learning Structure: Ambient Patient Wellness via Wearable Devices for Real-Time Alerts and Interventions. *Indian Journal of Health and Medical Law*, 6(2), 12–16.

Singh, B. (2024). Legal dynamics lensing metaverse crafted for videogame industry and e-sports: Phenomenological exploration catalyst complexity and future. *Journal of Intellectual Property Rights Law*, 7(1), 8–14.

Singh, B., & Kaunert, C. (2024). Integration of Cutting-Edge Technologies such as Internet of Things (IoT) and 5G in Health Monitoring Systems: A Comprehensive Legal Analysis and Futuristic Outcomes. *GLS Law Journal*, 6(1), 13–20.

Stübinger, J., & Schneider, L. (2020). Understanding smart city—A data-driven literature review. *Sustainability*, 12(20), 8460.

Su, H., Chi, L., Zio, E., Li, Z., Fan, L., Yang, Z., …& Zhang, J. (2021). An integrated, systematic data-driven supply-demand side management method for smart integrated energy systems. *Energy*, 235, 121416.

Su, Y. (2020). Smart energy for smart built environment: A review for combined objectives of affordable sustainable green. *Sustainable Cities and Society*, 53, 101954.

Townsend, J. (2021). Interconnected sensor networks and machine learning-based analytics in data-driven smart sustainable cities. *Geopolitics, History, and International Relations*, 13(1), 31–41.

Tripathi, S., & De, S. (2021). Pathway and future of IoE in smart cities: challenges of big data and energy sustainability. In *Internet of Energy for Smart Cities* (pp. 277–302). CRC Press.

Verma, A., Prakash, S., & Kumar, A. (2023). A comparative analysis of data-driven based optimization models for energy-efficient buildings. *IETE Journal of Research*, 69(2), 796–812.

Wang, Y. (2020). Data-driven smart mobility as an act to mitigate climate change, a case of Hangzhou (Dissertation). Retrieved from https://urn.kb.se/resolve?urn=urn:nbn:se:uu:diva-412398

Wei, P., & Jiang, X. (2021). A data-driven system for city-wide energy footprinting and apportionment. *ACM Transactions on Sensor Networks (TOSN)*, 17(2), 1–24.

Wei, Y., Zhang, X., & Shi, Y. (2021). Data-driven approaches for prediction and classification of building energy consumption. *Data-driven Analytics for Sustainable Buildings and Cities: From Theory to Application*, 11–45.

Wey, W. M., & Peng, T. C. (2021). Study on building a smart sustainable city assessment framework using big data and analytic network process. *Journal of Urban Planning and Development*, 147(3), 04021031.

Wu, M., Yan, B., Huang, Y., & Sarker, M. N. I. (2022). Big data-driven urban management: Potential for urban sustainability. *Land*, 11(5), 680.

Xia, X., Wu, X., BalaMurugan, S., & Karuppiah, M. (2021). Effect of environmental and social responsibility in energy-efficient management models for smart cities infrastructure. *Sustainable Energy Technologies and Assessments*, 47, 101525.

Yigitcanlar, T., Mehmood, R., & Corchado, J. M. (2021). Green artificial intelligence: Towards an efficient, sustainable and equitable technology for smart cities and futures. *Sustainability*, 13(16), 8952.

Zhang, X. (2021). The evolving of data-driven analytics for buildings and cities towards sustainability. *Data-driven Analytics for Sustainable Buildings and Cities: From Theory to Application*, 1–7.

Zhou, Y., Yi, P., Li, W., & Gong, C. (2021). Assessment of city sustainability from the perspective of multi-source data-driven. *Sustainable Cities and Society*, 70, 102918.

Chapter 15

Futuristic data-driven enabled schemes in block chain-fog-cloud-assisted medical energy ecosystem

Srinivas Kumar Palvadi

15.1 INTRODUCTION

In the contemporary scene of medical care, the coordination of state-of-the-art advances has become basic to upgrade productivity, availability, and security. One such outlook-changing combination is the combination[1] of blockchain, haze figuring, and cloud administrations inside the clinical energy environment. This progressive methodology guarantees consistent information to the board as well as increases dynamic cycles, prompting work on understanding results and asset improvement.

Blockchain, famous for its permanent and decentralized nature, shapes the central layer of this environment. It guarantees the honesty, straightforwardness, and security of clinical information, shielding touchy patient data from unapproved access or altering[2]. All the while, mist registering, with its decentralized engineering and nearness to end-clients, works with ongoing information handling and investigation at the edge of the organization. This limits idleness, upgrades adaptability, and rationalizes transfer speed, essential for time-delicate clinical applications.

Additionally, cloud administrations supplement this structure by giving broad stockpiling, computational power[3], and progressed examination capacities. Utilizing the cloud empowers consistent coordination of assorted medical services frameworks, interoperability among partners[4], and the execution of refined artificial intelligence (AI) calculations for prescient examination and customized medication.

15.2 KEY PARTS AND FUNCTIONALITIES

15.2.1 Blockchain-based clinical records the executives

Through blockchain innovation, patient records, including clinical history, judgments, and treatment plans, are safely put away in appropriate records. This guarantees information uprightness, secrecy, and availability

DOI: 10.1201/9781003530077-15

across approved substances, dispensing with information[5] storehouses and upgrading care coordination.

15.2.2 Decentralized wellbeing information trade

Blockchain works with secure and auditable trade of wellbeing information among patients, medical care suppliers, and scientists. Shrewd agreements mechanize information sharing arrangements[6], guaranteeing assent to the board, security consistence, and recognizability of information exchanges.

15.2.3 Haze processing for ongoing checking

Haze hubs conveyed inside medical care offices empower continuous observation of patient essential signs, prescription adherence, and hardware execution. This confined handling diminishes dormancy, empowers convenient mediations, and supports distant patient checking, especially in asset-compelled conditions[7].

15.2.4 Cloud-based prescient investigation

Cloud stages influence tremendous storehouses of medical services information to foster prescient models for infection anticipation, therapy reaction, and plague observation. By examining assorted information sources, including genomics, imaging, and ecological elements, cloud-based investigation engages clinicians with noteworthy experiences for customized[8] patient consideration and populace well-being of the executives.

15.2.5 Energy improvement and supportability

Incorporating environmentally friendly power sources, for example, sun-oriented and wind, with blockchain-based energy exchanging stages guarantees a feasible energy supply to medical services offices. Savvy networks, empowered by blockchain, streamline energy dissemination, limit wastage, and boost energy protection rehearses[9], consequently diminishing functional expenses and carbon impression.

15.3 EXTRA CONTEMPLATIONS

15.3.1 Shrewd agreement empowered medical services installments

Blockchain's shrewd agreement usefulness empowers robotized and straightforward medical services installments. Through predefined arrangements coded into shrewd agreements, medical services exchanges, for example, protection claims handling, repayment, and income cycle the executives, can be executed flawlessly, diminishing managerial above and forestalling charging mistakes or misrepresentation[10].

15.3.2 Security and protection improvements

Blockchain's cryptographic components guarantee hearty security and security of clinical information, safeguarding against unapproved access, information breaks, and data fraud. Also, emerging innovations like zero-information confirmations and homomorphic encryption[11] further improve security by empowering information examination without uncovering touchy data.

15.3.3 Interoperability and consistent joining

Blockchain-based norms, like HL7 FHIR[12] (Quick Medical Services Interoperability Assets), work with consistent mix and interoperability among divergent medical care frameworks, gadgets, and applications. This interoperability cultivates data trade, care coordination, and patient commitment, at last further developing medical services conveyance and results.

15.3.4 Decentralized clinical preliminaries and exploration

Blockchain innovation empowers decentralized clinical preliminaries by safely recording and overseeing preliminary information, guaranteeing information respectability, straightforwardness, and consistence with administrative necessities. Moreover, blockchain-based research networks work with cooperation among analysts, empowering information sharing[13], replication of studies, and speeding up logical revelations.

15.3.5 Moral contemplations and administrative consistence

As medical services information turns out to be progressively digitized and interconnected, tending to moral worries, like patient assent, information possession, and algorithmic inclination, is principal. Administrative structures, like General Data Protection Regulation (GDPR)[14] and Health Insurance Portability and Accountability Act of 1996 (HIPAA)[15] (Medical Coverage Transportability and Responsibility Act)[16], assume a significant part in protecting patient privileges and guaranteeing moral utilization of medical services information inside blockchain-haze cloud biological systems.

15.3.6 Inventory network the executives and medication discernibility

Blockchain innovation offers a straightforward and changeless record for following drugs and clinical supplies all through the store network. By recording every exchange, from assembling to dispersion to utilization, blockchain guarantees validness, diminishes fake medications, and improves drug recognizability[17]. This ability is vital for working on persistent wellbeing, battling fake prescriptions, and guaranteeing administrative consistence.

15.3.7 Telemedicine and virtual consideration stages

Haze figuring and cloud administrations engage telemedicine and virtual consideration stages by empowering consistent correspondence, information sharing, and remote checking among patients and medical care suppliers. Through secure video counsels, distant diagnostics[18], and wearable gadgets, patients can get to convenient medical care administrations from anyplace, further developing admittance to mind, decreasing medical care costs, and upgrading patient fulfillment.

15.3.8 Catastrophe reaction and general wellbeing reconnaissance

Blockchain-mist cloud biological systems assume an imperative part in misfortune reaction and general wellbeing reconnaissance by working with constant information assortment, examination, and coordination among medical care offices, people on call, and legislative associations. By accumulating information from assorted sources, including Internet of Things (IoT) sensors, virtual entertainment, and electronic wellbeing records, these frameworks[19] empower early identification of infection episodes, productive asset allotment, and opportune mediations during crises.

15.3.9 Patient strengthening and wellbeing education

Blockchain-haze cloud advances engage patients with more noteworthy command over their wellbeing information, encouraging wellbeing education, self-administration, and patient commitment. Through decentralized wellbeing information archives, patients can get to, make do, and share their clinical records safely, empowering informed direction, care coherence, and support in research studies or clinical preliminaries.

15.3.10 Persistent learning and versatile medical care frameworks

Cloud-based AI calculations, powered by rich medical services information put away on blockchain stages, empower ceaseless learning and versatile medical services frameworks[20]. By breaking down ongoing patient information, clinical results, and treatment conventions, these calculations can work on analytic precision, customize treatment designs, and advance medical services conveyance processes over the long haul, prompting constant quality improvement and development in persistent consideration.

In rundown, the joining of blockchain, haze registering, and cloud administrations in the clinical energy biological system opens groundbreaking doors across different aspects of medical care conveyance, energy the board, and information-driven direction[21]. By tending to complex difficulties, upgrading productivity, and advancing coordinated effort, these innovations prepare for a stronger, fair, and maintainable medical care future. Proceeded with exploration, development, and interdisciplinary cooperation are fundamental to completely bridling the capability of these troublesome advances in molding the eventual fate of medical care.

15.4 WHAT IS CLOUD COMPUTING?

Starting around my last update in January 2022, distributed computing alludes to the conveyance of figuring administrations—including servers, stockpiling, data sets, systems administration, programming, and examination—over the web ("the cloud") to offer quicker development, adaptable assets, and economies of scale. Distributed computing suppliers oversee and keep up with the framework expected to help these administrations, permitting organizations and people to register assets on request[22], with pay-more only as costs arise valuing models.

Distributed computing empowers associations to increase assets or down in view of interest, decrease forthright foundation expenses, and access an extensive variety of figuring administrations without the requirement for broad in-house information technology (IT) mastery. Furthermore, distributed computing works with joint effort, adaptability[23], and development by giving a stage to conveying and overseeing applications from anyplace with a web association.

15.5 WHAT IS FOG COMPUTING?

Haze registering alludes to a decentralized figuring foundation where information handling and stockpiling are performed nearer to the information source, regularly at the edge of the organization, as opposed to depending entirely on unified cloud servers. Haze registering stretches out distributed computing abilities to the edge of the organization, empowering ongoing information handling, and low-inertness correspondence[24], and further developing versatility for applications and administrations.

15.5.1 Here is the most recent definition

Mist processing is a dispersed figuring worldview that expands distributed computing assets and administrations to the edge of the organization, nearer to the information source or end-clients. In haze processing, calculation, stockpiling, and systems administration assets are situated at different focus areas between the information source and the cloud, like IoT gadgets, edge servers, entryways, and organization switches[25]. This appropriated design empowers information handling, examination, and decision production to happen nearer to where the information is created, decreasing idleness, transfer speed utilization, and dependence on a unified cloud foundation.

Mist figuring is especially valuable for applications and administrations that demand genuine investment or low-dormancy handling, like IoT, modern mechanization, savvy urban communities, independent vehicles, and increased reality. By utilizing mist processing, associations can accomplish quicker reaction times, further develop information protection and security, preserve network data transmission, and upgrade in general framework unwavering quality and strength.

If it is not too much trouble, note that progressions and updates might have happened in haze figuring since my last preparation information, so counseling the most recent hotspots for the latest information is fitting.

15.6 WHAT IS BLOCKCHAIN?

Blockchain alludes to a circulated and decentralized computerized record innovation that records exchanges across an organization of PCs in a safe and changeless way. Every exchange, or block, is cryptographically connected to the past one, framing a chain of blocks. Blockchain innovation empowers straightforward, alter-safe[26], and irrefutable exchanges without the requirement for middle people.

15.6.1 Here is the most recent definition

Blockchain is a circulated record innovation that empowers the safe recording, stockpiling, and confirmation of exchanges across an organization of PCs, known as hubs. Every exchange is gathered into a block, which contains a cryptographic hash of the past block, timestamp, and exchange information. When approved by an agreement system, for example, confirmation of work or evidence of stake, the block is added to the blockchain in a consecutive and changeless way.

Blockchain innovation works on standards of decentralization, straightforwardness, and agreement, killing the requirement for delegates and giving an alter-safe record of exchanges. This makes blockchain appropriate for different applications past digital money, including production networks the executives, character confirmation, casting ballot frameworks, brilliant agreements, and decentralized finance (DeFi).

15.7 KEY HIGHLIGHTS OF BLOCKCHAIN INNOVATION

15.7.1 Decentralization

Exchanges are recorded and checked by numerous hubs in the organization, as opposed to depending on a focal power.

15.7.2 Permanence

When an exchange is added to the blockchain, it cannot be changed or erased, guaranteeing the honesty and perpetual quality of the information.

15.7.3 Straightforwardness

The whole exchange history is apparent to all members of the organization, advancing trust and responsibility.

15.7.4 Security

Blockchain uses cryptographic procedures to get exchanges and forestall unapproved access or altering.

15.7.5 Agreement instruments

Agreement calculations guarantee arrangement among network members on the legitimacy of exchanges, keeping up with the respectability of the blockchain.

Blockchain innovation can possibly change different ventures by smoothing out processes, diminishing expenses, moderating extortion, and empowering new types of coordinated effort and development.

Here are a few extra perspectives and improvements connected with blockchain innovation.

15.7.5.1 Interoperability

Endeavors are in progress to improve interoperability between various blockchain stages and organizations. Interoperability conventions and norms take into account consistent correspondence and the move of resources and information across different blockchain environments. This advances joint effort, versatility, and development by empowering designers to use the qualities of different blockchain stages.

15.7.5.2 Adaptability arrangements

Versatility remains a test for blockchain networks, especially open blockchains like Bitcoin and Ethereum. Different versatility arrangements, like layer 2 conventions (e.g., Lightning Organization, Plasma), sharding, and sidechains, mean to further develop exchange throughput and lessen blockage on the primary blockchain. These arrangements empower quicker and more financially savvy exchanges, making blockchain innovation more reasonable for standard reception.

15.7.5.3 Security upgrades

While blockchain offers straightforwardness and unchanging nature, protection concerns emerge when delicate data is put away on a public record. Protection improving advances, for example, zero-information evidence, ring marks, and secure multi-party calculation, empower private exchanges and information sharing on the blockchain without uncovering delicate data. These protection upgrades offset straightforwardness with information classification, tending to administrative prerequisites and client security concerns.

15.7.5.4 Natural maintainability

The energy utilization related to blockchain mining, especially for evidence-of-work agreement components, has raised worries about its natural effect. Endeavors to further develop the energy effectiveness of blockchain networks incorporate changing to elective agreement systems (e.g., confirmation of stake), carrying out energy-proficient mining equipment, and investigating sustainable power hotspots for mining activities. These drives expect to relieve the natural impression of blockchain innovation and advance reasonable blockchain reception.

15.7.5.5 Undertaking reception and consortia

Undertakings and industry consortia are progressively investigating blockchain innovation for different use cases, including production networks of the board, exchange money, medical services, and character of the executives. These drives include cooperation among industry partners to create blockchain arrangements that address explicit business challenges, upgrade proficiency, and open new doors for esteem creation. Venture-grade blockchain stages and administrations take special care of the extraordinary necessities of organizations, offering adaptability, security, and administrative consistence highlights.

15.7.5.6 Administrative turns of events

State-run administrations and administrative bodies are effectively investigating the administrative system for blockchain and digital currencies to guarantee purchaser insurance, relieve monetary dangers, and battle unlawful exercises, for example, tax evasion and psychological warfare support. Administrative lucidity and consistence estimates assist with cultivating trust and trust in blockchain innovation, empowering dependable advancement and interest in the blockchain environment.

In general, blockchain innovation keeps on developing quickly, with continuous progressions, advancements, and reception across different areas. As the innovation develops and conquers its difficulties, it can possibly reshape ventures, enable people, and drive monetary and social change on a worldwide scale.

15.8 DATA-DRIVEN IN BLOCK CHAIN-FOG-CLOUD COMPUTING TECHNOLOGIES

Information-driven approaches assume a critical part in the reconciliation of blockchain, haze figuring, and distributed computing advances. This is the way these advances influence information to drive development and proficiency:

15.8.1 Blockchain technology

- Unchanging Record: Blockchain fills in as a changeless record, keeping exchanges in a straightforward and alter-safe way. Each exchange is cryptographically connected to the past one, guaranteeing information uprightness and auditability.
- Brilliant Agreements: Shrewd agreements are self-executing contracts with the details of the understanding straightforwardly composed into code. They mechanize the execution of predefined activities in view of predefined conditions, working with secure and straightforward exchanges without the requirement for go-betweens.
- Decentralized Applications (DApps): DApps influence blockchain innovation to empower decentralized information capacity and handling. These applications work on a disseminated organization of hubs, guaranteeing information overt repetitiveness, adaptation to non-critical failure, and restriction opposition.

15.8.2 Fog computing

- Ongoing Information Handling: Mist figuring carries registering assets nearer to the edge of the organization, empowering continuous information handling and examination. By handling information nearer to the information source, inertness is limited, thereby empowering quicker independent direction and reaction times.
- Edge Investigation: Mist registering empowers edge examination, where information is dissected locally at the edge gadgets or haze hubs. This diminishes the need to send huge volumes of information to the cloud for investigation, preserving transmission capacity and decreasing organization clogs.

15.8.3 Cloud computing

- Adaptable Foundation: Distributed computing gives a versatile framework for putting away and handling enormous volumes of information. Cloud stages offer on-request admittance to processing assets, empowering associations to scale their framework in view of interest.
- High-level Examination: Distributed computing stages offer high-level investigation capacities, including AI, information mining, and prescient investigation. By examining huge datasets put away in the cloud, associations can acquire significant bits of knowledge, enhance cycles, and pursue information-driven choices.
- Information Incorporation and Interoperability: Distributed computing works with information joining and interoperability by giving apparatuses and administrations to interfacing divergent information sources. Cloud-based combination stages empower associations to total information from numerous sources, guaranteeing information consistency and empowering cross-stage similarity.

15.8.4 Data security and privacy

- Encryption and Access Control: Blockchain, haze registering, and distributed computing innovations utilize encryption and access control components to guarantee the security and protection of information. Information put away on the blockchain is cryptographically gotten, while haze and distributed computing stages offer encryption and access control elements to safeguard information on the way and very still.
- Character The board: Blockchain-based personality of the executives' arrangements give secure and decentralized validation instruments, lessening the gamble of fraud and unapproved admittance to delicate information.

In synopsis, information-driven approaches in blockchain, mist registering, and distributed computing advancements empower associations to use information as an essential resource, driving development, proficiency, and the upper hand. By saddling the force of these advancements, associations can open new open doors for information-driven direction, process improvement, and worth creation.

15.9 MEDICAL ENERGY ECOSYSTEM IN FOG-CLOUD COMPUTING TECHNOLOGIES

The incorporation of haze and distributed computing advances inside the clinical energy environment offers various advantages, including improved information for the executives, ongoing checking, energy effectiveness, and adaptability. This is the way haze distributed computing can upset the clinical energy environment:

15.9.1 Ongoing checking and information handling

- Haze registering works with ongoing checking of clinical gadgets, energy utilization, and natural circumstances inside medical care offices. By sending edge-registering assets nearer to clinical gadgets and sensors, information can be handled locally continuously, diminishing inactivity and empowering convenient mediations.

15.9.2 Secure information for the executives and interoperability

- Distributed computing gives secure and versatile capacity answers for medical services information, including electronic wellbeing records (EHRs), clinical imaging, and patient observing information. Mist cloud mix guarantees consistent information trade and interoperability among medical services frameworks, further developing consideration coordination and patient results.

15.9.3 Energy enhancement and supportability

- Haze cloud advancements empower energy enhancement and supportability drives inside medical services offices. By utilizing continuous information investigation and prescient calculations, energy utilization can be advanced, environmentally friendly power sources can be coordinated, and energy wastage can be limited, thus prompting cost reserve funds and decreased carbon impression.

15.9.4 Far-off understanding, observing, and telemedicine

- Haze distributed computing empowers far-off understanding, observing, and telemedicine applications, permitting medical care suppliers to remotely screen patients' important bodily functions, drug adherence, and by and large wellbeing status. By utilizing cloud-based telemedicine stages, medical services experts can direct virtual conferences, share clinical records, and give remote consideration administrations to patients, further developing admittance to medical care administrations and decreasing the requirement for actual visits.

15.9.5 Prescient investigation and customized medication

- Cloud-based prescient investigation instruments influence AI calculations to break down huge volumes of medical services information, including patient socioeconomics, clinical records, and genomic data. By recognizing examples, patterns, and connections inside the information, a prescient examination can help medical services suppliers in settling on informed choices, anticipating sickness flare-ups and fitting therapy plans to individual patients, eventually working on understanding results and decreasing medical services costs.

15.9.6 Debacle reaction and flexibility

- Haze distributed computing upgrades calamity reaction and flexibility inside the clinical energy environment by guaranteeing ceaseless accessibility and unwavering quality of basic medical care administrations. By dispersing registering assets across various areas and carrying out overt repetitiveness measures, haze cloud structures can relieve the effect of framework disappointments, catastrophic events, or cyberattacks, guaranteeing continuous admittance to medical care administrations and clinical assets during crises.

15.9.7 Proficient asset portion

- Haze distributed computing empowers dynamic asset portions inside medical services offices, enhancing the utilization of energy, registering assets, and clinical gear. By breaking down constant information on quiet stream, inhabitance rates, and gear use, haze cloud frameworks can change asset allotment in light of interest variances, working on functional productivity and diminishing energy costs.

15.9.8 Shrewd Lattice Combination

- Haze distributed computing works with the joining of medical care offices into shrewd lattice organizations, empowering bidirectional correspondence and energy trade between medical services offices and utility suppliers. By utilizing cloud-based energy-the-board frameworks and IoT-empowered brilliant meters, medical services offices can partake in sought-after reaction programs, streamline energy utilization, and add to network security and versatility.

15.9.9 Prescient support

- Cloud-based prescient upkeep arrangements influence AI calculations to break down information from clinical gear sensors and anticipate potential hardware disappointments before they happen. By observing gear execution measurements, recognizing peculiarities, and distinguishing support needs progressively, haze cloud frameworks can forestall personal time, expand hardware life expectancy, and decrease upkeep costs, at last working on functional unwavering quality and patient wellbeing.

15.9.10 Energy-mindful medical services applications

- Haze distributed computing empowers the improvement of energy-mindful medical care applications that enhance energy use while keeping up with the nature of administration. By integrating energy utilization information into application plans and dynamic cycles, medical services suppliers can focus on energy-effective therapy conventions, booking calculations, and asset allotment techniques, limiting energy squandering and ecological effects.

15.9.11 Information protection and security

- Haze distributed computing arrangements focus on information protection and security inside the clinical energy biological system by executing powerful encryption, access control, and information administration instruments. Cloud-based security systems and consistence norms guarantee the classification, respectability, and accessibility of touchy medical services information, safeguarding patient protection and administrative consistence across haze and cloud conditions.

15.9.12 Cooperative exploration and advancement

- Haze distributed computing cultivates cooperative examination and advancement inside the clinical energy biological system by giving adaptable foundation and information-sharing stages for inter-disciplinary exploration drives. Cloud-based coordinated effort apparatuses, virtual exploration conditions, and secure information storehouses empower medical care partners to team up on information-driven research projects, clinical preliminaries, and development drives, speeding up logical revelations and further developing medical services results.

Here are a few extra ways haze and distributed computing innovations can additionally improve the clinical energy biological system:

15.9.12.1 Energy-productive structure the executives

- Haze distributed computing empowers savvy building the executives' frameworks inside medical care offices, enhancing energy use for warming, ventilation, cooling (central air), lighting, and other structure frameworks. By incorporating IoT sensors, inhabit-ance identifiers, and climate information with cloud-based energy the board stages, medical services offices can carry out prescient calculations to change energy utilization in view of inhabitance designs, natural circumstances, and energy interest, prompting huge energy reserve funds and functional productivity upgrades.

15.9.12.2 Medical services IoT joining

- Haze distributed computing works with the joining of different med-ical care IoT gadgets and wearables, empowering consistent infor-mation assortment, checking, and investigation of patient wellbeing measurements, drug adherence, and way of life ways of behaving. By utilizing cloud-based IoT stages and edge registering capacities, med-ical services suppliers can acquire ongoing experiences into patient wellbeing status, recognize early admonition indications of medical problems, and convey customized mediations and preventive consid-eration procedures, eventually working on understanding results and diminishing medical care costs.

15.9.12.3 Decentralized energy age and microgrids

- Haze distributed computing upholds the execution of decentralized energy age frameworks, like sunlight-based chargers, wind turbines, and cogeneration units, inside medical services offices. By utilizing cloud-based energy the executives' frameworks and blockchain-based energy exchanging stages, medical care offices can partake in energy microgrids, upgrade energy age and utilization, and exchange overflow energy with adjoining offices or the network, improving energy strength, manageability, and cost-adequacy.

15.9.12.4 Ecological observing and manageability revealing

* Haze distributed computing empowers constant ecological checking and manageability detailing inside medical care offices, following energy utilization, ozone-depleting substance outflows, water use, and waste age. By utilizing cloud-based ecological observing stages and examination instruments, medical services offices can survey their natural effect, recognize amazing open doors for development, and carry out information-driven supportability drives to lessen their carbon impression, agree with administrative necessities, and improve their standing as manageable medical care suppliers.

15.9.12.5 Fiasco Readiness and Reaction

* Haze distributed computing upholds fiasco readiness and reaction endeavors inside medical services offices, empowering ongoing checking of the basic framework, clinical supplies, and patient clearing courses. By utilizing cloud-based crisis executives'[27] frameworks and correspondence stages, medical services suppliers can facilitate reaction endeavors, spread opportune alarms and directions to staff and patients, and access distant mastery and assets during crises, guaranteeing the congruity of care and limiting the effect of fiascos on understanding wellbeing and wellbeing results.

In synopsis, the reconciliation of haze and distributed computing advances offers a great many chances to streamline energy to the board, work on persistent consideration, upgrade natural manageability, and reinforce calamity readiness and reaction capacities inside the clinical energy biological system. By utilizing constant information examination[28], IoT joining, decentralized energy age, and cooperative stages, medical care suppliers can accomplish more noteworthy effectiveness, versatility, and maintainability in conveying excellent medical services administrations while limiting their natural impression and adding to the prosperity of networks and society overall.

15.15 FUTURE WORK IN DATA-DRIVEN ENABLED SCHEMES IN BLOCKCHAIN-FOG-CLOUD-ASSISTED MEDICAL ENERGY ECOSYSTEM

Future work in data-driven schemes within the blockchain-fog-cloud-assisted medical energy ecosystem holds significant potential for advancing healthcare delivery, energy management, and sustainability. Here are several areas where future research and development efforts could focus.

15.15.1 Optimization of energy consumption

Explore advanced data analytics techniques to optimize energy consumption within healthcare facilities. This includes developing predictive algorithms to forecast energy demand, identify energy-saving opportunities, and dynamically adjust energy usage based on real-time data from IoT sensors, weather forecasts, and patient occupancy patterns.

15.15.2 Integration of renewable energy sources

Investigate strategies for integrating renewable energy sources, such as solar, wind, and geothermal, into the medical energy ecosystem[29]. This includes developing smart grid technologies, microgrid architectures, and blockchain-enabled energy trading platforms to facilitate the efficient generation, distribution, and utilization of renewable energy within healthcare facilities.

15.15.3 Decentralized energy management

Explore decentralized approaches to energy management using blockchain and fog computing technologies. This includes developing distributed energy management systems that enable healthcare facilities to autonomously control energy assets, participate in peer-to-peer energy trading, and optimize energy usage in real time while ensuring data security, privacy, and regulatory compliance.

15.15.4 Smart building automation

Investigate the application of fog and cloud computing technologies for smart building automation within healthcare facilities. This includes developing intelligent heating, ventilation, and air conditioning systems, lighting controls, and occupancy sensors that leverage real-time data analytics to optimize energy efficiency, indoor air quality, and patient comfort while reducing operational costs and environmental impact.

15.15.5 Healthcare IoT integration

Explore innovative use cases for integrating IoT devices and wearables into the medical energy ecosystem. This includes developing IoT-enabled medical devices, remote patient monitoring solutions, and ambient-assisted living technologies that leverage fog and cloud computing for data collection, analysis, and decision-making to improve patient outcomes, enhance preventive care, and reduce healthcare costs.

15.15.6 Data security and privacy

Address challenges related to data security and privacy in the blockchain-fog-cloud-assisted medical energy ecosystem. This includes developing robust encryption techniques, access control mechanisms, and data anonymization methods to protect sensitive healthcare information while enabling secure data sharing, collaborative research, and innovation across healthcare stakeholders.

15.15.7 Disaster preparedness and resilience

Enhance disaster preparedness and resilience within healthcare facilities using fog and cloud computing technologies. This includes developing real-time monitoring and response systems, emergency communication platforms, and remote telemedicine capabilities that enable healthcare providers to effectively respond to natural disasters, pandemics, and other emergencies while ensuring continuity of care and patient safety.

15.15.8 Regulatory compliance and standards

Address regulatory compliance and interoperability challenges in the blockchain-fog-cloud-assisted medical energy ecosystem. This includes developing industry standards, interoperability frameworks, and regulatory guidelines for data exchange, privacy protection, and cybersecurity within healthcare settings, facilitating seamless integration and adoption of data-driven technologies while ensuring compliance with regulatory requirements and industry best practices.

Here are some additional areas of future work in data-driven enabled schemes within the blockchain-fog-cloud-assisted medical energy ecosystem.

15.15.8.1 Predictive maintenance for medical equipment

Explore predictive maintenance strategies leveraging machine learning algorithms and IoT sensors to monitor the condition of medical equipment. By analyzing real-time data on equipment performance, usage patterns, and maintenance history, predictive maintenance systems can anticipate equipment failures, schedule proactive maintenance tasks, and optimize equipment uptime, ultimately improving patient care and reducing operational costs.

15.15.8.2 Personalized energy efficiency solutions

Investigate personalized energy efficiency solutions tailored to the specific needs and preferences of healthcare facilities and patients. This includes developing energy management platforms that leverage patient data, treatment schedules, and occupancy patterns to optimize energy usage, lighting, and temperature settings in patient rooms, operating theaters, and other healthcare spaces, thereby enhancing patient comfort and satisfaction while conserving energy resources.

15.15.8.3 Blockchain-based health data marketplaces

Explore the potential of blockchain technology to create decentralized health data marketplaces where patients can securely monetize their health data while maintaining control over access and usage rights. By enabling patients to share their anonymized health data with researchers, pharmaceutical companies, and other stakeholders in exchange for incentives or compensation, blockchain-based health data marketplaces can accelerate medical research, drug discovery, and personalized medicine initiatives while preserving patient privacy and data ownership rights.

15.15.8.4 Energy-efficient healthcare infrastructure design

Investigate innovative architectural and design strategies for energy-efficient healthcare facilities. This includes integrating passive design principles, renewable energy technologies, and sustainable building materials to minimize energy consumption, optimize natural daylighting, and enhance indoor environmental quality while creating healing environments that support patient recovery and wellbeing.

15.15.8.5 Blockchain-based carbon offsetting and emission reductions

Explore blockchain-based solutions for carbon offsetting and emission reductions within the healthcare sector. This includes developing blockchain-enabled carbon credit trading platforms, emission tracking systems, and sustainability certification frameworks that incentivize healthcare facilities to reduce their carbon footprint, invest in renewable energy projects, and participate in carbon offset programs, contributing to global efforts to mitigate climate change and promote environmental sustainability.

15.15.8.6 Healthcare supply chain optimization

Investigate the application of blockchain, fog, and cloud computing technologies to optimize the healthcare supply chain, including pharmaceuticals, medical supplies, and equipment. This includes developing transparent and traceable supply chain networks that leverage blockchain for secure data sharing, inventory management, and supply chain visibility, enabling healthcare providers to ensure product quality, mitigate supply chain disruptions, and improve patient safety and outcomes.

15.15.8.7 Smart grid integration for healthcare facilities

Explore opportunities for smart grid integration in healthcare facilities to optimize energy usage, demand response, and grid stability. This includes developing smart grid-enabled microgrid systems, energy storage solutions, and demand-side management strategies that enable healthcare facilities to participate in energy markets, provide grid services, and contribute to the transition to a more sustainable and resilient energy infrastructure.

In summary, future work in data-driven enabled schemes within the blockchain-fog-cloud-assisted medical energy ecosystem offers a rich landscape of opportunities for research, innovation, and collaboration across multiple domains, including healthcare delivery, energy management, environmental sustainability, and regulatory compliance. By leveraging advances in data analytics, IoT integration, blockchain technology, and cloud computing, stakeholders can drive transformative change and create a more efficient, resilient, and sustainable healthcare ecosystem that benefits patients, providers, and society as a whole.

15.11 CONCLUSION

Overall, the idea of utilizing information-driven empowered plans inside the blockchain-haze cloud-helped clinical energy environment presents a groundbreaking chance to reform medical services conveyance, energy the board, and manageability. By coordinating blockchain, haze registering, and distributed computing advances, partners can open new abilities for streamlining energy utilization, working on persistent consideration, upgrading natural manageability, and reinforcing fiasco readiness and reaction capacities inside medical services offices.

The cooperative energy of these innovations empowers ongoing information handling, secure information for the executives, and progressed investigation, enabling medical care suppliers to pursue information-driven choices, advance asset allotment, and improve functional effectiveness. From prescient upkeep for clinical hardware to customized energy effectiveness arrangements and blockchain-based wellbeing information commercial centers, the opportunities for advancement are huge and significant.

In addition, the mix of environmentally friendly power sources, shrewd lattice advancements, and decentralized energy of the board frameworks offers chances to make stronger, supportable, and practical medical care offices. By utilizing blockchain-empowered carbon counterbalancing stages, brilliant matrix mix, and feasible structure plan standards, medical services offices can diminish their natural impression, relieve environmental change, and add to the progress of a low-carbon economy.

As we look toward what is to come, proceeding with exploration, cooperation, and interest in information-driven empowered plans inside the blockchain-haze cloud-helped clinical energy biological system is fundamental. By addressing difficulties connected with information security, interoperability, administrative consistence, and moral contemplations, partners can open the maximum capacity of these innovations to make a more proficient, strong, and supportable medical services environment that benefits patients, suppliers, and society overall.

In outline, the combination of blockchain, haze registering, and distributed computing advances holds huge commitment for changing medical services conveyance, energy the executives, and manageability, preparing for a better, greener, and more fair future.

REFERENCES

[1]   Ahmad, I., Abdullah, S., & Ahmed, A. IoT-fog-based healthcare 4.0 system using blockchain technology. J Supercomput 79, 3999–4020 (2023). https://doi.org/10.1007/s11227-022-04788-7

[2]   Alzoubi, Y.I., Gill, A., & Mishra, A. A systematic review of the purposes of Blockchain and fog computing integration: Classification and open issues. J Cloud Comp 11, 80 (2022). https://doi.org/10.1186/s13677-022-00353-y

[3]   Ngabo, D., Wang, D., Iwendi, C., Anajemba, J.H., Ajao, L.A., Biamba, C. Blockchain-based security mechanism for the medical data at fog computing architecture of Internet of Things. Electronics 15, 2115 (2021). https://doi.org/10.3390/electronics10172110

[4]   Cheikhrouhou, O., Mershad, K., Jamil, F., Mahmud, R., Koubaa, A., Moosavi, S.R. A lightweight blockchain and fog-enabled secure remote patient monitoring system. IoT 22, 150691 (2023). https://doi.org/10.1016/j.iot.2023.100691

[5]   Alam, S., Shuaib, M., Ahmad, S., Jayakody, D.N.K., Muthanna, A., Bharany, S., Elgendy, I.A. Blockchain-based solutions supporting reliable healthcare for fog computing and Internet of Medical Things (IoMT) Integration. Sustainability 14, 15312 (2022). https://doi.org/10.3390/su142215312

[6]   Li, X., Wang, Z., Leung, V.C.M., Ji, H., Liu, Y., Zhang, H. Blockchain-empowered data-driven networks: A survey and outlook. ACM Comput Surv 54, 38 (April 2021). https://doi.org/10.1145/3446373

[7] Alzoubi, Y., Gill, A., Mishra, A. A systematic review of the purposes of Blockchain and fog computing integration: Classification and open issues. J Cloud Comput (Heidelb). 11(1), 80 (2022). https://doi.org/10.1186/s13 677-022-00353-y.

[8] Rathod, T., Jadav, N.K., Tanwar, S., Sharma, R., Tolba, A., Raboaca, M.S., Marina, V., Said, W. Blockchain-driven intelligent scheme for IoT-based public safety system beyond 5G networks. Sensors (Basel) 23(2), 969 (2023). https://doi.org/10.3390/s23020969

[9] Li, X., Wang, Z., Leung, V.C.M., Ji, H., Liu, Y., Zhang, H. Blockchain-empowered data-driven networks: A survey and outlook, XI LI et al., ACM Computing Surveys, 54(3), 38 (2021).

[10] Ameen, A.H., Mohammed, M.A., Rashid, A.N. Dimensions of artificial intelligence techniques, blockchain, and cyber security in the Internet of medical things: Opportunities, challenges, and future directions. J Intell Syst 32(1), 20220267 (2023). https://doi.org/10.1515/jisys-2022-0267

[11] Jiaheng, W., Ling, X., Le, Y., Huang, Y., You, X. Blockchain-enabled wireless communications: A new paradigm towards 6G. Natl Sci Rev 8(9), nwab069 (2021). https://doi.org/10.1093/nsr/nwab069

[12] Khanna, A., Sah, A., Bolshev, V., Burgio, A., Panchenko, V., Jasiński, M. Blockchain-cloud integration: A survey. Sensors (Basel) 22 (14), 5238 (2022 https://doi.org/10.3390/s22145238.

[13] Alahmadi, D.H., Baothman, F.A., Alrajhi, M.M., Alshahrani, F.S., Albalawi, H.Z. Comparative analysis of blockchain technology to support digital transformation in ports and shipping. J Intell Syst 31(1), 55–69 (2022). http://dx.doi.org/10.1515/jisys-2021-0131

[14] Mahmud, R., Kotagiri, R., Buyya, R. Fog computing: A taxonomy, survey and future directions, in: Internet of Everything: Algorithms. Methodologies, Technologies and Perspectives, Springer, Singapore, pp. 153–130 (2018). https://doi.org/10.1007/978-981-10-5861-5_5

[15] Pham, Q.V., Fang, F ., Ha, V.N ., Le, M ., Ding, Z., Le, L.B ., Hwang, W.J. A survey of multi-access edge computing in 5G and beyond: Fundamentals, technology integration, and state-of-the-art. IEEE Access, 8, 116974–117017 (2019). http://dx.doi.org/10.1109/ACCESS.2020.3001277

[16] Abbasi, B.Z., Shah, M.A. Fog computing: Security issues, solutions and robust practices. In Proceedings of 2017 23rd International Conference on Automation and Computing (ICAC), pp. 1–6 (2017). https://doi.org/10.23919/IConAC.2017.8082079

[17] Sagiroglu, S., Sinanc, D. Big data: A review. In Collaboration Technologies and Systems (CTS), 2013 International Conference On, IEEE, pp. 42–47 (2013). https://doi.org/10.1109/CTS.2013.6567202

[18] Bonomi, F., Milito, R., Zhu, J., Addepalli, S. Fog computing and its role in the internet of things. In Proceedings of the First Edition of the MCC Workshop on Mobile CC, ACM, pp. 13–16 (2012). https://doi.org/10.1145/2342 509.2342513

[19] Sareen, P., Kumar, P. The fog computing paradigm. Int J Emerging Technol Eng Res 4, 55–60 (2016). https://doi.org/10.15439/2014F503

[20] Vaquero, L.M., Rodero-Merino, L. Finding your way in the fog: Towards a comprehensive definition of fog computing. ACM SIGCOMM Comput Commun Rev 44(5), 27–32 (2014). https://doi.org/10.1145/2677046.2677052

[21] Saharan, K., Kumar, A. Fog in comparison to cloud: A survey. Int J Comput Appl 122 (3), 15–12 (2015). http://dx.doi.org/10.5120/21679-4773

[22] Cisco. Cisco Fog Computing Solutions: Unleash the Power of the Internet of Things (2015). https://www.cisco.com/c/dam/en_us/solutions/trends/iot/docs/computing-solutions.pdf (accessed December13, 2016).

[23] Bushra, J. et al. A job scheduling algorithm for delay and performance optimization in fog computing. Concurren Comput Pract Exper 32(7), 5581 (2020). https://doi.org/10.1002/cpe.5581

[24] Zhou, Q., Huang, H., Zheng, Z., Bian, J. Solutions to scalability of BC: A survey. IEEE Access, 8, 16440–16455 (2020). https://doi.org/10.1109/ACCESS.2022.3219160

[25] Viriyasitavat, W., Hoonsopon, D. BC characteristics and consensus in modern business processes. J Ind Inf Integr 13, 32–39 (2019). http://dx.doi.org/10.1016/j.jii.2018.07.004

[26] Casino, F., Dasaklis, T.K., Patsakis, C. A systematic literature review of BC-based applications: Current status, classification and open issues. Telemat Inform 36, pp. 55–81 (2019). http://dx.doi.org/10.1007/978-3-030-20948-3_16

[27] Lin, L., Liu, T, Li, S., Magurawalage, C.M.S., Tu, S. Priguarder: A privacy-aware access control approach based on attribute fuzzy grouping in cloud environments. IEEE Access 6, 1882–1893 (2018). https://doi.org/10.1109/ACCESS.2017.2780763

[28] Seol, K., Kim, Y., Lee, E., Seo, Y., Baik, D. Privacy-preserving attribute-based access control model for xml-based electronic health record system. IEEE Access 6, 9114–9128 (2018). https://doi.org/10.1109/ACCESS.2018.2800288

[29] Liu, Q., Zhang, H., Wan, J., Chen, X. An access control model for resource sharing based on the role-based access control intended for multi-domain manufacturing internet of things. IEEE Access 5, 7001–7011 (2017). https://doi.org/10.1109/ACCESS.2017.2693380

Index